Biotechnological Innovations in Health Care

THE UNIVERSIT

HAR LD C

BOOKS IN THE BIOTOL SERIES

The Fabric of Cells
Infrastructure and Activities of Cells

Biomolecules - Extraction and Measurement
Biomolecules - Purification Strategies
Biomolecules - Analysis and Properties

Principles of Cell Energetics
Energy Resource Utilisation in Cells
Biosynthesis and the Integration of Cell Metabolism

Organisation of Genetic Information
Regulation of Gene Expression

Crop Physiology and Productivity

Functional Physiology
Defence Mechanisms

Bioprocess Technology: Modelling and Transport Phenomena
Operational Modes of Bioreactors

Microbial Growth and Cultivation
Plant Cell and Tissue Cultivation
Animal Cell Cultivation

Bioreactor Design and Product Yield
Product Recovery in Bioprocess Technology

Techniques for Engineering Genes
Strategies for Engineering Organisms

Technological Applications of Biocatalysts
Technological Applications of Immunochemicals

Biotechnological Innovations in Health Care

Biotechnological Innovations in Crop Improvement
Biotechnological Innovations in Animal Productivity

Biotechnological Innovations in Waste Treatment and Energy Resources

Biotechnological Innovations in Chemical Synthesis

Biotechnological Innovations in Food Processing

Biotechnology Source Book: Safety, Good Practice and Regulatory Affairs

BIOTOL

BIOTECHNOLOGY BY OPEN LEARNING

Biotechnological Innovations in Health Care

PUBLISHED ON BEHALF OF:

Open universiteit   and   **Thames Polytechnic**

Valkenburgerweg 167   Avery Hill Road
6401 DL Heerlen   Eltham, London SE9 2HB
Nederland   United Kingdom

Butterworth-Heinemann

Butterworth–Heinemann Ltd
Halley Court, Jordan Hill, Oxford OX2 8EJ

PART OF REED INTERNATIONAL P.L.C.

OXFORD LONDON GUILDFORD BOSTON
MUNICH NEW DELHI SINGAPORE SYDNEY
TOKYO TORONTO WELLINGTON

First published 1991

British Library Cataloguing in Publication Data

Biotechnological innovations in healthcare.
  1. Medicine. Applications of biotechnology
I. Turnock, G.   II. Series
610.28

  ISBN 0-7506-1497-8
  ISBN 0-7506-1493-5 pbk

Library of Congress Cataloging-in-Publication Data

Biotechnological innovations in healthcare.
     p. cm.–(Biotechnology by Open Learning)
    Authors, M.J.H.I. Beekman and others: editor, G. Turnock.
    "Published on behalf of Open Universiteit and Thames Polytechnic."
    Includes bibliographical references and index.
    ISBN 0-7506-1493-5 (soft cover) :   ISBN 0-7506-1497-8 (hard cover) :
    1. Pharmaceutical biotechnology.   I. Beekman, M. J. H. I.
II. Turnock, G.   III. Open Universiteit (Heerlen, Netherlands)
IV. Thames Polytechnic.   V. BIOTOL (Project)   VI. Series:
Biotechnology by Open Learning (Series)
    [DNLM: 1. Biological Products.   2. Biotechnology.   OW 800 B6143]
R5380.B56   1990
615'.36–dc20
DNLM/DLC   90-15122
for Library of Congress   CIP

Composition by Thames Polytechnic
Printed and bound by Hartnolls Ltd, Bodmin, Cornwall

The Biotol Project

The BIOTOL team

OPEN UNIVERSITEIT, NETHERLANDS
Dr M. C. E. van Dam-Mieras
Professor W. H. de Jeu
Professor J. de Vries

THAMES POLYTECHNIC, UK
Professor B. R. Currell
Dr J. W. James
Dr C. K. Leach
Mr R. A. Patmore

This series of books has been developed through a collaboration between the Open universiteit of the Netherlands and Thames Polytechnic to provide a whole library of advanced level flexible learning materials including books, computer and video programmes. The series will be of particular value to those working in the chemical, pharmaceutical, health care, food and drinks, agriculture, and environmental, manufacturing and service industries. These industries will be increasingly faced with training problems as the use of biologically based techniques replaces or enhances chemical ones or indeed allows the development of products previously impossible.

The BIOTOL books may be studied privately, but specifically they provide a cost-effective major resource for in-house company training and are the basis for a wider range of courses (open, distance or traditional) from universities which, with practical and tutorial support, lead to recognised qualifications. There is a developing network of institutions throughout Europe to offer tutorial and practical support and courses based on BIOTOL both for those newly entering the field of biotechnology and for graduates looking for more advanced training. BIOTOL is for any one wishing to know about and use the principles and techniques of modern biotechnology whether they are technicians needing further education, new graduates wishing to extend their knowledge, mature staff faced with changing work or a new career, managers unfamiliar with the new technology or those returning to work after a career break.

Our learning texts, written in an informal and friendly style, embody the best characteristics of both open and distance learning to provide a flexible resource for individuals, training organisations, polytechnics and universities, and professional bodies. The content of each book has been carefully worked out between teachers and industry to lead students through a programme of work so that they may achieve clearly stated learning objectives. There are activities and exercises throughout the books, and self assessment questions that allow students to check their own progress and receive any necessary remedial help.

The books, within the series, are modular allowing students to select their own entry point depending on their knowledge and previous experience. These texts therefore remove the necessity for students to attend institution based lectures at specific times and places, bringing a new freedom to study their chosen subject at the time they need it and a pace and place to suit them. This same freedom is highly beneficial to industry since staff can receive training without spending significant periods away from the workplace attending lectures and courses, and without altering work patterns.

Contributors

AUTHORS

Dr M. J. H. I. Beekman, Eli Lilly, Nieuwegein, The Netherlands.

Dr H. J. M. van de Donk, National Institute of Public Health and Environmental Protection, Bilthoven, The Netherlands.

Dr R. A. Drost, Centocor b.v., Leiden, The Netherlands.

Dr G. Eestermans, Cilag Benelux n.v., Herentals, Belgium.

Dr G. J. Gamelkoorn, Merck Sharp and Dohme b.v., Haarlem, The Netherlands.

Dr Ir J. A. R. Keus, Duphar b.v., Weesp, The Netherlands.

Dr J. Lister-James, Centocor, Malvern, USA.

EDITOR

Dr G. Turnock, Leicester University, Leicester, UK

SCIENTIFIC AND COURSE ADVISORS

Dr M. C. E. van Dam-Mieras, Open universiteit, Heerlen, The Netherlands

Dr C. K. Leach, Leicester Polytechnic, Leicester, UK.

ACKNOWLEDGEMENTS

Grateful thanks are extended, not only to the authors, editors and course advisors, but to all those who have contributed to the development and production of this book. They include Dr N. Chadwick, Dr P. A. Kirschner, Professor E. H. Houwink, Professor R. Spier, Dr G. Lawrence, Mrs M. Wyatt and Miss J. Skelton. The development of BIOTOL has been funded by COMETT, The European Community Action programme for Education and Training for Technology, by the Open universiteit of the Netherlands and by Thames Polytechnic. Thanks are also due to Gist-brocades for financial support and to all those other companies who contributed case studies and generally encouraged the development of the product.

Contents

How to use an open learning text   viii

Preface   ix

1   **Some perspectives in medical biotechnology,**
G. Turnock   1

2   **Development stages for health care products,**
C . K. Leach   9

3   **Regulatory affairs,**
H. J. M. van de Donk   29

4   **Case Study: Human insulin of rDNA origin,**
M. J. H. I. Beekman   55

5   **Case Study: Erythropoietin - a growth factor
produced by rDNA technology,**
G. Eestermans   101

6   **Case Study: Hepatitis B vaccines: a product of
rDNA techniques,**
G. J. Gamelkoorn   141

7   **Case Study: Orthoclone OKT3 - A therapeutic
antibody produced by hybridoma technology,**
G. Eestermans   165

8   **Case study: A vaccine for Aujeszky's disease,**
J. A. R. Keus   199

9   **Case Study: Myoscint - A monoclonal antibody
preparation used for cardiac imaging**
J. Lister-James and R. A. Drost   225

Responses to SAQs   249

Appendices   283

How to use an open learning text

An open learning text presents to you a very carefully thought out programme of study to achieve stated learning objectives, just as a lecturer does. Rather than just listening to a lecture once, and trying to make notes at the same time, you can with a BIOTOL text study it at your own pace, go back over bits you are unsure about and study wherever you choose. Of great importance are the self assessment questions (SAQs) which challenge your understanding and progress and the responses which provide some help if you have had difficulty. These SAQs are carefully thought out to check that you are indeed achieving the set objectives and therefore are a very important part of your study. Every so often in the text you will find the symbol Π , our open door to learning, which indicates an activity for you to do. You will probably find that this participation is a great help to learning so it is important not to skip it.

Whilst you can, as a open learner, study where and when you want, do try to find a place where you can work without disturbance. Most students aim to study a certain number of hours each day or each weekend. If you decide to study for several hours at once, take short breaks of five to ten minutes regularly as it helps to maintain a higher level of overall concentration.

Before you begin a detailed reading of the text, familiarise yourself with the general layout of the material. Have a look at the contents of the various chapters and flip through the pages to get a general impression of the way the subject is dealt with. Forget the old taboo of not writing in books. There is room for your comments, notes and answers; use it and make the book your own personal study record for future revision and reference.

At intervals you will find a summary and list of objectives. The summary will emphasise the important points covered by the material that you have read and the objectives will give you a check list of the things you should then be able to achieve. There are notes in the left hand margin, to help orientate you and emphasise new and important messages.

BIOTOL will be used by universities, polytechnics and colleges as well as industrial training organisations and professional bodies. The texts will form a basis for flexible courses of all types leading to certificates, diplomas and degrees often through credit accumulation and transfer arrangements. In future there will be additional resources available including videos and computer based training programmes.

Preface

Medicinal products are important in the prevention, diagnosis and treatment of diseases and considerable resources are invested in their invention and development. The advent of new biochemical techniques has greatly increased our understanding of the molecular processes taking place in living organisms and has also created the possibility for their manipulation. These developments have already led to the production of many novel medicinal products and many more are being developed. This text is designed to provide insight and information in how modern biotechnology may be applied to produce useful medicinal products by drawing on industry's experiences to date.

There are two major issues in the application of biotechnology to health care. The first is of course biotechnology itself. We shall examine the biotechnological strategies and techniques aimed at the research and development of medicines. There is however another important issue. When a new drug or vaccine is developed, a marketing authorisation or licence must be obtained by the manufacturer or its representatives. The manufacturer has to submit for this purpose a licencing file to the licencing authorities. The licencing file must describe the clinical application of a product and its method of manufacture. The file must also list all the control procedures and provide an evaluation of the chemical, pharmacological, toxicological and therapeutic properties of the product. Licencing files and inspection reports are the tools used by licencing authorities to evaluate a product and to decide whether a licence can be granted. In recent years, the size of licencing files has increased substantially in response to the changing concepts of safety, efficacy and quality.

It will be evident that in the development of new medicinal products both the process of development and licencing procedures play an important role, and therefore both aspects will be dealt with in this text. We outline the whole route from the initiation of the development of a new product to the granting of a marketing licence. The text has been divided into two parts; an introductory part and a series of case studies.

We start with a general introduction to modern biotechnology, summarising its historical development, the possible role of modern biotechnology in the health care business and its impact on society. This is followed by a chapter describing the stages involved in developing and producing a new medicinal product. The introductory part concludes with a chapter that describes the general elements that a licencing file of a biotechnological medicinal product should contain.

The introductory part is followed by a number of case studies. These case studies have been selected to provide the reader with a breadth of experience. The structure of each case study is based upon the sequences of activities undertaken to produce a new product. Each begins with the initial concept by examining the need for the product and leads, through the development work, into the areas of product evaluation by discussion of the pre-clinical and clinical trials. Despite this commonality of structure, a different emphasis is given to each study. In one emphasis may be placed on the choice of host system, in another the emphasis is in the gene manipulations or the pre-clinical trials. A particular importance has been attached to key decision points in the projects under study. In the development of a new product, there is often more than one possible choice of route in the design of the production strategy and in quality control. The reasons that led to the final decisions in the case studies are explained.

The case studies include therapeutics (insulin, erythropoietin and Orthoclone OKT3), vaccines (hepatitis B, Aujeszky's disease) and diagnostics (Myoscint, Aujesky's disease).

The concepts and issues discussed in this text are quite advanced. The emphasis is on the application of biotechnology, rather than on the principles of the methods of biotechnology. It is assumed that the reader has a basic understanding of restriction enzymes and genetic manipulation procedures and of the principles underpinning cell fusion (hybridoma) technology. Since many of the products discussed are derived from or have an effect on the immune system, a basic knowledge of the processes of immunity will be helpful. Self assessment questions within this text assume a working knowledge of these techniques. Specialised techniques, however, are explained in appropriate places. If the reader is unfamiliar with these areas, advantage should be taken of studying other texts in the BIOTOL series.

Scientific and Course Advisors: Dr M. C. E. van Dam-Mieras
                                    Dr C. K. Leach

1

Some perspectives in medical biotechnology

1.1. Introduction   2

1.2. Historical   2

1.3. The key new developments in biotechnology   4

1.4. Biotechnology now and in the future   5

1.5. Impact on society   7

1.6. Regulation and patenting   8

Some perspectives in medical biotechnology

1.1. Introduction

Our increasing knowledge of the molecular mechanisms that underlie the basic functions of living cells is stimulating novel approaches in many areas of medicine. Some of these have already had an impact on the medical - pharmaceutical industry, and there is potential for substantial growth in this section of biotechnology. The historical development of biotechnology is surveyed briefly in this chapter, with particular emphasis being given to recent changes that have seen new products becoming available for both the diagnosis and treatment of disease.

The design and implementation of new products dependent, for example, on genetic engineering involves considerations not only of economics but also of safety and regulations. These are presented as a necessary part of the process of innovation in medical - pharmaceutical biotechnology, together with broader ethical issues that involve society as a whole.

1.2. Historical

1.2.1. Pre-Pasteur

Ways of producing alcoholic beverages such as beer and wine from fermentation have been known for thousands of years and cheese and bread making also go back to antiquity. In this sense biotechnology is very old. However, it was only in the nineteenth century that the process of fermentation began to be understood in scientific terms. A decisive step was the demonstration by Louis Pasteur in 1863 that living microbes are the active agents of fermentation.

1.2.2. Traditional biotechnology to antibiotics

The research into fermentation carried out by Pasteur initiated the move from descriptive biology towards a real understanding in terms of chemical reactions. His discoveries led in the first instance to improvements in the production of wine and vinegar. Such processes gradually became more industrial in character, whilst the demands of the First World War led to the development of the fermentative production of butanol, acetone and glycerol. However, the technology was still largely based on empirical knowledge.

biochemistry   Steady progress in the discipline that had become known as biochemistry, and in microbiology, helped to change this situation. The nature of proteins, especially enzymes, was defined and the chemical interconversions that make up the metabolism of microbial, animal and plant cells unravelled. An important principle that emerged at a relatively early stage was that basic metabolism is similar in all types of organisms.

The antibiotic, penicillin, was discovered as a result of a chance observation by Alexander Fleming in 1928. However, the challenge to produce it on a commercial scale was only met in response to the need to combat bacterial infections on the battlefields of the Second World War. Many problems had to be overcome, from aseptic cultivation of the producing-organism, *Penicillium chrysogenum*, on a sufficient scale and under appropriate conditions to the isolation and purification of penicillin from the culture medium.

process
engineering

The production of penicillin remains one of the most technically-demanding processes, because of the large scale on which it is carried out. Process engineering that is concerned with obtaining a product from the primary fermentation is now identified as a discrete discipline, and experience gained in the manufacture of penicillin has been applied to many other systems, for example the aseptic cultivation of animal and plant cells. The pharmaceutical industry has learned how to produce not only antibiotics but also vitamins, viral vaccines and steroid hormones.

1.2.3. Industrial biotechnology

Increasingly detailed knowledge of microbial metabolism allowed the systematic exploitation of microorganisms to produce a variety of metabolites and enzymes. This may be exemplified by the use of mutants to increase yields of amino acids, organic acids and antibiotics.

immobilised
enzymes

Further developments included the mass production of enzymes for use in detergents and the large-scale enzymic transformation of glucose to fructose. The latter process employed a new technique, an immobilised enzyme, in which the biological catalyst is used to effect a chemical change in the same way as conventional catalysts in the chemical industry. The advent of these new processes, together with continual improvements being made in established areas of production, emphasises the multi-disciplinary nature of biotechnology. It requires co-operation and the development of understanding between people trained in biochemistry, microbiology and engineering. Indeed, the European Federation of Biotechnology has defined biotechnology as: 'the integrated use of biochemistry, microbiology and engineering sciences in order to achieve technological (industrial) application of the capabilities of microorganisms, cultured tissue cells and parts there of'. An alternative way of looking at this new technology is to visualise it as a challenge in which the production of chemicals takes place in a biological environment, the fragility of which has to be accommodated by process engineering.

The rapid growth of the oil-based industries in the 1960s led to the use of mineral oil fractions for the mass cultivation of microorganisms for use as animal feed, so-called single cell protein (SCP). Some public concern was raised about possible risks to health, but it was the sharp rise in the price of oil in 1974 that made SCP uneconomic. Nevertheless the technological advances accompanying the development of SCP processes have found applications in other areas.

In a more prosaic area of life, the study of natural, mixed anaerobic fermentations has revealed opportunities for dealing with domestic and industrial waste water. Costs can be defrayed by using the methane ('biogas') that is produced and there is also the advantage that anaerobic bioreactors are considerably smaller than the equivalent aerobic treatment plants.

1.3. The key new developments in biotechnology

Biotechnology, as so far described, exploited the chemical processes and products of a range of organisms discovered empirically. The challenge to biologists, as more and more metabolic processes and products were described and their interrelationships defined, was to understand how the synthesis of the myriads of enzymes involved in metabolism may be inhereted and regulated. This challenge has to a large extent been met. It is molecular genetics that is providing powerful, new tools for biotechnology, tools that are particularly important for application in the medical - pharmaceutical industry.

universal
genetic code

Key discoveries were the experiments of Avery and his co-workers in 1943 that suggested that DNA is the genetic material, the model for the structure of DNA proposed by Watson and Crick in 1953 and the elucidation of the genetic code, with its universal applicability in all types of organisms. It is this last discovery, the universality of the genetic code, that fortuitously makes possible sophisticated experiments to investigate the regulation of gene expression and genetic engineering to facilitate the production of particular proteins.

1.3.1. Genetic engineering

cloning

recombinant
DNA

The characterisation of restriction endonucleases, enzymes that cut DNA at specific sites, allowed the development, in the mid-1970s, of DNA cloning and recombinant DNA technology. DNA molecules from different sources can be cut and rejoined to give new combinations of genetic material; hence the term, genetic engineering.

In order to be expressed, the DNA must be introduced into an appropriate cell. This was first achieved with bacteria but it is now also possible with animal and plant cells. Both in principle, and increasingly in practice, the isolation of genes and their manipulation to permit insertion into other types of cells provide elegant strategies for directing and controlling the synthesis of many potentially valuable products. There are several examples of this in the case studies in this volume, namely the production of human insulin by bacteria, human erythropoietin by mouse cells in culture, a hepatitis B antigen by yeast and the deliberate engineering of deletion mutants of pseudorabies virus.

1.3.2. Hybridoma technology

hybridoma

This technique, also developed in the 1970s, makes use of the fact that it is possible to fuse together different types of animal cells. A hybridoma is created by fusing an established cancer cell line, which can be grown *in vitro* indefinitely, with cells of the immune system, which cannot be grown *in vitro*. From the process of fusion, cells can be selected which combine the ability to make a specific antibody with that of long-term growth in culture. In this way, unlimited quantities of any desired antibody may be produced. One type of cell only synthesises one type of antibody; hence the description, monoclonal antibodies. They can be selected for predetermined applications by appropriate screening strategies and monoclonal antibodies are now finding wide use as reagents in specific diagnostic tasks. In another application, a monoclonal antibody linked to an appropriate column matrix, is used to bind its corresponding antigen (say, for example, interferon), thereby effecting initial purification. One of the case studies chosen to illustrate the production and use of a monoclonal antibody is orthoclone OKT3. This is an antibody directed towards a T-cell antigen and it is used therapeutically to reduce rejection after organ transplantation.

1.3.3. Modern bio-process technology

bioreactors

Considerable advances have been made in the design of bioreactors. Increasing emphasis is laid on the development of mathematical models of the fermentation process, with the parameters known to be important in controlling metabolic pathways

computer controlled

used to calculate optimal conditions for the particular application. In conjunction with continuous monitoring, computer controlled bioreactors then become a realistic proposition.

Downstream processing, (the isolation and purification of the desired product), is still largely empirical in character, requiring specific, individual procedures. Nevertheless, biotechnologists now have a whole range of techniques available to them and new techniques are being added all the time. For example, a novel approach uses genetic engineering to construct a gene that codes not only for the protein that is required but also for an additional sequence that facilitates recovery by affinity chromatography. At a later stage the additional sequence is removed by a specific cleavage reaction.

1.4. Biotechnology now and in the future

This brief historical review has illustrated the diversity of biotechnology and the increasing scale of industrial developments. However, it is worth noting the following comparisons:

Finance: Beer and wine, the classical products of biotechnology are still responsible for 60% of the turnover of all biotechnological products.

Volume: Application of biotechnology to environmental issues such as water reclaimation represents by far the largest application of biotechnology processing.

Innovation: This is most prominent in the application of biotechnology to medicine.

innovation

transgenic organisms

It is, of course, the excitement of innovation derived from the sophistication of modern biological sciences that makes the future of biotechnology so promising. It includes the introduction of new genes into plants and animals to create transgenic organisms. This is often discussed in the context of 'improving' agricultural yields, resistance to disease and so on, but very recently it has been shown that the genes for immunoglobulins can be introduced into and expressed in plants, thereby creating a new vehicle for the production of these valuable proteins. New, and often unexpected, ways of doing things will continue to be the hallmark of biotechnology, emphasising the importance of a broad training in basic sciences for anyone who wishes to pursue a career in biotechnology. The same point will become clear from the case studies in the second half of this volume.

1.4.1. The commercial promise of biotechnology in the production of therapeutics

There will undoubtedly be many applications of biotechnology to health care. These applications will be in the areas of diagnostics (monoclonal antibodies, DNA probes etc), in vaccines and in therapeutics.

Considering therapeutics alone, there is an enormous potential. Like any projections into the future, exact figures are difficult to predict, but they will certainly be in terms of thousands of millions of dollars per annum. Although the relatively long time scales needed to develop and test drugs means that the products derived from the new biotechnology did not reach the market until the mid 1980s, each year an increasing number of such projects and products reaches fruition. Table 1.1 lists a variety of products which have either reached the market or are under active development. An indication of the size of the market for each is also given. These figures are for the North America market alone, world figures may be at least three to four times those quoted.

Compound	Use	Market Size[1] ($ million per annum)
Atrial natriuretic factor	Diuretic	60-100
Epidermal growth factor	Wound recovery	160
Erythropoietin	Anaemias (kidney failure) blood enrichment	300
Factor VIII	Haemophilia	200
Follicle stimulatory hormone	Infertility	180
Human growth factor	Stature correction	250
Interferon α	Cancer	70
Interferon β	Cancer/infections	20
Interferon γ	Cancer/arthritis	50
Interleukin-2	Cancer//infection	300-500
Ripocortin	Anti-inflamatory agent	50
Monoclonal Antibodies (2)	Cancer therapy/infection Prophylaxis	1000
Tissue plasminogen activator	Degradation of blood clots	400-800

Table 1.1 Therapeutics derived from contemporary biotechnology
(1) Calculated from a combination of surveys for the American market at 1990 prices
(2) The figures quoted do not include the use of Monoclonal antibodies (MoAb) in research or as diagnostics.

The list provided is not complete and undoubtedly human ingenuity will greatly expand the variety of compounds that will find therapeutic application. If we add to these, the use of the new technology to produce safe vaccines and the development of diagnostic tools, it should be self evident that the contribution of biotechnology to the health care industry will be enormous.

1.4.2. Quality control in the new biotechnology industries

quality control

This is the major topic of chapter 3. Here we point out that quality control for the new generation of biotechnology products emphasises:

- validation of a process by 'spiking' studies to prove that 'unknown' contaminants can be eliminated effectively during purification;

- tumorigenicity studies on cell lines used in production;

- characterisation of master cell banks and the manufacturer's cell banks;

- pre-clinical safety tests have to be devised.

This is still being done on a case-by-case basis, because of the novel aspects of so many therapeutic agents obtained, say, by genetic engineering. The latest 'Notes to Applicants' of the European Commission recognises four groups of products: hormones, cytokines or other regulatory factors, blood products, monoclonal antibodies and vaccines. On biochemical criteria, each group is divided into three categories: proteins identical to-, closely related to- and distantly/unrelated to human proteins. How these work in practice is illustrated in the case studies.

Good Manufacturing Practice

Good Manufacturing Practice (GMP), as far as new biotechnology is concerned, draws heavily on traditional processes, in particular the production of viral vaccines. The prevention of microbial cross contamination is emphasised and separate production facilities for different steps in the overall process are generally required. Any substance that is used by injection must be free of pyrogens (endotoxins), necessitating aseptic techniques throughout with clean air and microbiological control of all materials and equipment. When genetically manipulated microorganisms or cell lines are involved, containment facilities appropriate for the protection of personnel and/or the environment must be employed.

1.5. Impact on society

All radically new technologies have dramatic effects on society and economists have attempted to analyse what happens as the dominance of one technology is replaced by another. Projecting this analysis into the future is not easy but a group within the European Commission postulated in 1980 that the pervasive effects of biotechnology in the broadest sense of the term could give rise to a 'biological society' in the 21st century. By this is meant, a society based on the conscious management of self-organising systems for the sustenance and enrichment of human life and purposes'.

biological society

economic criteria

This philosophical view has within it implicit assumptions concerning the economics of new biotechnological process, as well as those of the more traditional variety. To what extent will they be profitable? What effects will they have on employment? Basic questions like these are easily formulated but less easily answered; it is important, therefore, to appreciate that a good innovative idea has to meet stringent economic criteria if it is to give rise to a marketable product. This should be borne in mind in looking at the case studies.

ethics

Finally, some aspects of biotechnology raise ethical issues, many of which have yet to be resolved. They range from the basic question as to whether it is right for us to manipulate creation in any way we choose to specific examples involving the construction, use and release of genetically engineered plants and animals. When it comes to human genetics (diagnosis of genetic diseases, gene therapy) there is overlap with other concerns such as *in vitro* fertilisation and abortion that are not directly related to biotechnology. Over the next few years, philosophers, theologians and biologists will need to collaborate in discussion of the ethical questions raised by the radical, biological techniques that are now possible to ensure that their application in research, biotechnology and medical practice is properly understood and regulated. We can, however, be reassured that existing and evolving regulations are formulated to protect mankind, animals and the environment even if certain ethical matters have not been resolved to the satisfaction of all actions of society. These matters are discussed more fully in chapter 3.

1.6. Regulation and patenting

The regulation of any biotechnological process is concerned in the first instance with Quality Control (QC), Good Manufacturing Practice (GMP) and Good Laboratory Practice (GLP) during production. Secondly, any product designed for clinical use has to be evaluated rigorously in pre-clinical and clinical trails. Chapter 3 describes the sequence of measures that need to be undertaken in order for a product to be given market authorisation. A key issue is quality assurance and control.

The protection of products by patents is of critical importance for high-technology industries to enable development costs, which can be as high as $100 - $150 million for a new drug, to be recouped. Biotechnology, when it involves 'new' or 'modified' organisms, whether microbial, plant or animal in origin, has created peculiar problems as far as patenting is concerned and different countries still operate their own rules, to which detailed reference must be made in each case.

2

Development stages for health care products

2.1. The stages   10

2.2. Scientific, technical, legal, economic and social influences on the
       development of a new product.   10

2.3. Stage 1: Concept evaluation   12

2.4. Stage II and III: Research and development, scale up and product
       recovery   18

2.5. Pre-clinical and clinical studies   22

2.6. Stage IV: Product formulation   23

2.7. Stage V: Production   24

2.8. Stage VI: Marketing and product diversification   25

Summary and Objectives   27

Development stages for health care products

Life is very precious. The markets for products which preserve and improve life and health are consequently, potentially large and profitable. The development of new health care products is not, however, a simple matter. Considerable resources are needed before a product can be placed on the market. The expenditure is required to establish the efficacy and safety of the product as well as developing the production process and subsequent marketing and distribution of the product.

The aim of this chapter is to provide the reader with an understanding of the sequence of activities to bring a product from its initial conception to the market place. It will provide a model upon which subsequent chapters will build. It is not the intention to provide details of specific procedures, but to develop an overview of the key decision points in the research, development, production and evaluation of products derived from biotechnology. The case studies will provide specific examples of issues and procedures involved.

2.1. The stages

The stages through which the development and production of the new medicines generated by modern biotechnological procedures are not dissimilar to those of more traditional technologies. Like conventional products, the sequence of stages in the development of such new products is not a simple linear one. Although an early decision has to be made whether or not to proceed with a particular project, the results of subsequent stages may cause a re-appraisal of earlier decisions. Thus, although a more-or-less linear model will be described here, it must be recognised that in the real world there is considerable overlap of these stages and that the individual stages should be regarded as merely components of an integrated whole.

Figure 2.1. provides a general overview of the main stages in the development of a new product.

∏   Examine Figure 2.1 carefully and see if you can explain why we have displayed quality control as a long box running in parallel with the main stages in the development of the product.

2.2. Scientific, technical, legal, economic and social influences on the development of a new product.

economic factors

The successful application of biotechnology to the production of marketable medical products depends not only on the scientific and technical issues of research and development and production but also upon economic, regulatory and social factors. Thus, in discussing the stages of development, production and marketing of new products, we must keep in mind that we are not solely concerned with the scientific and technical issues. At each stage, economic factors are of great importance and the need to consider if the chosen strategy is the most effective solution is extremely important.

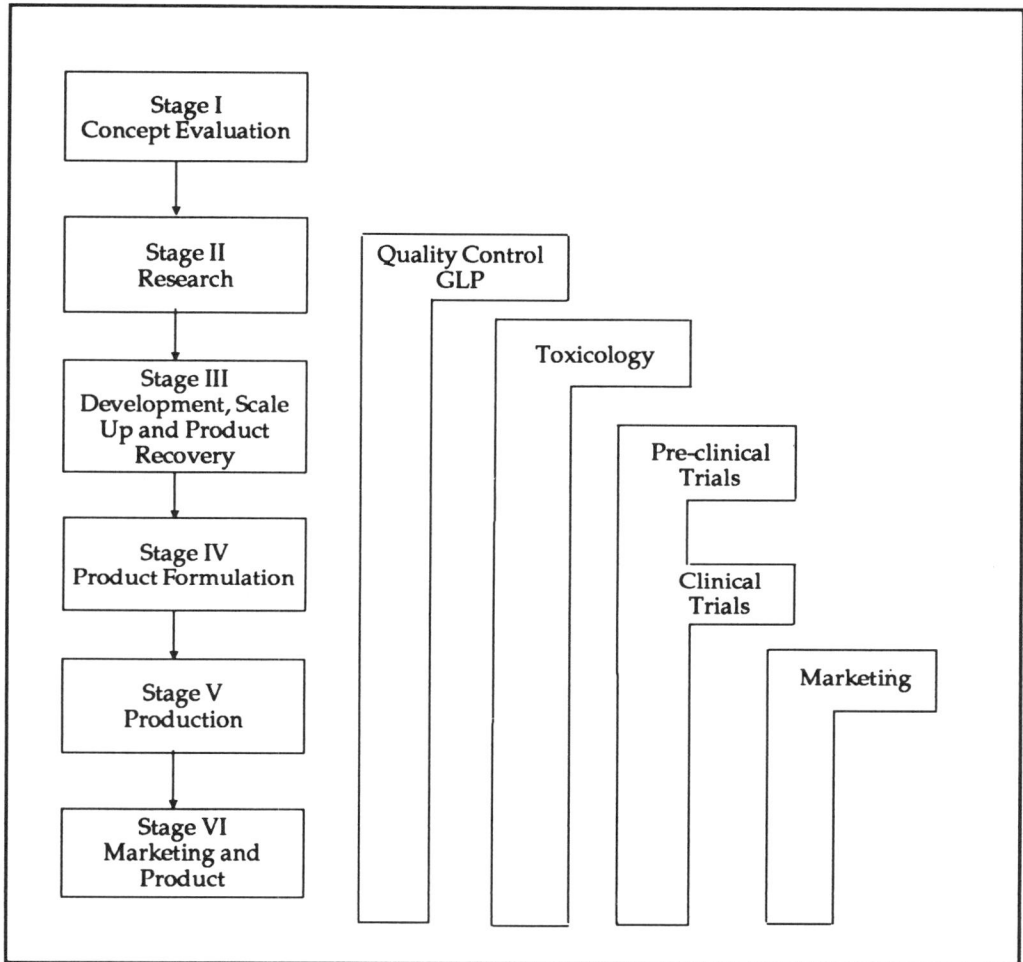

Figure 2.1. The main stages in the development of a new product

This may mean that the cheapest route may not be the best if it does not lead to a product of sufficient quality.

We would anticipate that much of the research and development activities will be directed towards optimising the production process. Nevertheless, this is not the only issue of the research and development phase. All medicines need to meet stringent regulatory criteria before they can be marketed. These regulatory issues focus on the need for the manufacturer to prove that the product is safe and effective. Thus much of the investment in research and development will involve chemical, pharmacological, toxicological tests that are needed to fulfil these regulatory requirements. Such tests are essential to demonstrate the safety of the product before human usage. Clinical trials are also essential to demonstrate the efficacy of the product. Thus the research programme must incorporate all the tests that are appropriate to the product and its usage. It is clearly irresponsible and unethical not to do so. The data gathered from these studies must be submitted to the regulatory authorities. It is important to realise that, when conducting safety and efficacy testing, the conditions under which the tests are

Regulatory requirements

safety and efficacy

undertaken must also conform to the regulations. By conforming to the most stringent regulations, a manufacturer can enhance the prospect of worldwide marketing of a product.

environmental issues
Regulatory requirements are not, however, solely focused on prospective customers. With the introduction of genetically engineered organisms, concern is expressed over the problems that may arise through the deliberate or unintentional releases of such organisms. This concern is built into the regulations which govern the use of genetically modified systems. The application of genetic engineering to the development and production of new products means that these must be conducted within the regulatory requirements. Regulatory issues are therefore of major significance in the development of new products. Chapter 3 considers the regulatory requirements in more detail. The case studies provides numerous examples of the tests that need to be undertaken to fulfil the regulatory criteria.

ethical issues
Some aspects of biotechnology raise ethical issues, many of which have yet to be resolved. They range from the basic question to whether it is right for us to manipulate living systems in any way we choose, to the use of laboratory animals in the pharmacological and toxicological evaluation of new products. These issues must remain a conscious component of the decision making process for they may well influence the acceptance or non-acceptance of the product. Such issues will, undoubtedly, come under increasing scrutiny and may well become subjected to further regulatory control. Thus, although the case studies will predominantly consider the technical aspects of applying biotechnology to the production of new medical products and with the regulatory criteria that must be responded to, the reader must recognise that the environment in which these activities operate may change quite significantly. Not only are changes anticipated in the development of techniques, but the legal and economic climate may also rapidly change.

With these considerations in mind, we will now examine the stages in the development of new products in a little more detail.

2.3. Stage 1: Concept evaluation

need and improvement
The development of a new product by the application of biotechnology stems mainly from two basic ideas:

- to respond to a medical situation for which diagnosis or treatment is required and is currently not adequately met by existing products;

- to manufacture an existing product more efficiently or economically.

The first stage in the development of the new process therefore involves the union of two major components namely the recognition of the potential of applying biotechnology to fulfil a need. The initial idea may come from the business sector or from a technical source, but, whatever its origin, both sectors have an important role to play in its evaluation. Here, we divide the discussion of the development and evaluation of the concept into two parts, the role of the biotechnologist and the role of the business manager. Since this text is primarily concerned with biotechnology, emphasis is placed on the technical aspect of the evaluation of a project. It is important however that those involved in a technical capacity are aware of the major issues which confront the business manager. Since it is fruitless to carry out a business evaluation unless a reasonably sound technical strategy can be developed, we will deal with the role of the biotechnologist first.

The biotechnologist's role

sequenced plan The key responsibility of the biotechnologist is to establish a scientific/technical rationale to develop the product. With this in mind, it is necessary to develop a sequenced plan of how to move from concept to the market place. This does not mean that all the answers need to be known at this stage but it does mean that most of the major unanswered questions need to be identified and plans laid for how answers may be achieved.

Π Before reading on examine Figure 2.1. again and see if you can write down a list of 5 or 6 major questions that the biotechnologist should address in order to develop a scientific rationale for the project.

The questions that you have written may be quite varied but what we hope your response would reflect the need to develop a sequenced plan. Let us see if we can give you help in developing such a plan. It is important to try to divide the project up into a number of stages.

The initial plan would invariably include strategies for obtaining suitable source materials. In biotechnology this predominantly means producing a cellular system capable of making the product. The kind of questions that require answers are:

* does a suitable cell system exist? If not can one be generated by genetic engineering, hybridoma technology or by some other means? There are of course, a lot of subsidiary questions. For example, if a suitable cell system exists, is it protected by patents? If genetic engineering is needed, are suitable vectors available, and is there a screening procedure which allows identification and isolation of the desired cell line also available. All of the case studies described in the second part of this text include discussions concerned with producing cell lines with the required characteristics. These case studies will therefore provide you with experience in evaluating alternative strategies for producing desirable cell lines so we will not dwell on specific cases here. You should however recognise that at an early stage in the planning of a project, it is essential to have a clear plan of how to obtain a suitable source materials/cell lines. Without such a plan, the project is almost certainly doomed to failure. Also important in this plan is an estimate of how long it will take to produce suitable cell lines.

* will such a system be stable and what strategies can be applied to maximise product yield? Once a route to obtaining suitable source materials/cells has been mapped out, the next set of questions relates to product yield and the need to produce a reliable manufacturing process. These questions are vital because they in turn relate to fulfilling market needs and to the economics of the process. It is of no value to develop a process which makes grossly insufficient quantities of products at an uneconomic price. These types of questions mark the transition between laboratory scale operation to the pilot plant/manufacture scale. We can ask such questions as, can the system be readily scaled up and are techniques available for the purification of the product? Important in this context is the need to obtain a product of sufficient purity to satisfy regulatory criteria.

* can the product be tested for efficacy and safety satisfactorily? This might seem a little surprising. Many products of biotechnology pose quite new problems.

Consider a proposal which involves production of a human protein using a yeast culture. The human protein may only be effective in humans and have different or modified, activities in animals. Almost certainly such a product would be immunogenic if administered to animals and cause an immunological (probably deleterious) effect. How is such a product to be tested if animal tests are unsatisfactory? We will meet with some excellent examples of this problem amongst the case studies, especially in the case study on erythropoietin. Although there are firm criteria for developing and marketing new medical products, many of the new products of biotechnology are being treated by the regulatory authorities on a case-by-case manner. Nevertheless the onus is on the manufacturer to devise and conduct appropriate tests to prove the efficacy and safety of the product and its applications. Thus even at an early stage in the development of a product, we need to consider such issues.

The key points we are trying to make are that before a project is undertaken it is necessary to develop a sequenced plan of how the product may be made and evaluated. This includes generating suitable source materials identifying a strategy for scale up and product recovery and devising suitable product evaluation to satisfy regulatory requirements.

The skill of the biotechnologist is not only in helping to identify such a pathway but also helping to make judgements concerning the likely outcome. The final result of this technical evaluation of a proposed project is not entirely in the hands of those conducting the technical evaluation. It is the business manager who claims the responsibility of using this technical input, together with other evaluations (eg economics) who will ultimately carry the responsibility of whether to proceed or not. Those who aspire to operate as biotechnologists in the medical and pharmaceutical industries will invariably have to interact with a variety of other technical staff (toxicologists, pharmacologists, clinicians) and with a business management team. It is important for the biotechnologist to recognise and respond to the concerns of the business section of his organisation. In the next section we turn our attention to their concerns.

2.3.2. The role of the business manager

cost v. gain   The business manager will attempt to balance two major factors, the costs of development and manufacture of the product against the probable commercial return. In reaching a decision on these issues, the manager will require a considerable amount of information. He will look to his technical staff for advice on details of the technical programme. The sorts of questions he will ask are:

* how long will the research and development work take?

* what equipment (research, pilot, production) will be required?

* how many staff will be needed for each stage?

* what technical issues need to be resolved and what is the likelihood that they can be resolved?

* what raw materials will be required?

* what are the costs of product evaluation?

In order to supply answers (or evaluating best estimates) to these, it is essential that the technical staff have developed a coherent plan for the whole process.

The questions raised above mainly represent the costs of development and manufacture. There are of course, other costs such as staff recruitment/retraining, taxation marketing and so on. The business manager will also have to consider other issues on the 'cost' side of the equation. For example, does the project offer reasonable security of employment to the workforce.

All of these factors need to be balanced by the commercial return of the project. The most obvious factors influencing this return are:

- the size of the potential market;

- the presence of rival products in the market;

- the value of the product;

- the accessibility of the market.

Although we do not intend to examine the business evaluation in detail, it is perhaps worthwhile to consider the factors above in a little more depth.

scale

We began section 2.3. by explaining that the first stage of development of a specific new product involves the interaction between recognising a need (i.e. identification of a market) and the possibility of applying biotechnology to fulfil this need. In the business area under consideration here, the need may be for a new or improved approach to disease diagnosis, prevention or treatment. For most organisations however, it is insufficient just to recognise a need for although this may offer commercial opportunity, what is critical is the scale of the need since this will govern the market size and therefore the profitability of the project. Clearly for a commercial organisation to invest millions of $ (or their equivalent) into the development of a product for which there are only a few potential customers, is a non-viable proposition. For non-commercial organisations, the situation is less clear cut.

market accessibility

barriers

In addition to identifying the size of the potential market, organisations will also need to consider where the market exists, whether it has access to this market and if that market can afford the product. We could, for example, envisage a disease condition in a region which was too poor to finance the treatment of the disease or in a country which was not accessible to the company because of trade or political barriers or a region in which the product developer has no distribution or marketing organisations. Such barriers may be surmountable by, for example, using other organisations to distribute the product or by changing the proposed site of production, but these may not always provide satisfactory alternatives. The use of other companies to market and distribute the product may reduce the control the producer has over the marketing of the product unless satisfactory contractual guarantees can be agreed. Moving production also has its limitations in terms of the cost of plant and raises questions about the availability of a workforce sufficiently skilled to operate the process. Such considerations are of great importance in reaching a decision as to whether or not to proceed with an innovative idea. Thus a perfectly sound technical idea might well be shelved for commercial reasons.

Also built into the equation of whether or not to proceed with a project is a consideration of possible rival products. The existence or likely existence of rival products may substantially influence the likely market penetration of the proposed product. The existence of a rival product, of course, raises a whole subset of questions relating to a comparison of quality (efficacy, side-effects, cost) and of the rivals marketing expertise.

high return
high risk

It is obvious that the development of a new medicine is an expensive business, involving considerable outlay before the product reaches the market. Indeed, even after this considerable outlay there is no guarantee of success. Innumerable products fail to meet regulatory requirements or do not have the anticipated therapeutic effects. Thus the business manager must also include the risk of failure into his equation. High risk projects are usually linked with high profits. This is of course understandable; it is not good business sense to take up a high risk option if the likely return is rather moderate.

Figure 2.2 provides an overview of this 'concept' evaluation stage. It must be remembered however that this is not a single activity. Each aspect will need to be kept under review as the project develops.

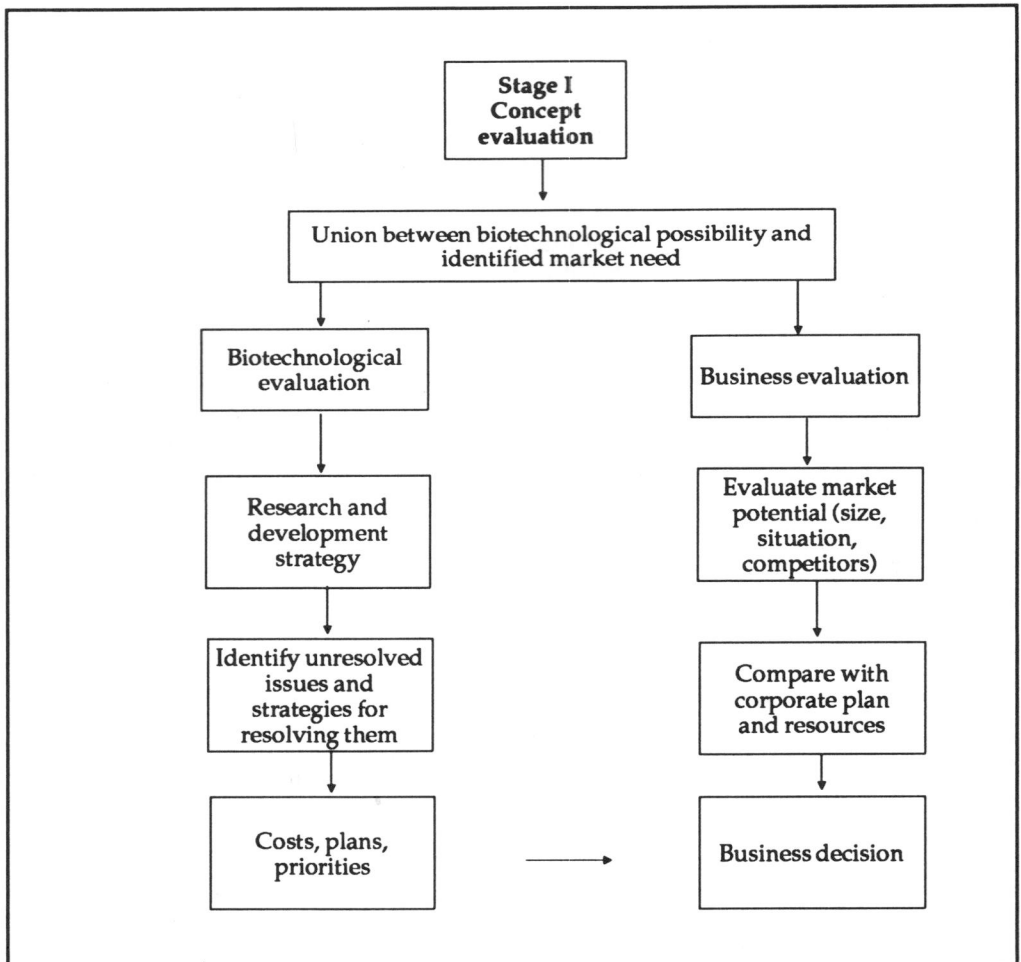

Figure 2.2. A representation drawing of the major factors involved in evaluating a project

We may summarise what we have learnt in this section by saying that the original concept for the type of product under consideration comes from recognising that modern biotechnology may provide novel solutions to medical needs. We have also learnt that an over-all technical strategy plan needs to be prepared and that the project must receive some detailed business evaluation.

The remaining stages of the project development are concerned with implementing the strategic planning. We re-emphasise that, in the light of technical data produced during the research phase or as a result of changes to the business and legal environment, the project has to be kept under constant review and must, if necessary, be adjusted accordingly.

SAQ 2.1.

Consider the following hypothetical circumstance.

WBC is a 'flu-like' illness thought to be caused by a DNA virus found only amongst a small Amazonian tribe in South America. Its symptoms are that it causes a slight fever, headache and, in severe cases, nausea and stomach cramp. Symptoms last for 2-12 days. The disease is reputedly spread by acts of cannibalism.

You are a member of a small team of biotechnologists, working for a small company based in Scotland, developing large scale fermentation processes, would you advise your company to become involved in producing a vaccine for this infection using genetically engineered bacteria to produce the coat protein? Yes or No?

Use an appropriate selection of the reasons given below to justify your answer.

1.   It is morally right to attempt to cure any disease.

2.   Although the disease is confined to a single tribe and therefore the market is small, this market is unlikely to attract competitors.

3.   The condition is likely to die out, as outside influences reduces the incidence of cannibalism, therefore it is pointless to develop a preventative reagent.

4.   There is little known about the disease, therefore a lot of costly work will need to be done before the possibility of raising a vaccine by genetically engineering bacteria can be realized.

5.   So little is known about the disease that competitors are unlikely to think up a way of producing a vaccine.

6.   The production of specific proteins by genetically engineered bacteria is at the very centre of the company's business. Therefore the development of this new project presents few difficulties in terms of equipment and staff skills.

7.   The market is in a remote place and unlikely to be subject to strict control or inspection. Therefore the vaccine could be produced and used without too much costly quality control providing care was taken to avoid inspection.

8.   The market for the vaccine is in such a remote place that it would be difficult for a rival to trial and distribute the product. Competitors would, therefore, be unlikely to enter this market, leaving it to your organisation.

SAQ 2.2.

In SAQ 2.1., it was suggested that a genetically engineered bacteria may be used to produce the rival coat protein. Below is a scheme to produce such an engineered bacteria. From the information supplied in SAQ 2.1., identify questions that need to be resolved (write a list of at least 10).

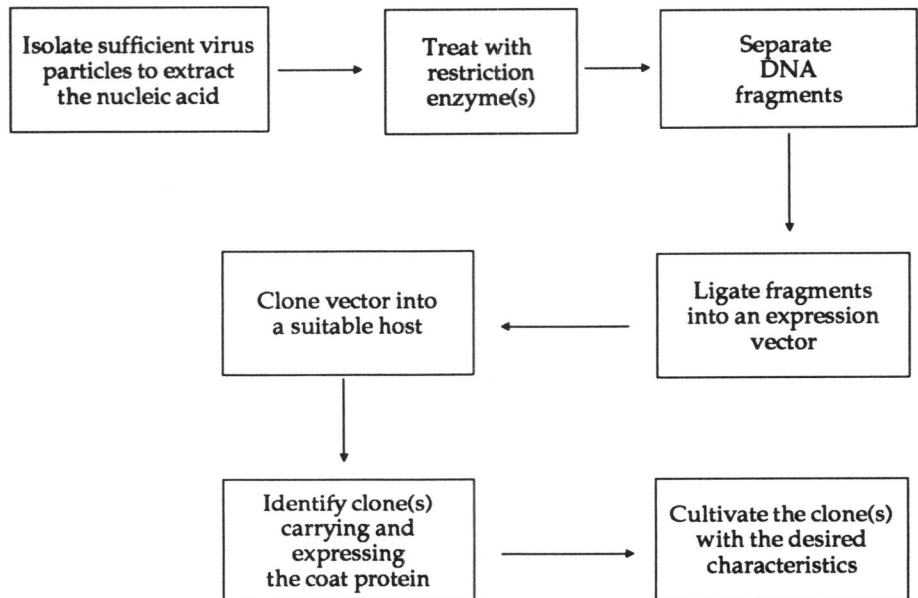

```
┌─────────────────────┐     ┌──────────────┐     ┌──────────────┐
│ Isolate sufficient  │     │ Treat with   │     │ Separate     │
│ virus particles to  │ ──▶ │ restriction  │ ──▶ │ DNA          │
│ extract the nucleic │     │ enzyme(s)    │     │ fragments    │
│ acid                │     │              │     │              │
└─────────────────────┘     └──────────────┘     └──────────────┘
                                                          │
                                                          ▼
┌──────────────────┐                          ┌──────────────────────┐
│ Clone vector into│                          │ Ligate fragments     │
│ a suitable host  │ ◀──────────────────────  │ into an expression   │
│                  │                          │ vector               │
└──────────────────┘                          └──────────────────────┘
        │
        ▼
┌──────────────────┐                          ┌──────────────────────┐
│ Identify clone(s)│                          │ Cultivate the clone(s)│
│ carrying and     │ ───────────────────────▶ │ with the desired     │
│ expressing       │                          │ characteristics      │
│ the coat protein │                          │                      │
└──────────────────┘                          └──────────────────────┘
```

2.4. Stage II and III: Research and development, scale up and product recovery

Stage I and II are, in many ways perhaps the most exciting phases for the biotechnologist since it is these stages which are concerned with implementing the research strategy formulated in Stage I. Of primary concern will be the development of the system to produce the product which might involve such activities as genetic manipulation using vector systems, production of hybrid cells using hybridoma technology, characterisation of the product and maximisation of product yield.

range of
objectives

This stage is also concerned with checking that the properties forecasted for the system are holding time. It should be anticipated that this stage involves some refinement of the original sequenced plan. It may be that unexpected difficulties are encountered or new opportunities present themselves. Work on the problem associated with scale up and product purification will also be encountered. This phase also marks the beginning of the experimental backup needed to support the validation of the process as demanded by the regulations (we will examine this in more detail in the next chapter).

We can therefore identify a range of objectives that need to be addressed. These include:

- development of suitable cell lines;

- evaluation of the performance of the cell lines in small and large scale cultures;

- the optimisation of yield;

- the development of appropriate downstream processing to achieve desired product purity;

- evaluation of the reliability of the process;

- validation of the process, to demonstrate that it works in a predictable manner and that the product is of the claimed composition and purity;

- development of specifications for the source materials, intermediates and products;

- development of quality control procedures to check that specification targets will be reached;

- production of sufficient quantities of the product firstly to undertake toxicological and pharmacological evaluation and, subsequently to begin pre-clinical testing.

In addition to these technical objectives, some effort will need to be directed towards ensuring that appropriate expertise is available not only for the research and development phase but also for the subsequent phases.

Let us examine, in turn, the objectives listed above.

selection of host system

The development of suitable cell lines is an early objective. In projects involving genetic manipulation it is important to select the best host system and to maximise the product yield. The following criteria might be applied to the selection of a suitable host:

- which of the possible host systems can be readily cultivated in bulk and what are the relative costs of cultivation of the hosts?

- do the various hosts present any particular risks to health to those working with them? Are they, for example, pathogens or might they carry oncogenes or viruses?

- which of the available hosts present least risk if they are unintentionally released into the environment?

- which of the available hosts are best understood in terms of the regulation of gene expression?

- for which of the hosts are a range of suitable gene vectors available?

- for which of the hosts can desired types be most easily cloned?

- which of the available host systems offers the greatest opportunity for protection of invention by patenting?

SAQ 2.3.

You plan to produce a mammalian peptide using genetically manipulated bacteria. Your initial plan was to introduce a gene coding for the mammalian peptide hormone into an attenuated strain of the bacterium *Escherichia coli* using a vector plasmid. Subsequent research shows that the host cell system is unsatisfactory because a variety of enzymes produced by the *E. coli*, degrade the hormone before its release. You decide to use an alternative host system. Use the criteria listed above to help you to select the most appropriate system from the following list.

1. A wild type strain of *Escherichia coli* (a bacterium)

2. A wild type strain of *Corynebacterium diphtheriae* (a bacterium)

3. *Saccharomyces cerevisiae* (a yeast)

4. *Clostridium sporogenes* (an anaerobic bacterium)

5. *Chlorella sp* (an alga)

6. A transformed mammalian cell line

7. A wild type strain of *Enterobacter aerogenes* (a gut bacterium)

In some of the case studies, we will address the question as to why a particular host systems used was selected. Why for example was *Escherichia coli* used to produce human insulin, while Chinese hamster ovary cells were used to produce erythropoietin? Why did *Saccharomyces* prove to be a better host system for producing hepatitis B vaccine than *Escherichia coli*?

maximising gene expression

The optimisation of yield is of course vitally important. This may involve the manipulation of culture media. Historically, penicillin production was improved several hundred fold mainly by modifications to the culture media. Increasingly however attention is focusing on genetic recombination as a device to improve yield. This may involve either generating multiple copies of the desired gene in each host cell or by selecting powerful promoters to increase the expression of the desired gene. Again we will meet examples of these in the case studies. For example, is the β galactosidase promoter of *E. coli* more or less productive than the tryptophan E (Tryp E) promoter? We will show you how to determine this in the insulin case study.

cost vs efficacy

Downstream processing needs to be cost-effective. The selection of unit operations is made partially on cost, and partially on efficacy. It is of no value to have a cheap process that does not achieve the desired purity. The case studies examine some aspects of downstream processing. Again we cite the insulin case study. You will learn of two ways of using *E. coli* to produce insulin. In one the two peptide chains of insulin are produced separately, in the other they are produced together. These alternatives pose different downstream problems in producing insulin of sufficient purity.

The remaining issues of the research and development phase mainly relate to fulfilling regulatory criteria. The regulations are, of course, predominantly concerned with the safety and efficacy of a product. They have been developed largely upon scientific evidence and should be regarded as a kind of audit of the safety and efficacy studies that needs to be done to protect the public and the environment. They are not an end in themselves and they should not be regarded as an imposition.

The studies conducted to achieve these objectives will become part of the dossier of information required to gain authorisation to market a new product. Chapter 3 is devoted to this aspect of product development but it is worthwhile here to point out that the research and development phase includes carrying out those studies required by the regulations in the manner described within the regulations. This does not mean that the manufacturer has no options. We will learn for example that insulin can be measured by bioassay or by an HPLC technique. New or alternative techniques must, of course, be validated. Important also is establishing the quality control procedures to be used in the production process.

We have included both research and development in this section because these two processes usually take place in tandem. The data derived from research is used directly in the development of the pilot (and ultimately the production) scale operation. Likewise many of the problems that may be encountered in the scale up may require further research to be undertaken.

The activities of the research and development phases therefore provide the data needed to make decisions concerning the production facilities, methods of operation and quality control procedures. We will illustrate this by conducting two short exercises. For this you will need to assume that you are developing a new medical product using genetically engineered bacteria.

∏ From your current knowledge, write down some possible ways that stocks of cell lines to be used in the process can be maintained. Then select what you think would be the best and give reasons for your selection.

cell line
maintenance

There are of course a number of possibilities such as freezing aliquots of master cell lines and using working stocks derived from them (a seed lot system); lyophilising (freeze drying) aliquots of master cell lines; maintaining master cell lines by continuous subculture. Which of these is the best will very much depend on the cell lines under development - only research will prove what is the best method.

The choice will be governed in part by cost, but mainly by the need to ensure cell line stability. The manufacturer does not want to lose valuable stocks which have cost considerable amounts to develop and will need to ensure product yield is predictable, if proper economic management of the process is to be maintained. Furthermore, it is a condition of the regulations governing the authorisation to market medical products derived by these processes that the quality of the product is guaranteed. Thus, along with developing a strategy for maintaining the cell lines, a set of criteria needs to be established to satisfy the requirement to demonstrate cell line stability. The choice of tests is partially governed by the nature of the cell lines used and partially by the need to fulfil regulatory obligations. Such tests may include nucleotide sequencing, demonstration of the absences of viruses, or contaminating nucleic acids or polypeptides. The regulatory requirements are discussed in more detail in the next chapter. During the exploration of the case studies make sure that you are able to identify the quality assurance measures offered with each product in terms of validation of the cell-lines used.

∏ Write down a list of items that need to be considered before selecting a bioreactor.

We would anticipate that you will have found several items including elements relating to the mode of operation, the size of the reactor and its construction. The selection of bioreactor depends upon many factors. We can describe these factors in the form of a series of questions that require answers:

- How big does the reactor need to be to produce the desired output?

- It is to be a batch or continuous cultivation system?

- Is it a single stage or multiple stage system?

- What parameters (eg O_2 pH etc) need monitoring and controlling?

- What form of agitation will be used?

- What are the prospective working lives of vessels made from different materials and how does this relate to cost?

Answers to these questions are based on the findings of research. They are however greatly influenced by the choice of host systems used for example, animal cell lines often are more susceptible to sheer forces than bacteria (bacteria are much smaller and coated by thick cell walls). Therefore in cultivating animal cells, considerably more attention has to be paid to the generation of stress forces in the vessel. Apart from satisfying the requirements of the cell lines in use, the choice of production system will also be governed by cost. Again we can illustrate this by an example. Monoclonal antibodies may be produced by hybridoma cells (animal cell lines derived by the fusion of two cells each carrying particular desired phenotypic characteristics). Such cells may be cultivated under aseptic conditions in a large volume bioreactor or by introducing them into the peritoneal cavity of an animal where they develop into ascites tumours. In terms of cost per unit of antibody produced, the ascites tumour route is by far the more efficient. The ascites tumour route, however, does raise ethical and product quality assurance issues not encountered with conventional bioreactors. Look for the reasoning behind the production system of choice in the case studies relating to myoscint and OKT3.

Now let us turn our attention to another aspect of the development of a product by asking at what stage should the pre-clinical and clinical studies be undertaken? We will give our thoughts on this in the next section.

2.5. Pre-clinical and clinical studies

Although we have separated pre-clinical and clinical studies from the research and development stages, the reader must not infer that these studies are tacked onto the end of the process. They are an integral part of the development of the process and product. It is important to the commercial producer to gain an indication of the likely pharmacological effects and the toxicological consequences of the proposed product as early as possible. Contra-indications at an early stage may save considerable resources that might otherwise be wasted. Delaying such studies, for example, until a large production plant had been built, is rather akin to playing commercial Russian roulette! If the product turns out to have unexpected toxic properties or proves to be ineffective, the high capital investment made may be wasted. Thus the pre-clinical testing should be carried out as early as possible. This usually depends on sufficient material being produced on a pilot plant scale. Furthermore, it makes commercial sense to carry out clinical trials before making a final commitment to the capital investment needed to finance the production plant. Look back to the flow diagram in Figure 2.1. you will see that we have represented the pre-clinical and clinical testing of the product as running in parallel with the development of the manufacturing process itself.

The pre-clinical and clinical studies should be regarded as part of the Research and Development effort.

pre-clinical pharmaco- kinetic and pharmaco- dynamic studies

The pre-clinical studies involve the examination of the pharmacological activities of the product. Important in this study is consideration of the rate of uptake and degradation/excretion of the product (ie pharmacodynamic and pharmacokinetic studies) since this will influence both the pharmacological activity of the product and its possible toxicological consequences. The pre-clinical studies also involve a toxicological evaluation of the product. Where possible, the activities and properties of the new product should be compared with existing products or with naturally occurring material (we will examine many examples of this in the case studies). Providing the product proves satisfactory at this stage, it may undergo clinical evaluation. Successful clinical trials may result in the acquisition of a licence enabling the product to go to the market.

These studies are also important in establishing the final formulation of the product.

2.6. Stage IV: Product formulation

We described product formulation as stage IV in the flow diagram of Figure 2.1. Developing product formulation is concerned with the form in which the product will be marketed and administered. The final formulation of the product is governed by two main concerns, namely to produce effective doses of the product in an acceptable form and to ensure that the product in its final formulation is safe to use and meets the specification regulations. By producing effective doses we mean that it has been proven to reach the target organs in appropriate concentration.

formulation

The former of these two requirements may mean that the primary product may require some "tailoring" before it becomes an effective reagent. More usually, however, the main concerns are with adding the product to a "cocktail" to ensure that it is administered in an acceptable and well controlled dose whose effects are predictable. Of great importance in the formulation are: the physical form of the product in relation to the route of administration (eg oral, injection, suppository); it's efficacy; its shelf-life.

Generally, orally presented medicines are more acceptable than injection. Additions of other compounds, themselves inactive, may enhance or inhibit the efficacy of a product. For example, polypeptide vaccines are usually much more immunogenic (i.e. they stimulate the production of greater quantities of antibodies) if they are administered in mineral oil or in the presence of certain types of particulate matter. Thus the presence of these latter two components (adjuvants) enhances the efficacy of such vaccines.

The final formulation of the product may however be greatly influenced by regulatory requirements. Of vital importance is the stability of the product. Much of the final formulation is aimed at improving the stability of the product. This is especially important with many of the products from biotechnology which are chemically labile. We shall learn in the next chapter that the testing of stability is of great importance and a 'shelf-life' (the time period for which a product can be stored and still meet specification standards) for a product needs to be determined. Decisions regarding final formulation of a product therefore depend upon the outcome of clinical and pre-clinical tests as well as upon the results of quality control measurements.

2.7. Stage V: Production

The decision to invest in large scale production facilities depends largely on the outcome of the research and development activities including the pre-clinical and clinical evaluation.

Π If you were a manager what would you need to know before you would agree to invest in large scale production facilities (make a list of four or five main items before reading on).

There are of course many questions that need answering but we can divide them into two broad groups. One group are concerned with the production process itself. For example:

- how reliable is the process?

- will it produce a predictable yield?

- what are the costs of production?

The other group concerns the product. For example:

- will the product satisfy regulatory requirements?

- what is the current and predicted value of the product?

facilities   The production process has two main facets. One relates to the production facilities. This includes the premises and equipment needed. These will in part be dictated by the outcome of the research and development activities. They are also influenced by the need to fulfil certain regulatory requirements. For example, if a genetically engineered cell line is used a particular level of containment will be required.

The second aspect of the production process concerns operational procedures. For the manufacture of medicines, these must conform to the standards demanded under Good

GMP   Manufacturing Practice (GMP). Thus in developing the production facility, it is insufficient merely to consider the design of the manufacture process. Considerable efforts is required to ensure that sound manufacturing practices are operative and monitored. The development of these practices is of course a component of the research

QA   and development phases of the project. Thus these stages will need to establish Quality Assurance procedures required for good science and for regulatory and commercial

GLP   reasons. We will learn in the next chapter that Quality Assurance (QA) includes Good Laboratory Practice (GLP), Good Manufacturing Practice (GMP) and Quality Control

QC   (QC).

In order to meet these objectives, it is essential that the appropriate level of expertise is available amongst operatives and that there is an accountable management system. The expertise required may be achieved by recruitment or by retraining.

There are many other factors to be considered in setting up the production facilities such as where are the facilities to be cited? In this text, we do not intend to consider these general issues. What we have sought to emphasis is the importance of the research and development activities in the design and operation of the production process.

2.8. Stage VI: Marketing and product diversification

It is largely outside of the scope of this text to consider marketing in any depth. Commercially, of course, successful marketing is of vital importance. For medical products marketing often depends on the data generated from the research and development activities. It is these activities which supply information concerning the efficacy and safety of the product. These two features are vital for the acceptance of a medicine. Long before a product reaches the market place, it makes commercial sense to consider the possibility of diversifying the product and to explore if markets exist which were not the original primary target for the product. Conventional antibiotics such as penicillin are now produced in a wide variety of forms, each with their own special properties (eg different disease targets, different shelf-life, different pharmacokinetic properties). The diversification of the product often extends the size of its market, thereby improving the commercial return on the original investment. Such diversification is not universally applicable and we will not dwell on this at this stage. However, it should be borne in mind that the skills and knowledge gained during the research, development and marketing of a new product may be directly transferable to secondary and subsequent projects.

SAQ 2.4.	Which of the following need to be considered before embarking upon the development of a new medical product based on a requirement to genetically manipulate cell lines?

1) The availability of suitable genetic vectors

2) The nature of the host cell line to be used

3) The availability of suitably trained staff

4) The projected number of patients who may be treated by the new product

5) The availability of alternative treatments of the disease

6) The number of animals to be used in determining potency and toxicity testing

7) The geographical distribution of the patients to be treated

8) The design and size of the bioreactor to be used in the production of the product

9) The other projects under consideration by the organisation.

SAQ 2.5.

Arrange the following activities into the form of a flow diagram to represent the development of a new product. (Do not refer to Figure 2.1.!!)

Research	Quality Control
Development scale up	Marketing strategy
Pre-clinical Testing	Clinical testing
Toxicology	Concept evaluation

Summary and Objectives

The primary objectives of this chapter have been to provide you with an understanding of the sequence of activities which brings a product from its initial conception to the market place. Special emphasis has been placed on the purposes of research and development in this sequence. The essential points to remember are that research and development encompasses not only the production process but also the evaluation of the safety and efficacy of the product. These two aspects of the development process are conducted in an integrated manner and in accordance with the methods prescribed by the regulations.

Now that you have completed this chapter you should be able to:

- explain, in general terms, why the development of a product to be used in medicine produced by the application of biotechnological procedures does not depend solely on the application of biotechnological experts;

- outline the technical stages of development of a new product before it can be placed on the market;

- list the main factors that must be taken into consideration in the evaluation of the concept before the project is undertaken;

- describe the main objectives of the research and development programme associated with the production of a new medicine;

- explain why the evaluation of efficacy and safety of the product should be carried out early in the process of product development.

3

Regulatory affairs

3.1. Introduction   30

3.2. Licencing authorities   31

3.3. What documents need to be complied with?   32

3.4. Regulation of genetic manipulation   37

3.5. GLP, GMP, and QC   39

3.6. Market authorisation   43

Summary and objectives   53

Regulatory affairs

3.1. Introduction

This chapter describes, in general terms, what a licencing file for a biotechnologically produced medicine should contain and how a licence can be acquired.

efficacy and safety

The relative recent introduction of genetic engineering and hybridoma techniques has led to changes in the regulations and it must be anticipated that further refinement of these regulations will take place. Nevertheless, the basic objectives of the regulations remain similar to those set for the marketing of medicinal products produced by more conventional technology. In simple terms, the manufacturer of a medicinal product has to demonstrate that it will contribute to the health of patients (or the public at large) and that it will be safe to use. This involves considerations of the safety of those manufacturing the product as well as the extensive testing of the product all the way through to trials with humans.

content and purpose of a licencing file

We can anticipate therefore that a licencing file will consist of descriptions of:

• the clinical use of the product;

• the method of manufacture;

• the quality control measures to be undertaken;

• the chemical and physical properties of the product;

• the results of pharmacological tests;

• the results of toxicity tests;

• the results of clinical studies.

It is also necessary for the manufacturing facilities to be licenced, a process normally accompanied by inspection. We will examine the contents of a licencing file in more detail in later sections. At this stage we wish to establish, in a general way, the contents of such files and their purpose. The purpose of a licencing file is to provide the licencing authority with the information it needs to decide if a licence should be granted.

confidentiality

A licencing file, therefore, represents a large dossier of information. In a commercially sensitive environment, the maintenance of confidentiality is important and presents a number of practical problems especially as many people need to have access to the file in order to evaluate the merits of the case being made. A product licence is granted only after full evaluation of the licencing file and the product.

Before we examine the content of licencing files in more detail, we will consider the question to whom should licencing files be submitted.

3.2. Licencing authorities

national
authorities
supernational
authorities

Historically, in Europe, licencing files were submitted to national authorities. Each national authority would apply national regulations in making a judgement on each application. Thus a product would need to be submitted for evaluation in several states if the intention was to manufacture or use the product in those states. The emergence of the European Community has led to several modifications in the procedure and a manufacturer now has some choice between submitting licencing files to national authorities or to supranational authorities such as to the EC Committee for Proprietary Medicinal Preparations (CPMP).

multi-state
application
supranational

Since 1977, as far as proprietary medicinal products are concerned, it has been possible to use such a supranational body to obtain marketing authorisation in EC member states. Instead of submitting individual applications, a common application to five (reduced to two in 1985) or more member states is made using the CPMP. This, however, can only be done after authorisation has been obtained in one member state. This first application is crucial because the decision, assuming it is favourable, must be taken into account by other member states as they become involved.

The sequence is therefore:-

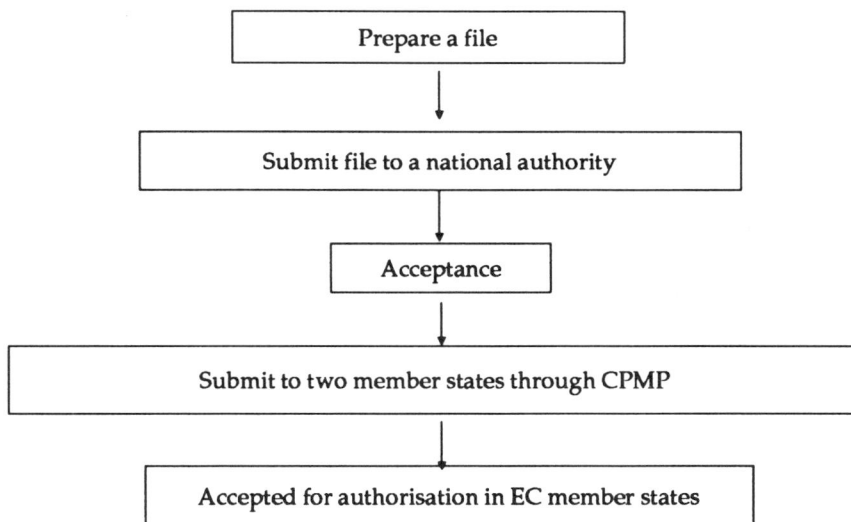

```
            ┌─────────────────────────────┐
            │        Prepare a file        │
            └─────────────────────────────┘
                          │
                          ▼
        ┌─────────────────────────────────────┐
        │  Submit file to a national authority  │
        └─────────────────────────────────────┘
                          │
                          ▼
              ┌───────────────────────┐
              │      Acceptance        │
              └───────────────────────┘
                          │
                          ▼
    ┌───────────────────────────────────────────────┐
    │   Submit to two member states through CPMP      │
    └───────────────────────────────────────────────┘
                          │
                          ▼
      ┌─────────────────────────────────────────────┐
      │  Accepted for authorisation in EC member states │
      └─────────────────────────────────────────────┘
```

In 1987 the above, multi-state procedure was extended to 'high tech' products (ie those obtained by genetic engineering and monoclonal antibodies) even when the intention was to submit to a single member state.

The objective was to enable questions relating to the quality, safety and efficiency of biotechnological medicinal products to be resolved within the EC by CPMP. This procedure would facilitate subsequent access to the markets of other member states and enable a co-ordinated and consistent stance to be taken within the EC member states. A list of addresses of the 'Competent Authorities' to whom application may be made is provided in Appendix 1.

USA and the FDA

In the USA the administrating body for the introduction of new medicines is the Food and Drug Administration (FDA). In the USA, the Codes of Federal Regulation (CRF) describes both administrative and specific requirements. The specific requirements are confined to drugs that have existed for at least a decade (eg vaccines, sera and blood derived drugs). For more recent products like monoclonal antibodies and those involving recombinant DNA, so called 'points to consider' have been drafted. There is particular concern to stimulate discussion within industry.

Japan and the PAB

In Japan, the body responsible for reviewing applications to licence new drugs is the Pharmaceutical Affairs Bureau (PAB) of the Ministry of Health and Welfare. In the main, the EC, FDA and PAB are attempting to harmonise their approach to licencing medicines. This is especially true in the approach to the licencing of medicines derived from biotechnology. It is not surprising that other licencing authorities are following the lead of the EC, FDA and PAB. Thus although some important differences still exist, there is a growing trend towards a global approach.

In order to maintain clarity we will focus on the situation in Europe.

3.3. What documents need to be complied with?

EC directive and guidance documents

The area of market authorisation for medical products has been particularly subject to a sequence of new EC directives and guidance documents. In the evolving European situation many directives and guidance notes have been issued. We can expect there will be many more. A substantial, but not complete list is provided here. It is our intention to explore with you the structure of the regulations and to provide some information on the EC approach to market authorisation. The documents that need to be complied with to gain market authorisation, fall into two general categories: broad ranging EC directives; and EC guidelines. The reader should note that EC directives do not bind the members of the European Community, but contain an obligation for the Member States to bring into force laws, regulations and administrative provisions necessary to comply with the directive concerned. Guidelines on the other hand provide help to applicants to provide the information that is required before authorisation can be given.

A list of the major community directives is given in Table 3.1. Note that these documents are published in the "Official Journal" (OJ) which is available in all Member States. Copies of the Official Journal are usually placed in main libraries or can be obtained from the offices of the Official Journal in each Member State .

The key directives are:

- Council Directive 65/65/EEC which provides the basic instructions required to make an application for market authorisation.

- Council Directive 75/319/EEC which provides details of how to make a multi-state application.

- Council Directive 87/22/EEC which extends multi-state application to high-technology products.

Many of the remaining directives are more specific. For example Council Directive 78/25/EEC relates to the addition of colouring matters to medicinal products.

Read through Table 3.1 and then attempt SAQ 3.1. It is sufficient at this stage that you know of their existence and have some idea of the issues they cover.

EEC Directives

-Council Directive 65/65/EEC of 26 January 1965 on the approximation of provisions laid down by law, regulation or administrative action relating to proprietary medicinal products (O. J. n° L 22 of 9.2.65).

-Council Directive 75/318/EEC of 20 May 1975 on the approximation of the laws of Member States relating to analytical, pharmaco-toxicological and clinical standards and protocols in respect of the testing of proprietary medicinal products (O. J. n° L 147 of 9.6.75).

-Council Directive 75/319/EEC of 20 May 1975 on the approximation of provisions laid down by law, regulation or administrative action relating to proprietary medicinal products.

-Council Decision 75/320/EEC of 20 May 1975 setting up a Pharmaceutical Committee (O. J. n° L 147 of 9.6.75)

-Council Directive 78/25/EEC of 12 December 1977 on the approximation of the laws of the Member States relating to the colouring matters which may be added to medicinal products (O. J. n° L 11 of 14.1.78).

-Council Directive 81/851/EEC of 28 September 1981 on the approximation of the laws of the Member States relating to veterinary medicinal products (O. J. n° L 317 of 6.11.81).

-Council Directive 81/852/EEC of 28 September 1981 on the approximation of the laws of the Member States relating to analytical, pharmaco-toxicological and clinical standards and protocols in respect of the testing of veterinary medicinal products (O. J. n° L 317 of 6.11.81).

-Commission Communication on parallel imports of proprietary medicinal products for which marketing authorizations have already been granted (O. J. n° C 115 of 6.5.82).

-Council Directive 83/189/EEC of 28 March 1983 laying down a procedure for the provision of information in the field of technical standards and regulations (O. J. n° L 109 of 26.4.83).

-Council Directive 83/570/EEC of 26 October 1983 amending Directives 65/65/EEC, 75/318/EEC and 75/319/EEC on the approximation of provisions laid down by law, regulation or administrative action relating to proprietary medicinal products (O. J. n° L 332 of 28.11.83).

Table 3.1. A library of community directives to medicinal products

EEC Directives (cont'd)

-Council Recommendation 83/571/EEC of 26 October 1983 concerning tests relating to the placing on the market of proprietary medicinal products (O. J. n° L 332 of 28.11.83).

-Commission Communication on the compatibility with Article 30 of the EEC Treaty of measures taken by Member States relating to price controls and reimbursement of medicinal products (O. J. n° C 310 of 4.12.86).

-Council Directive 86/609/EEC of 24 November 1986 on the approximation of laws, regulations and administrative provisions of the Member States regarding the protection of animals used for experimental and other scientific purposes (O. J. n° L 358 of 18.12.86).

-Council Directive 87/18/EEC of 18 December 1986 on the harmonization of laws, regulations or administrative provisions relating to the application of the principles of good laboratory practice and the verification of their applications for tests on chemical substances (O. J. n° L 15 of 17.1.87).

-Council Directive 87/19/EEC of 22 December 1986 amending Directive 75/318/EEC on the approximation of the laws of the Member States relating to analytical, pharmaco-toxicological and clinical standards and protocols in respect of the testing of proprietary medicinal products (O. J. n° L 15 of 17.1.87).

-Council Directive 87/20/EEC of 22 December 1986 amending Directive 81/852/EEC on the approximation of the laws of the Member States relating to analytical, pharmaco-toxicological and clinical standards and protocols in respect of the testing of veterinary medicinal products (O. J. n° L 15 of 17.1.87).

-Council Directive 87/21/EEC of 22 December 1986 amending Directive 65/65/EEC on the approximation of provisions laid down by law, regulations or administrative action relating to proprietary medicinal products (O. J. n° L 15 of 17.1.87).

-Council Directive 87/22/EEC of 22 December 1986 on the approximation of national measures relating to the placing on the market of high technology medicinal products, particularly those derived from biotechnology (O. J. n° L 15 of 17.1.87).

-Council Recommendation 87/176/EEC of 9 February 1987 concerning tests relating to the placing on the market of proprietary medicinal products (O. J. n° L 73 of 16.3.87).

-Council Directive 88/182/EEC of 22 March 1988 amending Directive 83/189/EEC laying down a procedure for the provision of information in the field of technical standards and regulations (O.J. n° L 81,26.3.88).

-Council Directive 88/320/EEC of 9 June 1988 on the inspection and verification of Good Laboratory Practice (GLP) (O. J. n° L 145 of 18.12.86).

Table 3.1. (cont'd) A library of community directives to medicinal products

| SAQ 3.1. | You will need to refer to Table 3.1 to answer the following questions. |

You will need to refer to Table 3.1 to answer the following questions.

1) Which of the Council's Directives specifically refer to the addition of colouring matter to medicinal products?

2) Which of the Directives relate to the analysis, pharmacological, toxicological and clinical evaluation of medicinal products?

3) Which of the Directives provides for the inspection and verification of Good Laboratory Practice?

In comparison with the directives, the guidelines, issued as a series of recommendations, offer a greater degree of flexibility.

∏ A list of recommendations is provided in Table 3.2. Read through this list and see if you can identify those which are concerned with a) evaluating the toxicity of a product; b) pharmacological studies; c) pre-clinical and clinical trials.

EEC Guidelines (O. J. = Offical journal)

a) Recommendation 83/571/EEC relates to the following five notes for guidance, published in the O. J. n° L 332 of 28.11.83:

-repeated dose toxicity;

-reproduction studies;

-carcinogenic potential;

-pharmaco-kinetics and metabolic studies in the safety evaluation of new drugs in animals;

-fixed-combination products.

b) Recommendation 87/176/EEC includes the following fourteen notes for guidance, published in the O. J. n° L 73 of 16.3.87:

-single dose toxicity;

-testing of medicinal products for their mutagenic potential;

-cardiac glycosides;

-clinical investigation of oral contraceptives;

-user information on oral contraceptives;

-data sheets for antimicrobial drugs;

-clinical testing requirements for drugs for long-term use;

-non-steroidal anti-inflammatory compounds for the treatment of chronic disorders;

-anti-epileptic/anticonvulsant drugs;

Table 3.2. Community guidelines prepared with the Committee for Proprietary Medicinal Products

EEC Guidelines (cont'd)

-investigation of bio-availability;

-clinical investigation of drugs for the treatment of chronic peripheral arterial diseases;

-pharmaco-kinetic studies in man;

-anti-anginal drugs;

-topical corticosteroids.

c) Other guidelines adopted or in preparation by CPMP.

-recommended basis for the conduct of clinical trials of medicinal products in the European Community;

-production and quality control of monoclonal antibodies of murine origin intended for use in man;

-production and quality control of medicinal products derived by recombinant DNA technology;

-chemistry of the active ingredient;

-stability tests on active substances and finished products;

-development pharmaceutics and process validation;

-preclinical biological safety testing;

-antidepressant drugs;

-testing medicinal products in the elderly;

-clinical trials in children;

-herbal remedies;

-trials of medicinal products in the treatment of cardiac failure;

-antiarrhythmic drugs;

-anti-cancer agents in man, III/699/88;

-studies of prolonged action forms in man, III/1962/87;

-production and quality control of cytokine products derived by modern biotechnological processes, III/3791/88;

-anti-epileptic/anticonvulsant drugs, III/3128//88: revision;

-analytical validation, III/844/87;

-general pharmacodynamics, III/480/87;

-production and quality control of monoclonal antibodies derived from human lymphocytes intended for use in man, III/3975/88;

-good clinical practices, III/3976/88;

-local toxicity and skin/eye toxicity, III/3979/88;

-control tests in the finished product, III/3978/88;

-mental deficiency in the aged, III/3977/88.

Table 3.2. (cont'd) Community guidelines prepared with the Committee for Proprietary Medicinal Products

generic and
specific
guidelines

From your efforts to carry out the last activity, you will have recognised that some guidelines could span more than one aspect of drug development, evaluation and use. Some refer to specific drugs or families of drugs while others are much more generic. For example the recommendations regarding single dose and repeated dose toxicity testing have wide ranging applications. On the other hand, the guidelines referring to cardiac glycosides are much more specific. Thus in preparing a licencing file, the applicant may need to respond to both wide ranging and specific guidelines.

We will be examining the development and production of specific health care products using biotechnological procedures in chapters 4 to 9. For each of these products a licencing file had to be prepared before production and marketing were authorised. You will learn by studying the kind of information that needs to be generated and the procedures that need to be followed in order to fulfil the obligation specified in the directives and guidelines. For example, data relating to single and repeated dose toxicity studies are presented in a number of the case studies. By implication, these studies will have been conducted within the guidelines established in recommendations 83/571/EEC and 87/176/EEC.

3.4. Regulation of genetic manipulation

contained use
and deliberate
release

Most of the regulatory issues discussed above relate to medicines produced by any route for example by chemical synthesis or extraction of natural material. The introduction of genetic manipulation has led to new approaches to producing medicines. In the 1970's, it was recognised that some risks to humans and to the environment may arise from the application of recombinant DNA technology.

Since many of the products discussed in the case studies are derived from genetically manipulated systems, it is important that you are alerted to the regulatory issues which govern the use of genetically modified organisms. The regulatory framework is provided by directive 88/160/EC. This directive has a wider application than just to the development, production and marketing of medicines. It applies to any activity involving genetically modified micro-organisms. This directive distinguishes between the contained use of genetically modified micro-organisms (GMMOs) and the deliberate release of genetic modified organisms (GMOs). Both these aspects are relevant to the case studies. We will learn, for example, that the production of insulin and erythropoietin both involve the contained use of genetically modified systems. On the other hand, the vaccines produced for Hepatitis B and Aujeszky's disease involve the deliberate release of such manipulated systems.

The directive relating to the use and release of genetically modified micro-organisms is quite an extensive document. Appendix 2 provides a summary of its main provisions. We recommend that you read this carefully before attempting the following SAQs. Our intention is to give you some idea of the content and structure of an EC directive. We do not intend that you learn it by heart. In other words, we are using this as an illustrative example. More extensive information about EC directives and biosafety are brought together in a source book to be published within the BIOTOL series.

SAQ 3.2.

Indicate by a tick those of the following technique which are exempt from the EC directive on the use of genetically modified micro-organisms.

1) mutagenesis

2) self cloning of certain non pathogenic, naturally occurring micro-organisms

3) recombinant DNA techniques using vectors

4) techniques involving the direct introduction of heritable material prepared outside of the micro-organism.

5) construction of somatic animal hybridoma cells.

SAQ 3.3.

Delete the inappropriate option in each of the following statements concerning the EC-directive on the use of genetically modified organisms.

1) Type [A] or [B] operations are operations used for teaching, research or non-commercial purposes.

2) Type [A] or [B] operations are operations that are conducted in a total volume of less than 10 litres.

3) Group [I] or [II] organisms have a long record of safe use and are considered safe.

4) The procedures for Group 1A operations are [less] or [more] stringent than those for Group IIB operations.

SAQ 3.4.

The EC directive on the deliberate release of genetically modified organisms is divided into four parts (A to D). Identify these parts from the following list.

Purpose   Exemptions

General Provisions   Confidentiality

Authorisation   R & D (except placing on the market)

Placing the product on the market   Notification

                                                 Final Provision

| SAQ 3.5. | Complete the following statements relating to the EC directive on the deliberate release of genetically modified organisms.

1) On receipt of a notification of an intention to place a genetically manipulated organism on the market, the competent authority shall respond within [   ] days.

2) Competent authorities have [   ] days to reach agreement over a proposal to place a genetically manipulated organism on the market. If agreement is not reached after this period, the commission shall take a decision in accordance to a specific procedure.

3) The competent authority must respond to anyone notifying them of their intention to release genetic ally manipulated organisms as part of a research and development activity within [   ] days.

4) The competent authority sends a summary of notification to release genetically modified organisms to the commission within [   ] days.

3.5. GLP, GMP, and QC

Many of the guidelines described above contain the requirements for testing medicines at the different stages of development and production. In addition to these requirements, the regulations also include general guides concerning laboratory and manufacturing practice.

3.5.1. GLP

GLP

Good Laboratory Practice (GLP) guides give a general description of the practice, procedures and conditions that should be maintained within laboratories used for the development and testing of medicines. The GLP guides cover such items as laboratory design, equipment and facilities, management structure, operating procedures, qualified persons, education and training of staff, record keeping and quality assurance and independent auditing of the activities of the laboratory and the data producer.

purposes of GLP

The main purposes of Good Laboratory Practice is to promote the development of quality data generation that is mutually acceptable to a wide range of countries and institutions. The application of agreed laboratory procedures should avoid the need for duplicative testing and reduce technical barriers to trade. They should also improve the protection of human health and the environment.

standard operating produces

GLP is concerned with all laboratory practice including the organisational process and the conditions under which laboratory studies are planned, performed, monitored, recorded and reported. The GLP guides define, for instance the responsibilities of individuals from the director downwards and they provide conditions that the laboratory facility should fulfil including the need to establish regular inspection of apparatus and equipment, its maintainance and calibration. All routine procedures and practices must be enshirned in standard operating procedures which effectively set the standards for the laboratory. Particular emphasis is given to the importance of generating quality data and maintaining its integrity during processing.

3.5.2. GMP

Good Manufacturing Practice (GMP) guides are similar to GLP guides except they relate to the process of manufacturing. The holder of a Manufacturing Authorisation must manufacture medicinal products so as to ensure that they are fit for their intended use and comply with the requirements of the Marketing Authorisation. GMP is concerned with both production and quality control (QC).

The basic requirements of GMP are:

- Manufacturing processes are clearly defined, systematically reviewed and shown to be capable of consistently manufacturing medicinal products of the required quality and complying with their specification.

- Critical steps of manufacturing processes and significant changes to the process are validated.

- All necessary facilities for GMP are provided (including qualified personnel, adequate space, suitable equipment and services, correct materials, containers and labels, suitable storage and transport).

- Instructions and procedures are written in a clear specific and unambiguous language.

- Operators are trained to carry out procedures correctly.

- Records are made (manually or by instrumentation) during manufacture which demonstrate that all the steps required by the defined procedures and instructions were taken and that the quantity and quality of the product was as expected. Any significant deviations are fully recorded and investigated.

- Records must be kept of manufacture and distribution which enable the complete history of a batch to be traced. Such records must be retained in a comprehensible and accessible form.

- The distribution of the product minimises any risk to its quality.

- A system be developed to recall any batch of product.

- Complaints about products must be examined. Any quality defects must be investigated and appropriate measures taken to prevent re-occurence.

Having read this list carefully you will notice that special emphasis is again placed on maintaining the quality of the product and on keeping records of the manufacturing process and the distribution of the product. GMP also includes some items which ensure the quality of the product after it has left the factory.

3.5.3. QC

Let us turn our attention to a specific aspect of GMP, namely Quality Control (QC). It is that part of GMP which is concerned with sampling, specification and testing. It encompasses the organisation documentation and release procedures which ensure that the necessary and relevant tests are carried out. It is designed to ensure that materials are not released for use until their quality has been judged to be satisfactory.

The requirements for QC are:

- Adequate facilities, trained personnel and approved procedures are available for sampling, inspecting and testing starting materials, packaging materials, intermediate, bulk and finished products. Where appropriate these must be adequate to monitor environmental conditions for GMP purposes.

- Samples for QC analysis are taken in an approved manner.

- Test methods are validated.

- Records are made which demonstrate that all required sampling, inspection and test procedures were actually carried out. Any deviations to be recorded and investigated.

- The finished product must contain active ingredients complying with the qualitative and quantitative composition of the Marketing Authorisation. It must be of the required purity, enclosed in its proper container and correctly labelled.

- Records are made of the results of inspection and testing and that these are formally assessed against specification.

- No batch of product is released for sale or supply prior to certification by a Qualified Person that is not in accordance with the requirements of the Marketing Authorisation.

- Suffient reference samples of the starting materials and products are retained to permit future examination of the product.

GMP also provides details of what should be included in the specifications used in the manufacture of a product. The reader should be alerted to the fact that specifications are included in the authorisation to market a product. Specifications have to be established for starting materials, packaging, intermediate, bulk and finished products.

3.5.4. Quality Assurance (QA)

Many people are confused by the terms Quality Assurance and Quality Control, so let us see if we can make the distinction clear. Quality Assurance (QA) is a wide ranging concept which covers **all** matters which individually and collectively influence the quality of a product. It is the total of all the arrangements made to ensure that a product is of the quality for its intended use. QA therefore incorporates GLP, GMP and QC.

We can therefore represent the relationship of QA, GLP, GMP and QC as shown in Figure 3.1.

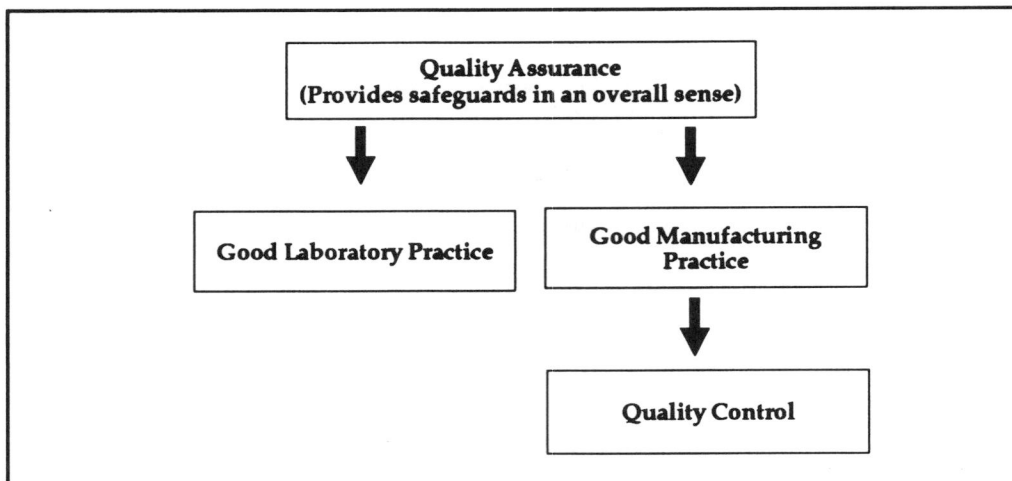

Figure 3.1. The relationship between QA, GLP, GMP and QC.

SAQ 3.6.

Use Tables 3.3. and 3.4. to decide at what stages of manufacture QA, QC and GMP play a part. Do this in the form of a diagram.

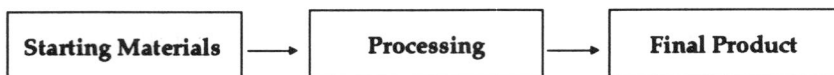

Starting Materials ⟶ Processing ⟶ Final Product

Try to indicate on your diagram when particular emphasis is placed on QC and on GMP.

Let us summarise what we have learnt about the regulatory framework governing the market authorisation of medicinal products. First we have learnt that in order to achieve authorisation a licencing document must be submitted to the appropriate authority. This may be done at a national level or by a multi-national mechanism. We have also learnt that the licencing file must cover descriptions of the clinical use of the product, the method of manufacture, the results of pharmacological, toxicological and clinical testing, as well as details of the production facilities. In the European Community these items must respond to the EC directives and guidelines (Tables 3.1. and 3.2.) while in the USA, the relevant Codes of Federal Regulations (eg Licencing, General Biological Products Standards) must be upheld.

We have also learnt that laboratory studies and manufacturing processes must be conducted within the framework of guides (GLP, GMP and QC) which provide product Quality Assurance (QA).

3.6. Market authorisation

At the beginning of this chapter we indicated that a large dossier of information needs to be submitted to the licencing authorities before authorisation to market a product could be gained. We learnt in subsequent sections some of the details of directives and guidelines that need to be followed in the development and production processes. In this section, we examine the contents of dossier that needs to be submitted to the licencing authorities.

Table 3.3. provides a summary of the kind of information which is included in a licence dossier. You will notice some of these items have already been discussed under the headings of GLP, GMP and QC. We do not wish to dwell on them but to think about the rationale behind the production of the dossier.

Contents	Remarks
Premises (i.e buildings)	These relate to GMP and to GLP and are mainly monitored by inspection
Personnel	
organisation	
education and training	
Process validation	
reduction of contaminants	
reproducibility of quality	
Source materials	These relate to GMP and are subject to QC.
chemicals	
cultures	
Production procedures	The evaluation of these are part of GMP and are subject to QC
bulk product	
final product	
filling lot	
Quality control	Subject to GLP
safety	
potency	
reproducibility	
Stability	Subject to QC
Reprocessing	Subject to QC
Pre-clinical testing	
Clinical testing	

Table 3.3. The information contained in a licence application dossier

The main objectives of the dossier are to provide the licencing authority with documentary evidence that the product fulfils the requirements of all of the relevant regulations concerning the efficacy and safety and that the procedures adopted for the production and evaluation of the product provide an assurance of quality. The onus therefore is placed on the manufacturer to produce a specification for the product and process which fulfils these requirements.

Let us examine each of the headings provided in Table 3.3 in turn to learn more of what is involved.

3.6.1. Premises

clean

We have learnt in our discussion of GMP that premises must be designed and constructed to facilitate effective cleaning and maintenance. Many biotechnological medicinal products cannot be autoclaved and so have to be manufactured under aseptic conditions, involving classified rooms with air of appropriate quality. Some care has to be taken that the product is not a vehicle for transmitting infection.

3.6.2. Personnel

education
training

A scheme defining the responsibilities for personnel in charge of production, quality control and quality assurance should be included in the licencing file. The same individual should not be responsible for both production and QC to avoid possible conflicts of interest.

There must be sufficient qualified personnel to carry out all the tasks involved in production. This implies a commitment to education and training on the part of manufacturers.

3.6.3. Process validation

Reduction of contaminants

A basic principle of GMP is that the maintenance of quality requires not only extensive testing of the final product redbut also high standards throughout the process of manufacture. This may be illustrated by reference to the problem of removal of bacteria by filtration. It can be very difficult to detect low levels of bacteriological contamination, as illustrated in the following table (Table 3.4.) and so it is essential to reduce to a minimum any possibility of contamination prior to the final filtration.

The data of Table 3.4. is based on testing 20 vials in each batch.

Number of contaminated vials	Batch Size		
	1000	10 000	100 000
1	2%	0.2%	0.02%
10	20%	2.2%	0.22%
100	88%	18.4%	2.0%

Table 3.4. Chance of detecting contamination in one vial if twenty vials are tested from each batch

The data shows that if only 1 vial is contaminated in every batch of 1000 vials and 20 vials are tested, then there is only a 2% chance of detecting the contamination. If the batch size is 100 000 vials, then the chances of detecting a single contaminated vial if 20 vials are tested is only 0.02%.

spiking
experiments

Eliminating pathogenic viruses, a potential risk when mammalian cells are used as a source of material, is even more difficult. Because of the relative insensitivity of many assays for viruses, the reduction factor in each purification step should be established for a number of viruses. This is done by trial experiments in which known quantities of virus are added before carrying out the purification scheme (ie 'spiking' experiments). When using source materials proven to be free of contaminants (like viruses), reduction factors should be in the range 10^5 to 10^{10}. The objective of the validation of the reduction is a precaution against as yet 'unknown' viruses that could conceivably be present. (We will examine these principles in detail in several of the case studies).

SAQ 3.7.

In the process validation specification submitted in a Licensing file, it is proposed to examine 40 vials out of every batch of 10 000 vials produced for contaminating bacteria.

What are the approximate chances of detecting contamination if:

1) 1 vial in every batch of 10 000 vials is contaminated

2) 100 vials in every batch of 10 000 vials are contaminated?

Reproducibility of quality

five
consecutive
batches

Chemical by-products arising from the growth of cells cover a wider range than those accompanying chemical synthesis. Sometimes they can only be detected by bioassays, and so establishing reproducibility of quality for a GMP can be quite difficult. Commonly, five consecutive batches are produced to demonstrate consistency before submitting a licencing file. The case studies deal with both chemical and bioassay procedures directed towards demonstrating the reproducibility of quality.

Here we have included only three types of process validation studies, two relating to removal of contaminants, one relating to reproducility of quality. The reader should recognise that the performance of such studies enables the manufacturer to make certain specifications for the product. For example it could be specified that the product would contain less than a certain number of viruses or that fewer than a specified proportion of the product would contain bacteria.

3.6.4. Source materials

It is obvious that source materials should be of high quality, using, when appropriate, guidance given by pharmacopoeias.

seed lot system

In a text on the biotechnological production of medicines, cells containing and expressing deliberately inserted or amplified genetic material require special consideration. We must prove the cultures' purity. Cultures are grown to a suitable density, dispensed into containers that allow a long period of storage (usually in liquid nitrogen) without jeopardising the quality. Subsequently, the cells are tested for contaminants such as bacteria, viruses and mycoplasma. Each time a new batch of product is required, one or more containers of the stock culture is used to seed the fermenter (hence 'seed lot'), ensuring as far as is possible consistency in the source of the biological material. This practice is widely accepted in the production of vaccines, monoclonal antibodies and recombinant DNA products.

3.6.5. Production procedures

batch
continuous

Products can be made one batch at a time or by continuous culture. Either way, the description of the method of cultivation should demonstrate clearly that contamination with other products made in the same premises cannot occur. Again this can be built into the specification of the product.

In batch production, cells are cultivated until reasons related to economy or formation of contaminants necessitate stopping the process and the cells are harvested. During continuous production, the cells remain in the fermenter while nutrients flow into and product out of the fermenter. A higher yield may be possible, but it is likely to be more difficult to prove that the product is consistent in quality.

Before we turn our attention to the tests that may be conducted during the production process, let us first make clear the main stages of product processing, after the cultivation stage. We can divide this into three main phases. The production of the bulk product, the final product and the filling lot. We define these as follows:-

bulk product

Bulk product. After harvesting, the cells are ruptured if the product is intracellular and cell debris is discarded. The preparation at this stage is referred to as the bulk product. It is appropriate to test for unwanted materials (eg viruses) for which relatively insensitive assay methods are available. If 'absent' from the bulk product, there is a further safety margin provided by subsequent purification and/or inactivation procedures.

final bulk

Final bulk. After purification and possible inactivation, the final - or purified - bulk product is obtained. It is at this stage that most of the tests such as potency, chemical and biological characterisation are performed. We will briefly deal with a few of the more common ones here.

filling lot

Filling lot. After dispensing into the final containers, the product is referred to as the final lot or filling lot. At this stage, tests for sterility, safety and potency are carried out. It is not our intention here to examine all of the tests that may be carried out on the various stages described above. You will learn of many of these in exploring the case studies in chapters 4 to 9. These tests will be predominately concerned with the physical and chemical identity and with the purity of the product at each stage. Thus for each stage, the quality of the product must be specified. If, at any time this specification is not achieved, the product must be rejected or reprocessed. The tests will also involve evaluating the potency of the product and in determining the level of contamination. With products generated by biotechnology particular concern is attached to the presence of residual DNA and viruses in the product. Thus at this stage we will briefly consider the potency, DNA and virus tests that may be conducted on the product. You will find mention of these tests in all of the case studies.

Potency test

Wherever possible, the potency of a product should be tested against an international reference preparation. The goal should be a product that differs from the standard only in terms of concentration, although this might require qualification depending on the stability of the standard during storage.

There are four categories of potency tests: animal, tissue culture, immunological and chemical.

challenge
potency tests

Animal tests, especially challenge potency tests, simulate the wanted effect relatively well. In the case of many vaccines, animals are immunised with dilutions of either the product or the reference preparation. The animals are subsequently challenged with the agent against which the vaccine should provide protection. The surving animals are counted and the potency is calculated using a probit distribution.

tissue cultures
and challenge
potency test

Tissue cultures can be used as an alternative to the challenge potency test. Animals are immunised with the vaccine and the resulting antibodies used to neutralise a fixed amount of the toxin. The mixture of antibodies and toxin is added to a tissue culture to establish the remaining cytotoxic effect. These tests use fewer animals, are easier to standardise and cause less harm to animals. Likewise, tissue culture tests can be used to establish the potency of live viral vaccines. The tissue culture is incubated with dilutions of the vaccine; the number of plaques of cells infected with the virus establishes the potency of the preparation.

immunological
tests

Immunological tests can involve immuno-electrophoresis, immuno-diffusion or ELISA (Enzyme Linked Immune Sorbant Assays). In immuno-electrophoretic and immuno-diffusion tests, the antiserum is added to the gel and during electrophoresis/diffusion of the product and the reference standard, precipitations occur at equimolar concentrations. The size of the peaks (in rocket electrophoresis) or rings is proportional to the concentration of the product.

Chemical potency tests make extensive use of HPLC, (High Performance Liquid Chromatography) and other biochemical separation techniques. A good example of this will be found in the insulin case study.

DNA test

The presence of DNA is of particular concern in many biotechnology products. The amount of residual DNA can be established by a nucleic acid hybridisation assay, the detection limit for a repetitive sequence being of the order of 10^{-12} g (1 pg). This is a sensitive indicator of the presence of a host cell DNA in a product derived from those cells. However, there is still concern whether it provides adequate 'protection' against putative oncogenic DNA sequences. Tests with oncogenic viral DNA in animals suggest that the risk of a tumorigenic event after receiving 100 pg of DNA is of the order of 1 in 5-20 million. There is considerable uncertainty in the assumptions made in the calculations. In general, it is advised that the DNA present per dose should be no more than 10-100 pg.

Virus tests

Several techniques are available for establishing that products are not contaminated with viruses. Commonly used are:

- inoculation, as appropriate to the virus, of chick embryos, suckling or adult mice or higher species

- inoculation as above and testing for the production of antibodies against specific viruses with ELISA or immuno-fluorescence assays. Amounts of viruses just sufficient to cause propagation of the virus can be detected

- cell cultures can be used to test for viruses that are cytopathogenic.

3.6.6. Quality control

A medicinal product should be free of toxic side-effects (i.e. be safe), effective (i.e. be potent) and be prepared consistently. The conditions of GMP, quality control and clinical testing should all support the above goals. Each filling lot, (occasionally the final bulk), should be evaluated by the following tests.

Safety

In a general test for safety, a relatively high dose is inoculated intraperitoneally into young guinea pigs and young mice. After one or two weeks, the animals should not show any ill effects such as loss of weight relative to control animals.

LAL test   Pyrogens, usually bacterial fragments, are detected by injecting three rabbits intravenously and recording any rise in temperature. An alternative assay uses an extract of amoebocytes from the horseshoe crab (*Limulus polyphemus*); called the LAL test. The extract forms a gel-clot with pyrogens. Not surprisingly, the rabbit and LAL tests for pyrogens do not always coincide and require careful validation.

Potency

reference to a   The potency of a product should preferably be expressed in units relative to the
standard   international standard preparation. Such a standard will not be available for a new preparation and in such a situation the batch that has been subjected to the most extensive clinical evaluation should be used as a reference preparation.

Reproducibility

trend analysis   The results of potency and other tests on consecutive batches should be submitted to show trend analysis. Any undue trend should be investigated and correcting action taken in accordance with contingency plans.

microbial   Other safety tests include screening for bacteria, viruses, mycoplasma and DNA. (These
contamination   tests greatly overlap with the tests conducted during the production process).

3.6.7. Stability

expiry term   Samples of several (preferably all) batches should be stored at the appropriate temperatures during the expiry term (the maximum time during which the product can be stored without serious loss of quality). The potency should not be lower than the claimed potency at the time of issue. Often additives are included in the final formulation of a product to improve stability. The reader should note however that these additives and potency are both included in the specification of the product.

If there is a possibility that toxic products can be formed during storage, it must be shown that the concentrations of these products do not exceed safety limits by the full expiry time.

3.6.8. Reprocessing

main stream   Reprocessing is a rescue operation used once a product (or intermediate stage in its
side stream   production) is out of specification. It may involve repeating a production step or applying an additional procedure in order to regain the specification. Reprocessing should only be allowed if it has been validated and described in the licencing file. In 'main stream' reprocessing a procedure is repeated on fractions of an intermediate product, whilst 'side stream' reprocessing involves handling fractions that would

normally be discarded. The latter carries more risk of changing the specification of the product.

∏ Examine Figure 3.2 and see if you can think of a reasons why 'side stream' reprocessing involves more risk of changing the specification than main stream reprocessing. (Do this before reading on).

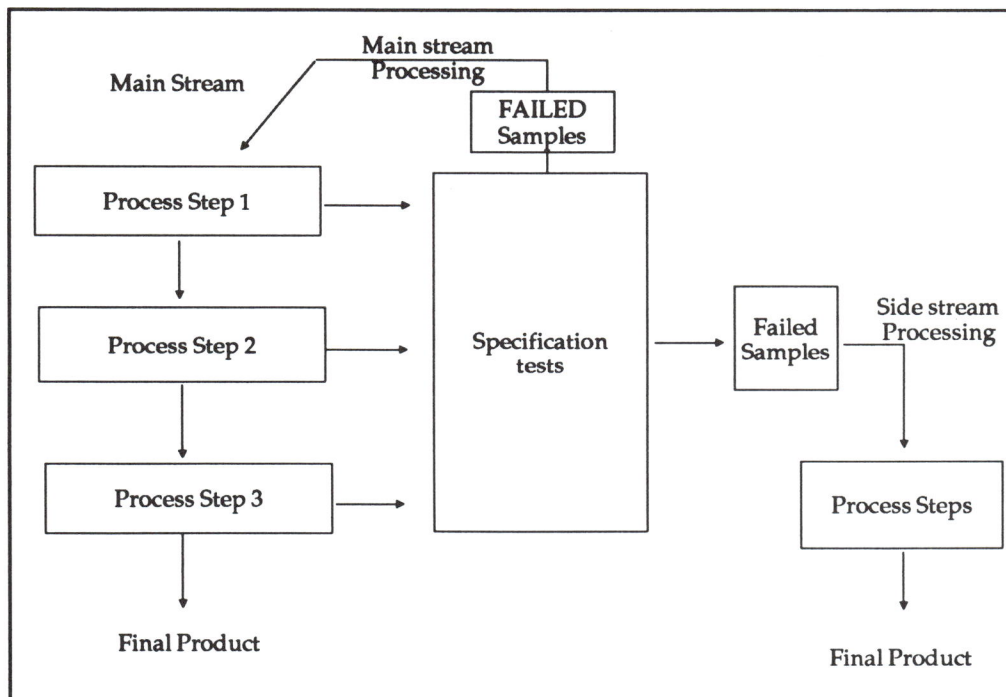

Figure 3.2. The two routes of reprocessing medicinal products

Side stream reprocessing often involves purification processes which are different from those conducted in the main manufacturing stream. These different processes could therefore make some alterations in the final product.

3.6.9. Pre-clinical testing

The general aim of the pre-clinical testing of any new medicinal product is to ascertain whether it has the potential to cause unexpected and undesirable effects. The EC document on pre-clinical trials of biotechnological medicinal products subdivides them into four groups: hormones, cytokines or other regulatory factors, blood products, monoclonal antibodies and vaccines (see Figure 3.3). Each group is subdivided into products **closely** related to human substances and products **distantly** related to human products. The subdivision implies an increasing need for pre- clinical testing (Figure 3.3.).

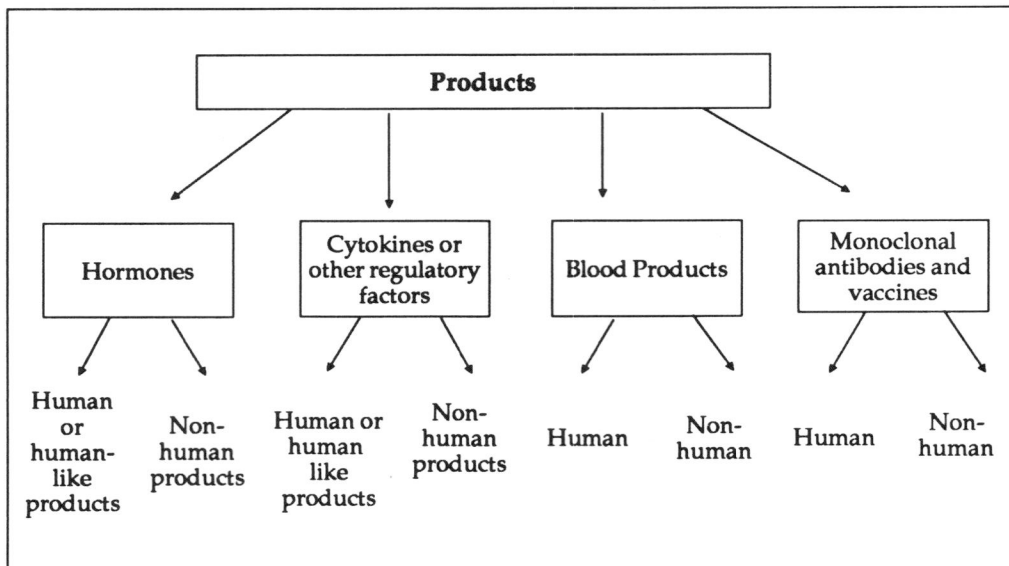

Figure 3.3 The subdivision of pre-clinical trials

Clearly such products may exhibit high pharmacological activity or influence the immune or other physiological system. Evaluation of safety will often begin with pharmacokinetic studies in which the pattern and time course of absorption, distribution, metabolism and elimination of the product are determined. If appropriate, test animals lacking, or with less effective immune systems can be used. The case studies provide examples of the types of pre-clinical testing that needs to be carried out. You will learn for example that the situation which applied to vaccines (eg Hepatitis B) is substantially different than for a therapeutic hormone (eg erythropoietin). For a hormone, absorption, distribution and metabolism are of great importance.

3.6.10. Clinical testing

A clinical trial involves:

- the administration of a medicinal product by a doctor or dentist to patients;

- producing evidence that the product may have effects beneficial to the patients;

- ascertaining whether, and to what extent, the product has other beneficial or harmful effects.

- exemption

Exemption with respect to the second requisite can be granted in the case of the products such as vaccines, administration of which may protect the subject in the future, hence allowing the participation of volunteers. In order to obtain an authorisation for a clinical trial in the UK, the manufacturer has to submit an application, including a description of the chemistry and pharmacy of the product, the result of any experimental tests with animals and of clinical work that may have already occurred.

In the USA it is only necessary to inform the regulatory agency, which has the authority to object. The manufacturer has to file a notice of Claimed Investigational Exemption for a new medicinal product (Investigational New Drug, IND). This must provide similar details to that cited for the UK. The Food and Drug Administration (FDA) may object to the notice within 30 days of receiving the IND.

placebo

double-blind

A clinical trial should preferably be randomised, one group of patients receiving the new product, another receiving a control preparation (a placebo or another medicinal product). It should also be carried out in a double-blind manner, neither the patient nor the investigator knowing which is used in any individual case.

A clinical trial can be divided into the following phases:

Phase I - Clinical pharmacology and toxicology. This phase is primarily concerned with safety and is usually performed on human volunteers, with dose ranging studies on 20-50 subjects.

Phase II - Initial clinical investigation for the effect of the treatment. This is intended to demonstrate efficacy and relative safety with 100-200 subjects.

Phase III - Full-scale evaluation of treatment. Trials in this phase are designed to gather additional information about safety, tolerance and definition of adverse effects.

Phase IV - Post-marketing surveillance (PMS). The aim here is to establish possible adverse effects occurring at a low frequency and to establish morbidity and mortality. In many countries, PMS is not considered to be part of a clinical trial and can be performed without involvement from the National Control Authority.

Before moving onto the case studies attempt the following SAQs concerning the production of a licencing file.

SAQ 3.8.

1) In a licensing file, procedures for controlling the quality of a protein product have been proposed. These procedures have been designed to detect DNA at a level of 1 ng/ml of final product. It is proposed to reject the product if it contains more than 1 ng/ml. It is also proposed to use the product in 0.1 ml portions and to administer 10 such doses to each patient to treat a medical condition.

   Estimate the likely risk of inducing a tumour in a patient receiving such treatment. Use the risk factor that the chance of a tumorigenic event is about 1 in 5-20 million after receiving 100 pg DNA (NB 1 ng = 10^{9} g, 1 pg = 10^{-12} g).

2) Your production process involves the use of the enteric micro-organism *Escherichia coli* to produce a mammalian hormone. Which of the following tests would you build into the analysis of your production process and final product?

   a) Limulus Amoebocyte Lysate assay; b) DNA assay; c) screening for bacteria other than *Escherichia coli*; d) immunoassay of the product; e) chemical assay of the product by HPLC; f) check for bacterial contamination of the final product.

| SAQ 3.9. | Put the following into the sequence in which they should be conducted. |

Put the following into the sequence in which they should be conducted.

a) Preliminary trials with human volunteers to establish absorption, distribution and metabolism of the product

b) Chemical characterization of product

c) Analysis of the absorption, distribution, metabolism and excretion of the product in animals

d) Trials using several thousand patients to get a fully validated evaluation of the product

e) Test the product on 20-50 human volunteers to gain data concerned with the dose size required

f) Conduct post-market surveillance of the product

If you have successfully answered SAQ 3.8. and 3.9. you have already grasped some of the information required by the licencing authorities and also the time sequence in which the various requirements may be undertaken.

Now turn to the case studies, to gain fuller experience of responding to the regulatory requirements at the same time as learning how recombinant DNA and hybridoma technology is being applied to produce new sources of medicinal products.

Summary and objectives

This chapter has provided a brief summary of the requirements that a licence dossier has to fulfil. Now that you have completed this chapter you should be able to:-

- explain why a licencing dossier is needed and what the overall objectives of the licencing dossier are;

- describe, in outline, the multi-state authorisation procedure which operates within the European Community;

- describe in outline major content areas of a licence file;

- explain the relationships between QA, GMP, QC and GLP;

- place into sequence the stages of assessment of the efficacy and safety of a medicinal product.

4

Case Study: Human insulin of rDNA origin

4.1. Introduction   56

4.2. Diabetes mellitus and insulin   56

4.3. Biotechnological development of rDNA human insulin   62

4.4. Characterisation of rDNA human insulin   80

4.5. Clinical studies   94

Summary and objectives   100

Case Study: Human insulin of rDNA origin

4.1. Introduction

4.1.1. Human insulin and recombinant DNA technology

The production of human insulin utilising recombinant DNA-technology marked a significant accomplishment in the field of molecular biology and it provides a secure source of insulin for the future treatment of insulin-dependent diabetics. The reason why insulin became one of the first products of recombinant DNA-technology was the knowledge of the amino acid sequence of insulin, which made it possible to construct a gene coding for the correct amino acid sequence.

This case study will review our knowledge of insulin and its use in the treatment of diabetes mellitus. It will also provide the rationale for utilising recombinant DNA-technology in the production of insulin. The procedures used to prepare highly purified human insulin are described, as are analytical tests demonstrating that human insulin produced using recombinant DNA-technology is of high purity and safe for use in humans. The study ends with a survey of the clinical studies carried out with this insulin.

4.2. Diabetes mellitus and insulin

4.2.1. History

pancreas

Diabetes mellitus is a disease that has been recognised since the time of the ancient Egyptians. Yet until the early 1920s, a diagnosis of diabetes was equal to a death sentence. The only method of treatment of diabetes was to put the patients on a diet which resembled starvation. Little was known about the disease, only that the pancreas played a role.

On July 30th, 1921, Dr. Frederick G. Banting and his assistant, Charles H. Best, both of the University of Toronto, injected a crude preparation from the pancreas of one dog into another from which the pancreas had been removed. Their excitement when they noted that the blood sugar of the diabetic animal fell to normal levels can easily be imagined. The first diabetic human being to receive Banting and Best's crude insulin preparation derived from canine pancreas was Leonard Thompson, a Canadian boy, who was surviving on a diet of 450 calories per day. As therapy continued, Leonard Thompson became able to eat a more normal diet, gained weight and, in time, returned to active life.

Although insulin was used by millions of diabetics from the early 1920s onwards, it was not until 1955 that Frederick Sanger at Cambridge University reported the chemical structure of insulin. We will deal with the structure of insulin later.

4.2.2. Diabetes mellitus

Definition

disorders

Diabetes mellitus (diabetes: passage; mellitus: sweet as honey) is a chronic, systemic disease characterised by disorders in:

- metabolism of carbohydrate, fat and protein and of insulin;

- the structure and function of blood vessels.

The principal early symptoms and signs are usually related to the metabolic defects; findings late in the disease are linked with complications resulting from vascular defects caused by the disease process.

Diabetes mellitus ordinarily appears as one of two recognised clinical conditions:

- the juvenile type (growth-onset; type I) with absolute insulin deficiency;

- the adult type (maturity-onset; type II) which is more often the result of a delayed release of endogenous insulin.

Glucose metabolism

Diabetes mellitus is the result of an absolute or relative shortage of insulin; insulin is necessary to control the use of glucose in the body. For a better understanding of the disease, the metabolism of glucose will be described.

Carbohydrates are the main sources of energy. In the body, carbohydrates are converted into monosaccharides, principally glucose; the main fuel for metabolism. When there is no, or too little, insulin, glucose cannot be taken up from the blood by many of the cells of the body and, in consequence, the blood glucose concentration rises resulting in hyperglycaemia (hyper: too high; glycaemia: blood glucose concentration).

hyperglycaemia

Another function of insulin is in the regulation of storage of glycogen (as a reserve-pool for glucose) if there is no insulin, no glucose can be stored, and so the rise in blood glucose concentration is accentuated.

glucosuria

When the blood glucose concentration exceeds a threshold value, the kidneys excrete glucose in the urine; in order for this process (glucosuria = presence of glucose in urine) to occur, the kidneys require extra water (osmotic diuresis). With the high concentration of glucose in urine, water is retained and the volume of urine increases. This process is known as osmotic diuresis. That is why thirst, excessive drinking and polyuria (increased urine excretion) are general symptoms of diabetes mellitus.

polyuria

Fat and protein metabolism

Because cells cannot take up glucose efficiently in the absence of insulin, fats and proteins are used as sources of energy; fats (free fatty acids) are oxidised in the mitochondria to acetyl-CoA. In the absence of sufficient oxaloacetate, ketone bodies (acetoacetate, acetone and hydroxybutyrate) are produced from the acetyl-CoA. The excretion of these substances impairs the acid-base balance and further increases loss of water in the urine.

Proteins, after hydrolysis to their constituent amino acids, are transported to the liver by the portal vein. Thirteen of the amino acids are considered glucogenic, because their metabolism results in net synthesis of pyruvate and oxaloacetate; two essential amino acids, leucine and lysine, are called ketogenic, as their carbon chains can be converted into ketone bodies but cannot contribute to a net synthesis of glucose. Five amino acids are both glucogenic and ketogenic. The brain is mainly restricted to use glucose as a fuel and it is poisoned by the products of the uncontrolled metabolism of proteins and fats; that is why coma is a constant threat for diabetics. The metabolic defects and their associated clinical symptoms of diabetes mellitus are summarised in Table 4.1.

	Metabolic Defects	Chemical Abnormalities	Clinical Correlates
A) Carbohydrate Metabolism	1) Diminished uptake of glucose by tissues such as muscle and adipose tissue	Hyperglycaemia	Polyuria Polydipsia Polyphagia Fatigue Muscle weakness Pruritus
	2) Overproduction of glucose (via glycogenolysis and gluconeogenesis) by the liver		Blurred vision Diminished mental alertness
B) Protein Metabolism	1) Diminished uptake of amino acids and diminished synthesis of protein	Negative nitrogen balance Evaluated levels of branched-chain amino acids Elevated blood urea nitrogen level	Loss of muscle mass Weakness
	2) Increased proteolysis	Elevated potassium level	
C) Fat Metabolism	1) Increased lipolysis	Elevated plasma fatty acid level Elevated plasma glycerol level	Loss of adipose tissue Nausea and vomiting
	2.) Decreased lipogenesis	Elevated plasma ketones	Abdominal pain Acetone on breath
	3) Increased production of triglycerides	Hypertriglyceridemia	Exudative xanthoma (skin lesions) Lipemia retinalis Pancreatitis (abdominal pain)
	4) Decreased removal of triglycerides	Metabolic acidosis	Hyperventilation Rapid breathing

Table 4.1. Clinical and biochemical correlates of the diabetic syndrome

The following SAQ will test your understanding of the causes and consequences of diabetes mellitus.

<table>
<tr>
<td>

SAQ 4.1.

</td>
<td>

Indicate which of the following statements are correct:

1) The urine produced in the untreated condition of diabetes mellitus is characterised by containing ketones such as acetone and acetoacetate.

2) The urine of the sufferers of diabetes mellitus contains high levels of glucose.

3) Coma is caused by excessive drinking in severe cases of diabetes mellitus.

4) The absolute deficiency of insulin is characteristic of type II (adult type) diabetes.

5) In diabetes mellitus there is a rapid incorporation of glucose into glycogen in muscles and the liver.

</td>
</tr>
</table>

4.2.3. Insulin

insulin

Insulin is an anti-diabetic hormone produced in the pancreas by β-cells of the *islets of Langerhans*. Insulin from bovine (derived from the cow) and porcine (derived from the pig) pancreata have been used extensively in the treatment of diabetes mellitus.

Since the structure of insulin was described in 1955 by F. Sanger, the way has been open for the development of an insulin which closely resembles the actual structure of human insulin.

In Figure 4.1, the amino acid sequence of human insulin is shown. Insulin consists of two chains of amino acids, an A- and a B-chain, which are interconnected by disulphide bonds; there is also a disulphide bond in the A-chain itself.

Bovine insulin differs in 3 amino acids from human insulin and porcine insulin in only 1 amino acid as shown in Figure 4.2.

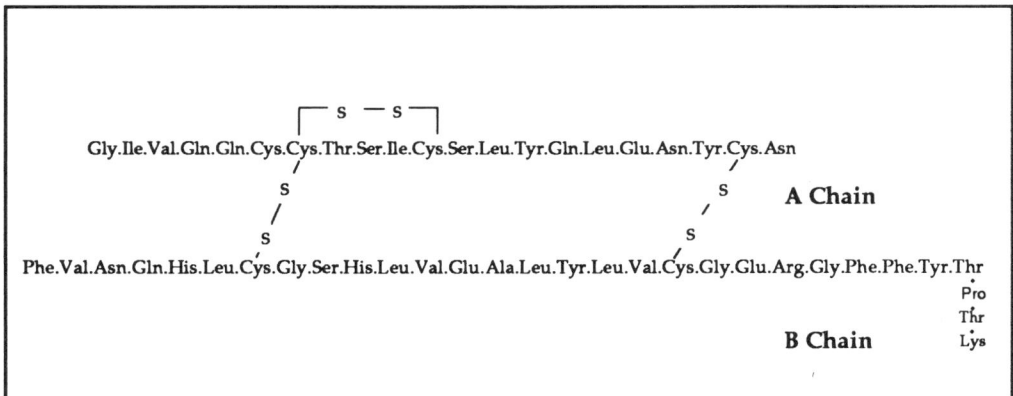

Figure 4.1. The amino acid sequence of human insulin

A-Chain			B-Chain
	8	10	30
Bovine	Ala	Val	Ala
Porcine	Thr	Ile	Ala
Human	Thr	Ile	Thr
All of the remaining amino acids are the same			

Figure 4.2.   Sequence differences between bovine, porcine and human insulin

∏   Can you think of any disadvantages of using bovine and porcine insulins for treating diabetics? (How do animal bodies react to foreign compounds?)

One disadvantage is that the differences in amino acid composition can result in the formation of antibodies; bovine insulin in particular is highly immunogenic. Furthermore, impurities like pro-insulin, can further stimulate antibody formation.

∏   What will be the effect of the presence of antibodies to insulin be for an insulin-dependent diabetic?

anti-insulin
antibodies

The presence of antibodies will lead to neutralisation of the insulin, so the diabetic will need more and more insulin to obtain the same hypoglycaemic effects.

insulin from
pancreata

Another disadvantage in using insulin of animal origin is that the pancreata must be removed fresh and must be frozen directly to avoid denaturation of insulin by proteolytic enzymes.

A number of reasons, therefore, call for the production of human insulin, identical in chemical structure to human pancreatic insulin and not containing antigenic impurities.

Possible routes available for the preparation of human insulin are:

• extraction from pancreata of human cadavers;

• full chemical synthesis;

• enzymic conversion of porcine insulin into an insulin with the same amino acid composition of that of human insulin;

• recombinant DNA technology.

availability of
cadaver

The availability of human cadaver pancreata is limited, and denaturation of insulin by proteolytic enzymes is rapid (post-mortem autodigestion), whilst fully-synthesised human insulin is too expensive for treatment of diabetes mellitus.

In the past, chemically synthesised human insulin was sometimes regarded as a potential source of commercial insulin; however, because of the complexity of the manufacturing process (more than 60 different steps!), it is unlikely that chemical synthesis will ever be a commercial source of insulin. In general, the manufacture of a drug may not exceed 15 steps; otherwise it is difficult to obtain commercial benefit. Thus only the last two possibilities can reasonably be considered.

semisynthetic
or humanised
insulin

Porcine insulin differs in only one amino acid from human insulin; the 30th amino acid in the B-chain is alanine in porcine insulin and threonine in human insulin. An enzymatic modification has been developed to replace alanine by threonine and this insulin is called semisynthetic or humanised insulin (Fig. 4.3.)

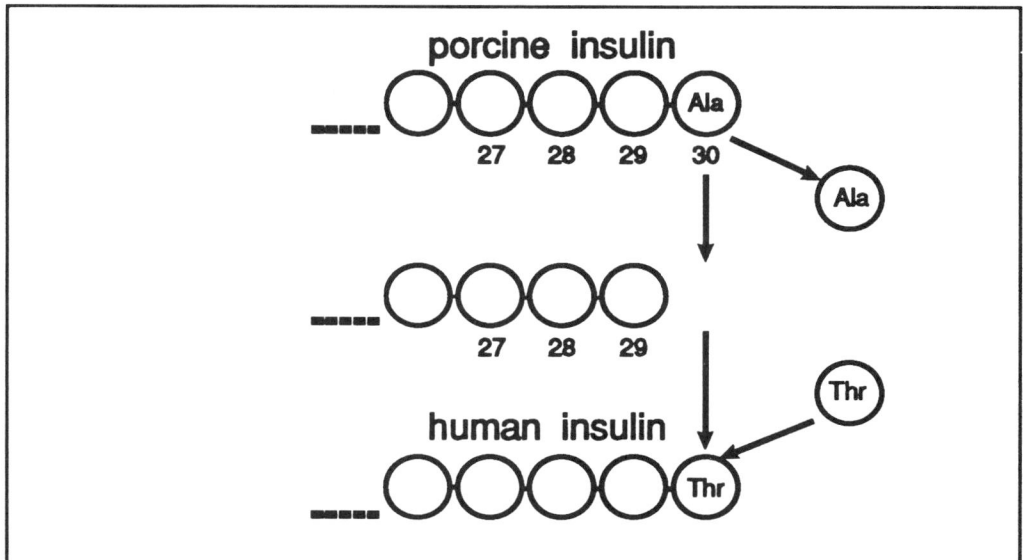

Figure 4.3. Conversion of porcine insulin into 'human' (semisynthetic) insulin

Although the total number of pigs and cattle being slaughtered continues to grow from year to year, and we do not by any means collect all of the pancreas glands for the production of insulin, the rate at which insulin is being used is still increasing more rapidly than the rate of increase in the availability of pancreas glands.

∏ The number of diabetics using insulin is expected to have increased by a factor of more than two towards the end of this century. Can you give any reasons why? If so, write them down before reading on.

There are, in fact, several reasons.

Due to a change in eating habits, an increase in type II diabetics is expected, and their treatment is changing in favour of use of insulin. In addition, diabetics live longer than they did a few decades ago and, the care of pregnant diabetic woman has much improved, making it safer for them to have children, so that the prevalence of diabetes in the general population is expected to increase.

We can add that improved methods of detection and a greater public awareness of diabetes in all developed countries have also increased the demand for a dependable supply of insulin.

insulin by recombinant DNA technology

In the light of all of this, Eli Lilly & Company came to the conclusion that the genetic alteration of microorganisms turning them into producers of insulin held greatest promise for a continuation of insulin-therapy in the future. An extra advantage was recognised that human insulin produced by recombinant DNA technology would be free of pancreatic peptides, such as proinsulin, glucagon, somatostatin, pancreatic polypeptides and vaso-active intestinal peptides that are present as impurities in insulin preparations derived from pancreatic extraction, because only the insulin sequence would be produced.

4.3. Biotechnological development of rDNA human insulin

4.3.1. Principles of biotechnology

Isolation of the insulin gene

To isolate the gene for insulin, human pancreas cells are used as a source for the messenger RNA molecules that have been copied from the gene as part of the process of protein synthesis.

reverse transcriptase

Some RNA viruses are able to convert the genetic information in mRNA into DNA by means of the enzyme reverse transcriptase. Such viruses (retroviruses) convert RNA nucleotide sequences into their DNA equivalents. (Figure 4.4.).

cDNA

The viral enzyme, reverse transcriptase, mixed with human mRNA *in vitro* can be used to synthesise the complementary or copy DNA (cDNA) - sequence of the required gene (the insulin gene) (Figure 4.5.).

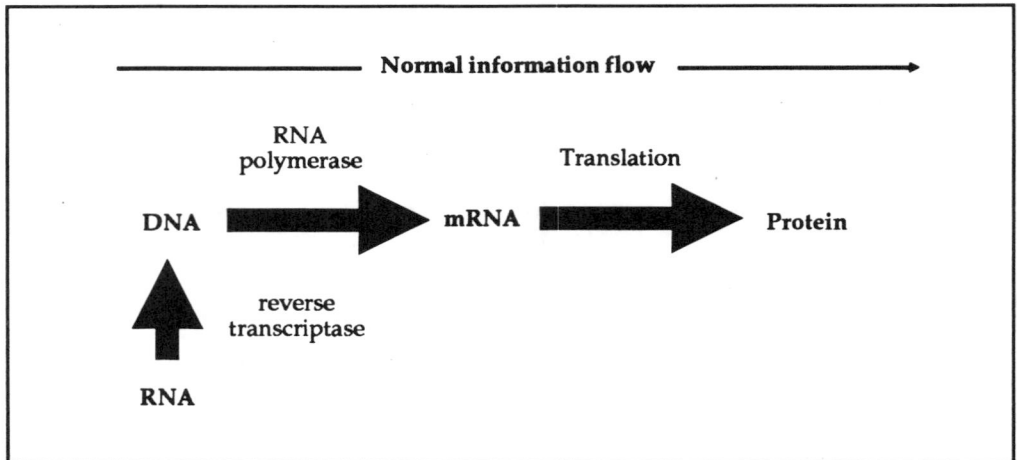

Figure 4.4.   Reverse transcriptase action in relation to the central dogma of molecular biology

Another way to obtain the desired DNA would be by synthetic nucleotide chemistry, but to do this demands knowledge of the nucleotide sequence required and quite complex chemical synthesis.

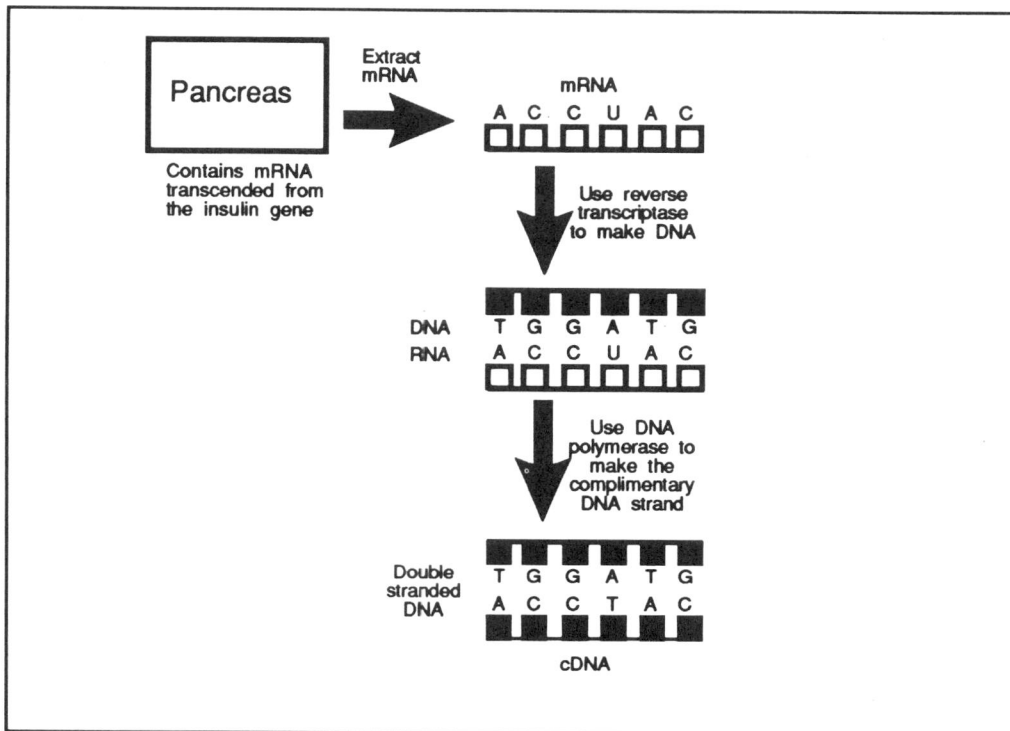

Figure 4.5. How to produce DNA containing the nucleotide sequence of the insulin gene

Plasmids

Once isolated, the DNA molecule coding for insulin must be introduced into an appropriate host cell. The insertion of genes can be facilitated by bacterial plasmids (small, circular, cytoplasmic DNAs). Insertion of the human cDNA-gene for insulin into a plasmid, facilitates its introduction into a microbial host for a biotechnological application. Plasmids used in this way are called vectors.

vectors

Host system

Π *Escherichia coli* was chosen to be the host, why do you think this was so? (Again write down as many reasons as you can think of before reading on).

attenuated

E.coli strain

There are many factors to be considered. The principle ones are concerned with the ease of cultivating the host and with issues of safety. For reasons of safety and containment Eli Lilly & Co. chose the K-12 strain of *E. coli* for use in recombinant DNA research and production. Because it is a weakened strain of *E. coli* it does not survive outside the very carefully controlled conditions of the laboratory and production process. It is important to emphasise this, because one of the fears that society initially had about this technology when it moved into large-scale production was what might happen if some of the material escaped into the environment. In reality, the organism used would not survive more than a few hours. In addition, the bioengineered product inside the *E. coli* is not biologically active until isolated, chemically modified and purified.

Figure 4.6. summarises the steps of the process by which the insulin gene is introduced into a bacterial plasmid. After isolation, the plasmid is cut with a restriction enzyme. Specific linker sequences are added to the ends of the insulin DNA so that these ends fit exactly into the opened plasmid. Finally, the insulin DNA is covalently attached to the plasmid with the enzyme, DNA ligase.

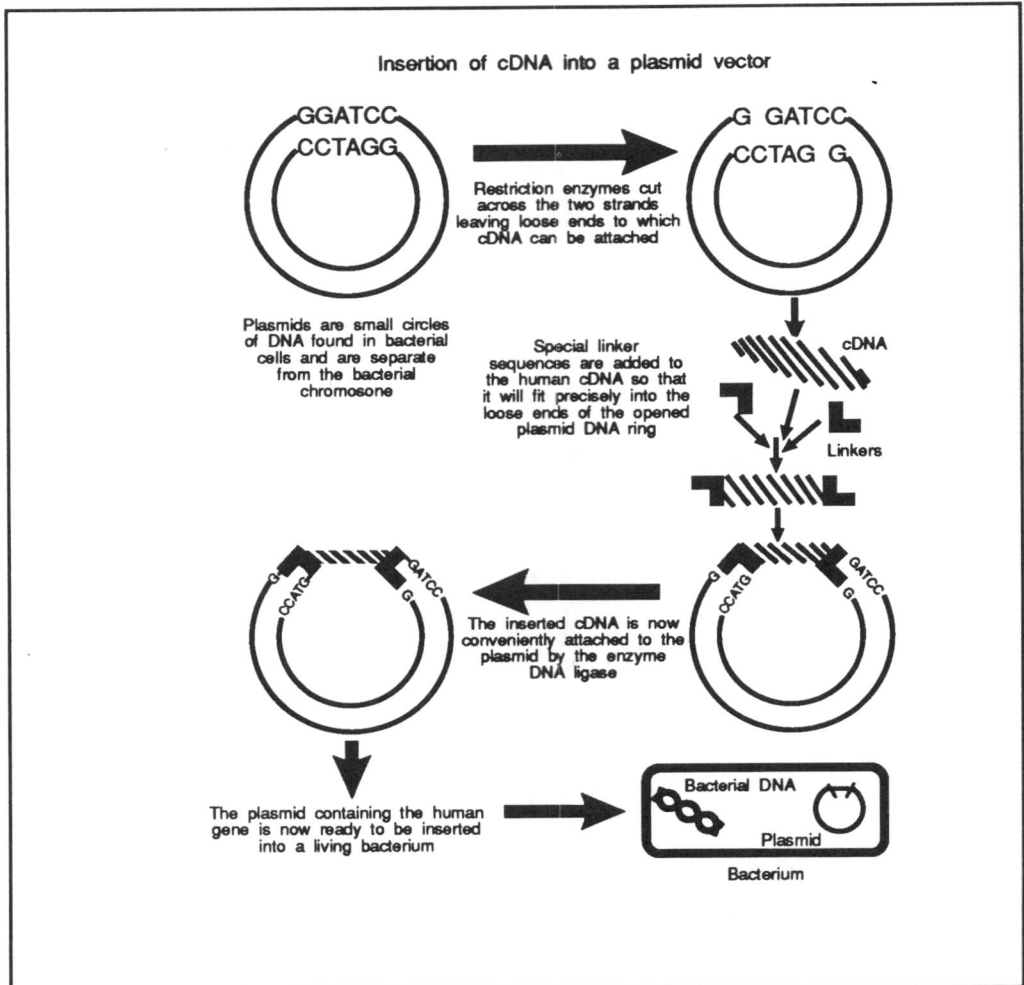

Figure 4.6. How the insulin gene is isolated

Let us examine a little more closely the insertion of the insulin gene into the plasmid.

When the plasmid is cut with a restriction enzyme it produces a linear molecule with short single strands of DNA at each end. In the example shown in Figure 4.6., these ends would have the structure:

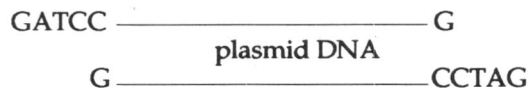

GATCC ———————————————— G
                  plasmid DNA
G ————————————————CCTAG

To make these ends specifically join up with the insulin gene, then we have to produce an insulin gene with the appropriate nucleotide sequence at its ends. We could, for example, add to each end of the insulin gene the nucleotide sequence CTAG, to produce a sequence with the structure:

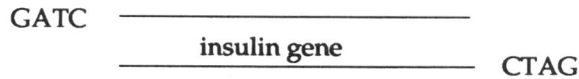

GATC ————————————————
                    insulin gene
              ———————————————— CTAG

If we now incubate the linear plasmid DNA with the insulin gene, then they will naturally align themselves in the manner shown in Figure 4.7.

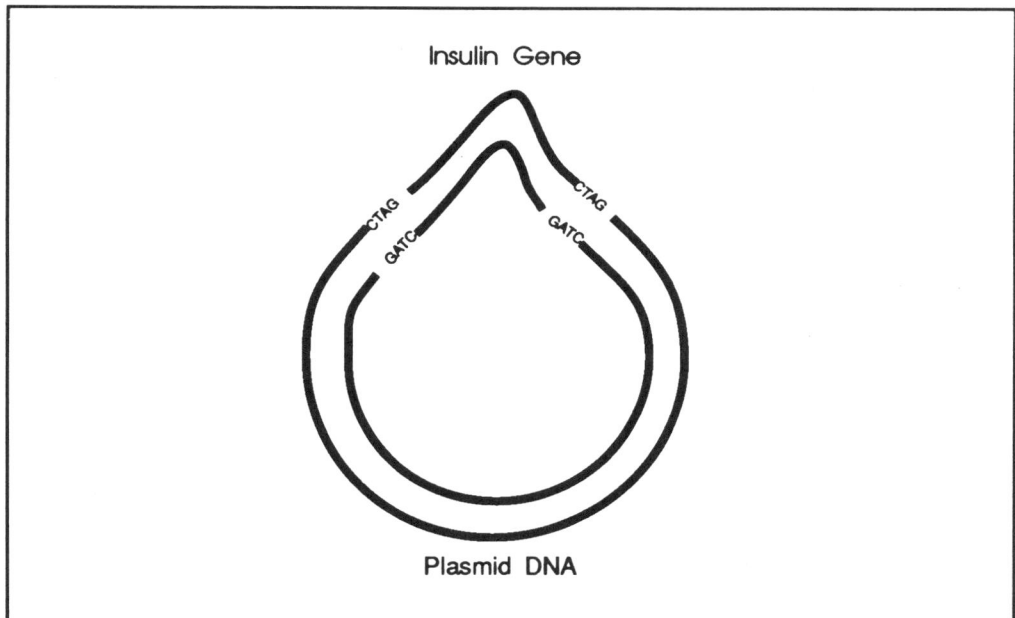

Figure 4.7.   The insertion of the insulin gene into plasmid DNA

linkers

A ligating enzyme will then join up the molecule. We can regard the CTAG - sequence added to the insulin gene as a "linking" sequence. We can define the function of such linking sequences (or "linkers" as they are often called) as ensuring that cDNA is compatible with the plasmid DNA into which they are to be inserted. There are a wide variety of linkers available which can be used for different restriction enzyme sites.

Clone selection

clone selection

Plasmid pBR322 is used extensively in genetic engineering. It has two genes of special interest. One codes for a protein that enables any host bacterium to resist the lethal effects of the antibiotic ampicillin and the other confers resistance to tetracycline. The latter contains a sequence of six bases recognised by the restriction enzyme *Bam* HI.

When a human insulin gene is inserted into this *Bam* HI restriction site, the tetracycline-resistant gene no longer functions as normal, so that a bacterial cell with the new gene incorporated is sensitive to tetracycline, whilst still being resistant to ampicillin. This is the basis of a simple test to check for incorporation of a new gene. (Figure 4.8.)

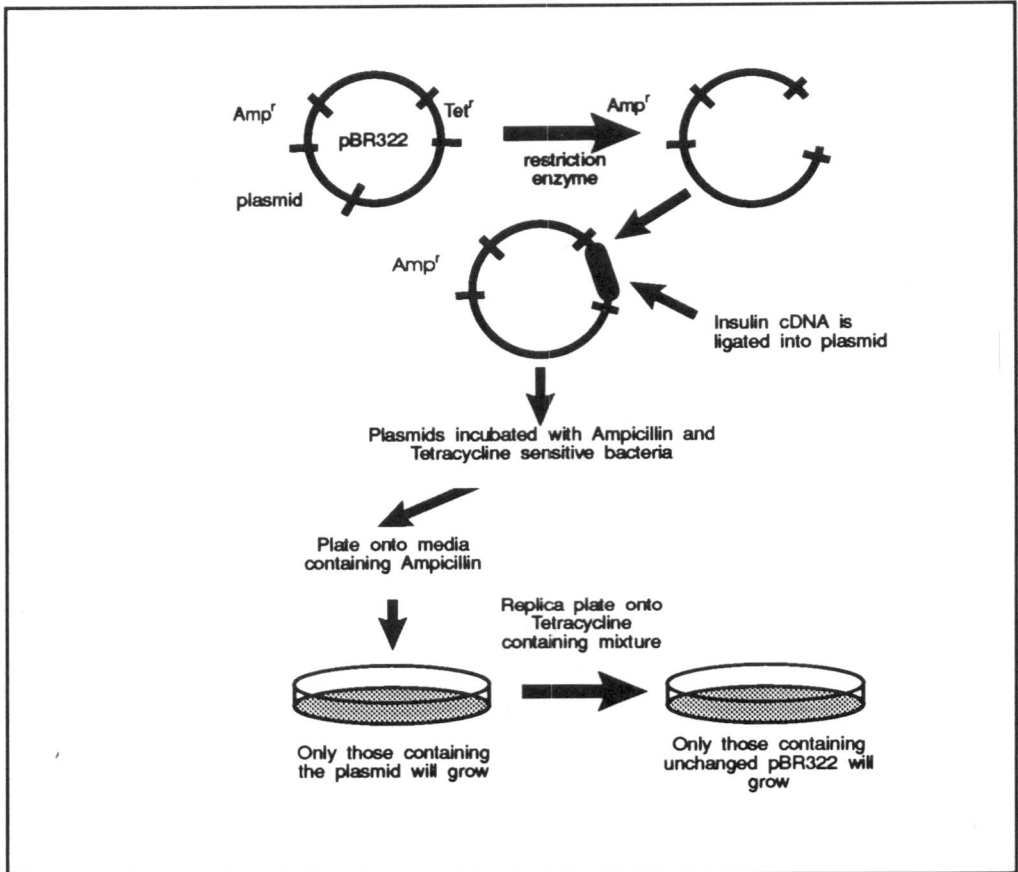

Figure 4.8.   Screening technique used to identify bacteria carrying plasmids with genetic material inserted in tetracycline resistance gene, Amp^r = Ampicillin resistant gene, Tet^r = Tetracycline resistant gene

A mixture of bacteria and plasmids are incubated and then spread onto nutrient plates containing ampicillin. Only those cells which have received a plasmid will grow. These are replica plated onto a medium containing tetracycline. Those cells which contain unmodified plasmids will grow. Those which have foreign DNA in the middle of the tetracycline gene will not. Thus these may be identified.

Another key test is to pinpoint the bacterial colonies that are manufacturing insulin. This is done by attaching insulin antibodies to a membrane which is placed on bacterial colonies grown from a culture transformed with the recombinant plasmid. (Figure 4.9.).

The membrane picks up any insulin molecules present and is then transferred to a dish containing radioactively-labelled insulin antibodies. The membrane containing antibody-insulin-radioactive antibody complexes is finally placed against a photographic film, and the resulting appearance of silver grains unmistakably maps the positions of insulin-manufacturing colonies.

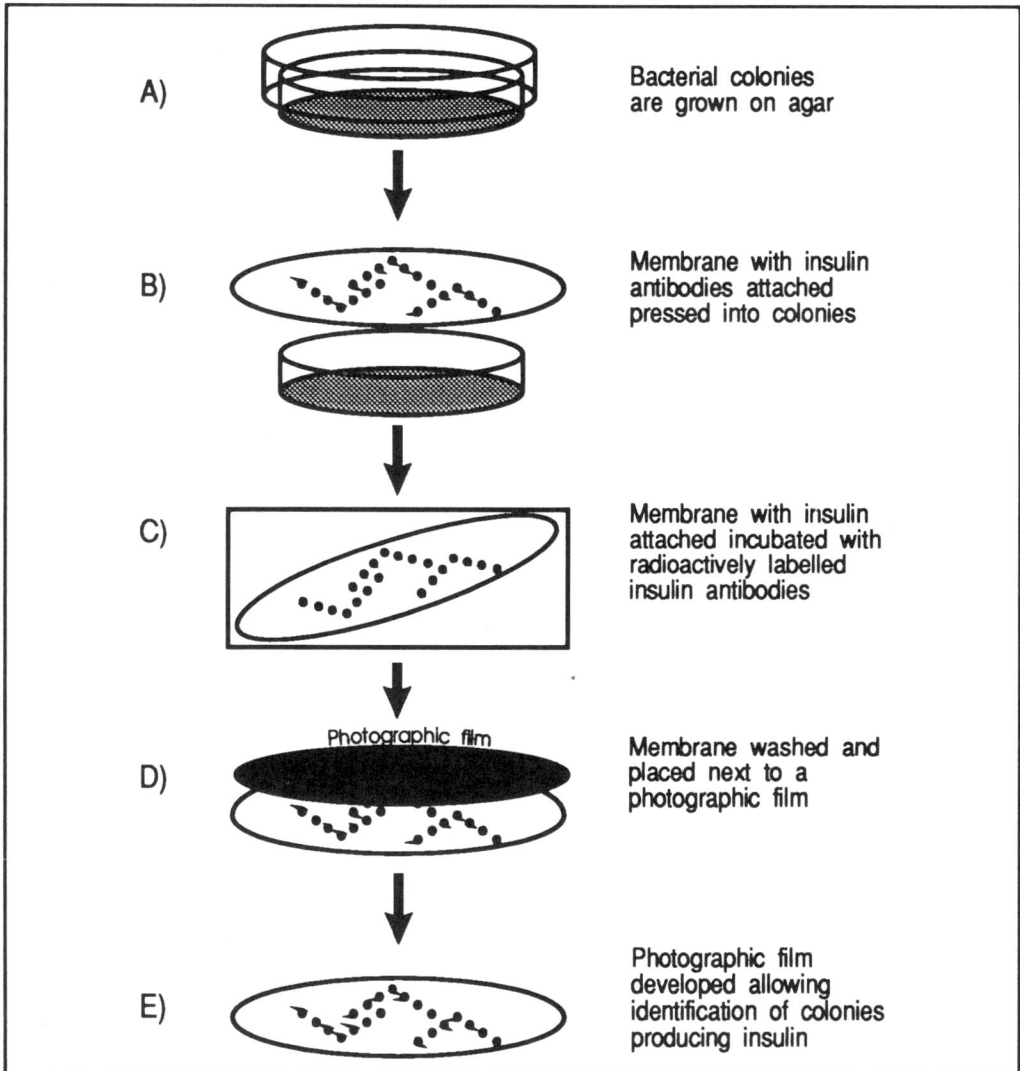

Figure 4.9. Detection of colonies producing insulin by blotting technique

Gene cloning

Plasmids can be replicated inside the host cell. Once inside the bacterium, the plasmid containing the human cDNA can multiply to yield several dozen replicas. With cells dividing rapidly (every 20 minutes) a bacterium containing human cDNA encoding for insulin will, in a relatively short time, produce many millions of similar cells (a clone) containing the same human insulin gene. (Figure 4.10.).

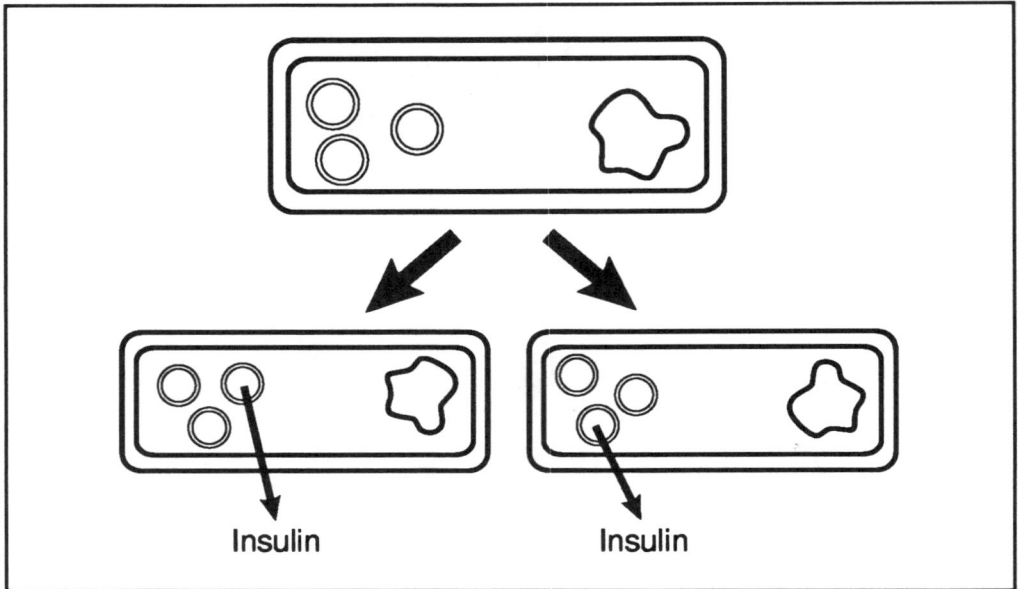

Figure 4.10. Cloning a human insulin gene leading to the production of a large number of cells which will produce insulin

| **SAQ 4.2.** | In attempting to produce clones of *Escherichia coli* capable of making human insulin, place the activities listed below in the order that they should be carried out. Thus, if you think that step A should be done first, write A in the first box and so on. (Omit those steps which you do not think should be carried out). |

Step

A) Lay a membrane onto a petri dish containing colonies of *Escherichia coli.*

B) Collect some Islet of Langerhans cells and extract their RNA.

C) Add some human DNA to a suspension of plasmid pBR322 and treat with a mixture of restriction enzymes.

D) Add reverse transcriptase to a mixture of mRNA isolated from pancreatic cells and deoxyribose nucleotide triphosphates, and, after incubation, purify the cDNA synthesized.

SAQ 4.2.
cont'd

E) Cut open plasmid pBR322 with a restriction enzyme.

F) Add pBR322 to a culture of *Escherichia coli*.

G) Add a DNA ligating system to a mixture of linear pBR322 and cDNA.

H) Place *Escherichia coli* culture which has been incubated with pBR322 which had previously been ligated with human cDNA onto nutrient media containing ampicillin.

1st	2nd	3rd	4th	5th	6th	7th	8th

4.3.2. Biotechnological production of human insulin

prepro-insulin

The mRNA for insulin in a human cell encodes four distinct regions- the A, B, C chains and the signal sequence. (See Figure 4.11.).When insulin mRNA is translated, prepro-insulin is formed which contains all four parts.To obtain active insulin, human pancreatic cells first remove the signal sequence, which contains an address directing the newly synthesised molecule to the appropriate part of the cell. This yields pro-insulin from which the C-chain is removed, leaving only linked A- and B-chains, which make up the functional hormone.

In the production of genetically engineered insulin, the DNA for proinsulin is inserted into a plasmid. This is then introduced into *E. coli* to be transcribed and translated to produce proinsulin; the proinsulin is extracted and the C-chain removed to form insulin (Figure 4.11.). We will examine this in more detail later.

promoter

In order to maximise the production of the desired protein in *E. coli*, the gene that is inserted into the plasmid also contains a so called promoter. This is a DNA sequence which determines the rate at which mRNA is formed; thus, with a strong promoter, more mRNA is formed and consequently there is greater production by the cell of the desired gene product.

For human insulin biosynthesis, two promoter systems have been used, originally β-galactosidase (β-gal) and now tryptophan synthetase (Tryp E). The latter yields more insulin as compared to the ß-galactosidase promoter system.

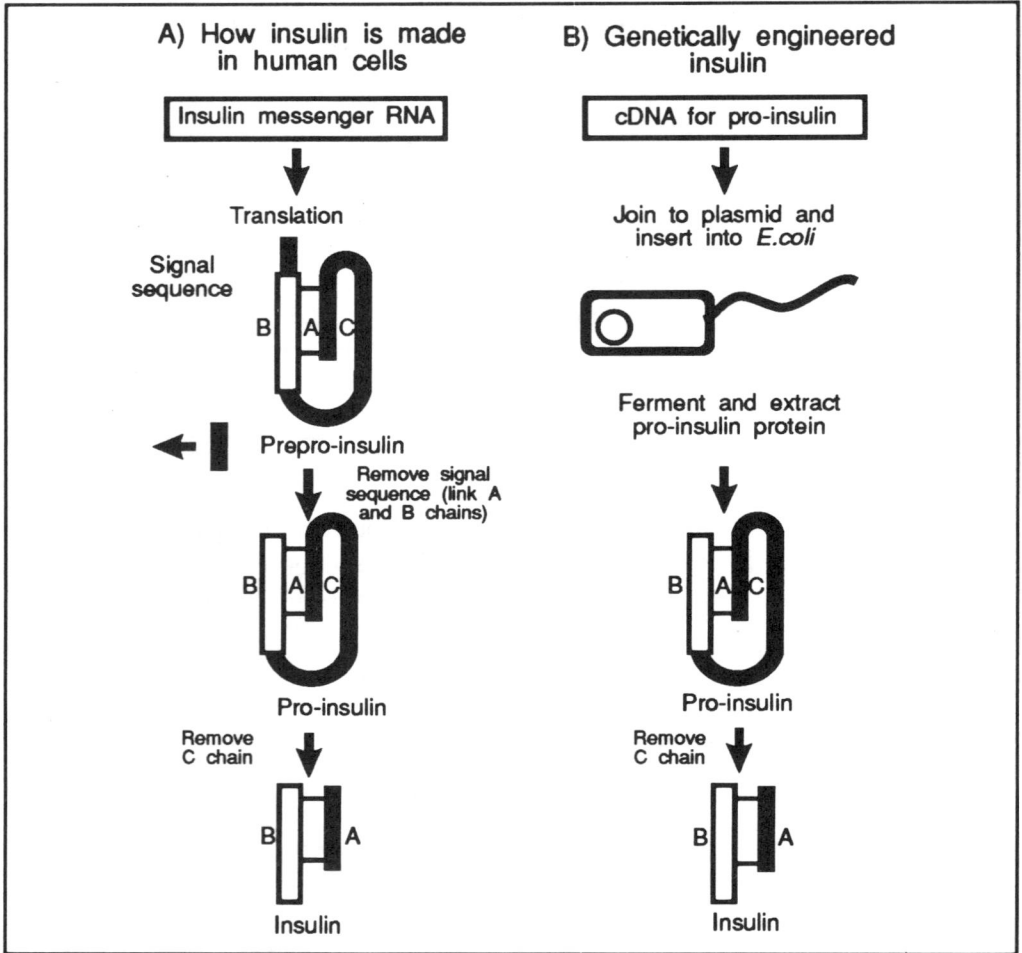

Figure 4.11. How insulin is produced (a) in a human cell and (b) by genetic engineering.

Let us consider these each in turn.

When human insulin gene is attached to β-galactosidase *gene we will have produced the gene sequence:*

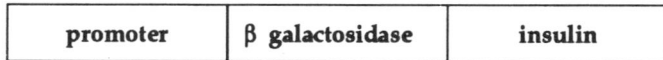

promoter	β galactosidase	insulin

chimeric protein

We can anticipate this gene would produce a hybrid protein namely β-*galactosidase insulin.* Such hybrid proteins are called *chimeric proteins.*

⊓ What would be the product if the insulin gene was attached to the Tryp E gene?

We would expect the gene to produce the chimeric protein Tryp E insulin. Figure 4.12. gives some more examples of chimeric proteins.

∏ Draw the gene combinations which would result in the production of the chimeric proteins shown in Figure 4.12.

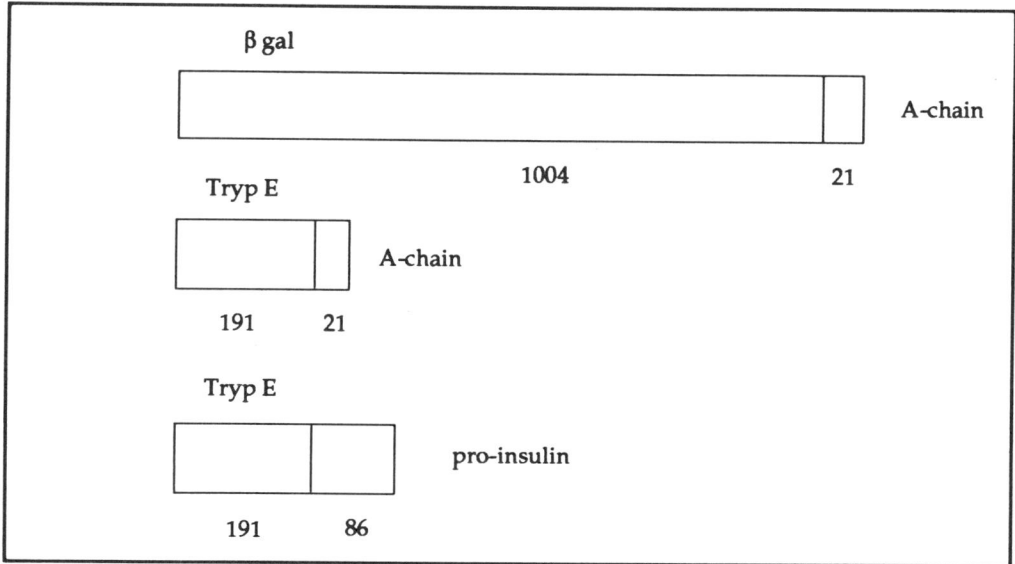

Figure 4.12. Chimeric proteins contain insulin amino acid sequences. (Numbers represent the number of amino acids in each part of the protein).

The gene combination which would produce the proteins shown in Figure 4.12. are:

β gal promoter	β gal structural gene	cDNA for insulin A gene portion

Tryp E promoter	Tryp E structural gene	cDNA for insulin A gene portion

Tryp E promoter	Tryp E structural gene	cDNA coding for pro-insulin

∏ Which of these gene combinations would be most efficient in terms of insulin production?

Examine Figure 4.12. again. The number of amino acids in each part of the chimeric proteins is reported.

Consider first the system using a β-galactosidase promoter system. The A chain of insulin is 21 amino acids long, the B chain is 30 amino acids long.

To make one A chain molecule, the cell will have to produce a protein chain which is 1004 + 21 amino acids long. To make one B chain molecule, the cell will have to produce a protein chain which is 1004 + 30 amino acids long. Therefore to produce one molecule of insulin (which is composed of 1 A chain and 1 B chain), it will have to join a total of 1004 + 21 + 1004 + 30 amino acids together, i.e. 1 insulin molecule might be produced per 2059 amino acids joined together.

If the tryptophan synthetase (Tryp E) promoter is used, then to make one A chain molecule 191 + 21 amino acids will have to be joined together and for one B chain, 191 + 30 amino acids will have to be joined together.

Thus in this case 191 + 21 + 191 + 30 amino acids will have to be joined together to produce 1 insulin molecule, i.e. 1 insulin molecule is produced for every 433 amino acids used.

We can report this as a ratio. For the β-galactosidase promoter we produce one (51 amino acids) insulin molecule for each 2059 amino acids joined together, i.e. a ratio of 51:2059 = 0.025. For the tryptophan synthetase promoter, this ratio is 51:433 = 0.12.

Thus a much higher proportion of the protein synthesis promoted via the Tryp E system is insulin compared with the protein synthesised promoted by the β-galactosidase system.

Thus the tryptophan synthetase promoter is a much more efficient system for producing insulin. When *E. coli* is producing the desired gene product, one can see electron dense bodies (representing secretion granules, produced by polyribosomes). Immunocytochemical techniques have shown these to be the A chain (or pro-insulin) linked to the bacterial protein.

We can visualise two routes for producing human insulin by recombinant DNA technology. One method is to make the A- and B-chains in separate *E. coli* strains, while the second route is to use a single *E. coli* to produce pro-insulin which is subsequently converted into insulin. It is important to recognise that both methods yield an equivalent preparation of human insulin. Since 1985, the pro-insulin route has been used in producing human insulin by Eli Lilly & Co. Figure 4.13. illustrates the identical HPLC-curves for human insulin produced by the two methods.

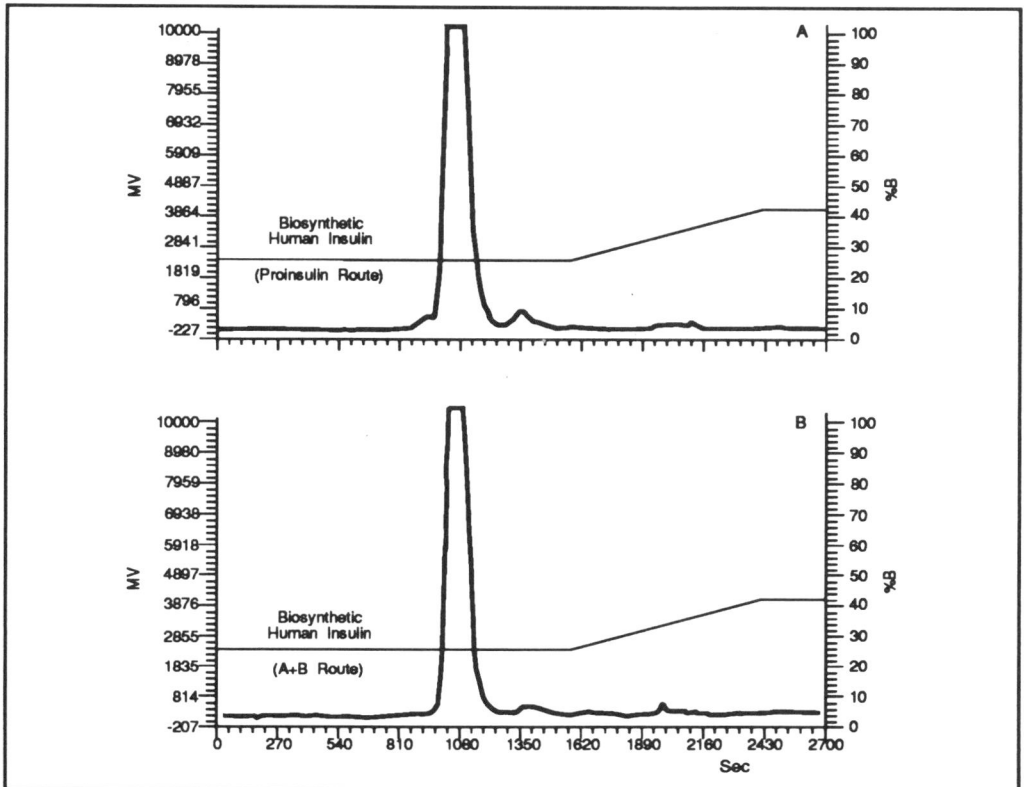

Figure 4.13. HPLC of human insulin (recombinant DNA) produced by way of pro-insulin (top) and A- and B-chain combination (bottom)

∏ Can you think of any advantages of the pro-insulin route? To give you a clue, attempt the following SAQ.

SAQ 4.3.

Using the data presented in Figure 4.12. and the sample calculation of the efficiency of the promoter systems, calculate the ratio of insulin produced against total protein synthesis in a system producing the Tryp E pro-insulin chimeric protein.

Your calculation should have shown that this is a more efficient route for producing insulin in terms of the proportion of insulin in the chimeric proteins produced. There are, however, other advantages. These are: that the pro-insulin route only requires a single bacterial fermentation while the production of two separate chains (A and B) requires two bacterial strains in separate fermentations; the pro-insulin route also leads to the production of the C peptide as well as pro-insulin and insulin. Thus although the primary objective is to produce insulin, the C peptide has interesting properties of its own and could be of value.

4.3.4. The two methods of producing rDNA insulin

Method I - The separate chain method

An outline of this method is described in Figure 4.14.

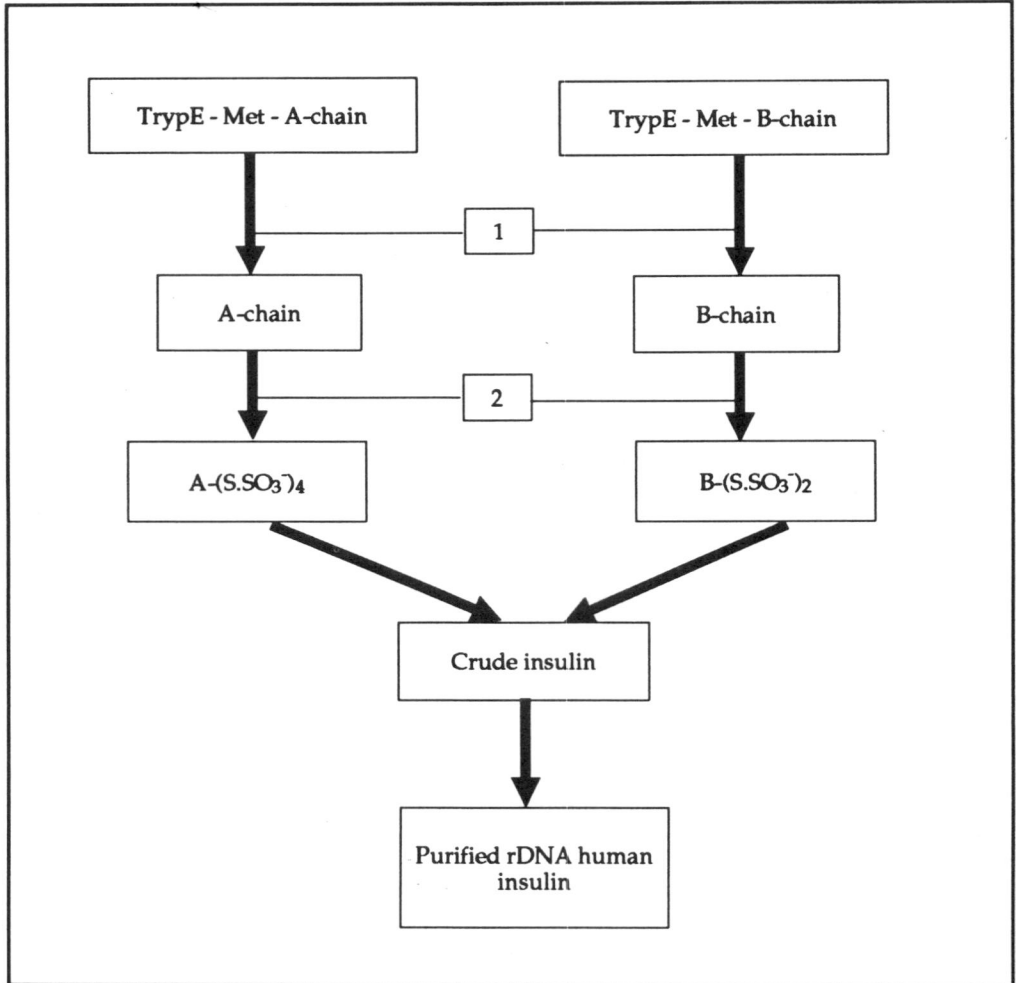

Figure 4.14. A chimeric product is represented schematically as Tryp-E-Met-A-chain
1) CNBr: cyanogen bromide (splits desired polypeptide from promoter protein)
2) Oxidative sulphitolysis

importance of the code for methionine in the linker

In constructing a linker for the incorporation of the human insulin gene into a plasmid, it was provided with the code for the amino acid methionine. The methionine (Met)-linkage provides a specific chemical cleavage site for release of the desired polypeptide from the promoter-protein, Tryp-E since methionine is sensitive to cyanogen bromide (CNBr). (Reaction 1 in the scheme described in Figure 4.14.).

Methionine does not occur elsewhere in pro-insulin and it is for this reason that a methionine residue has been deliberately introduced at the junctions between Tryp-E and the insulin sequences. Thus for the insulin A chain we could write:

Tryp E   insulin A   Tryp E   insulin A
| CNBr
Met

Similar for the insulin B chain we could write:

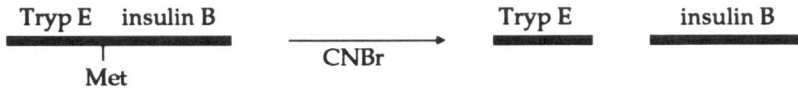

Tryp E   insulin B   Tryp E   insulin B
| CNBr
Met

After the CNBr-cleavage, the A- and B-chains are converted to the stable S-sulphonate derivatives, purified and chemically combined to yield insulin. The insulin is then purified by gel-filtration and ion-exchange chromatographic procedures.

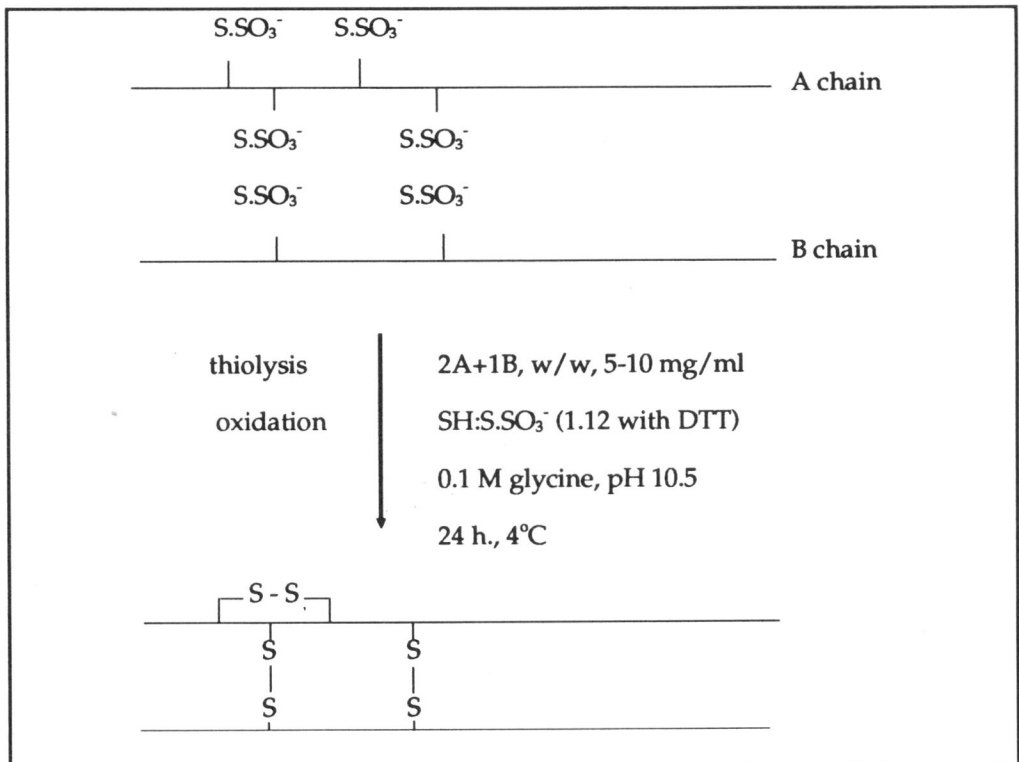

Figure 4.15. Scheme for combining the separate chains of insulin

The optimal conditions that have been established for preparing human insulin by the chain combination route are outlined in Figure 4.15. A predetermined amount of dithiothreitol (DTT) to give an SH (sulphydryl):SSO₃ (S-sulphonate) molar ratio of approximately 1:2 is dissolved in a 0.1 M glycine buffer, pH 10.5 and quickly added to a chilled solution of A and B chains (5-10 mg/ml), containing a 2:1, A:B ratio, by weight, of S-sulphonated chains in the glycine buffer. The resulting solution is stirred for 24 hours at 4°C in an open vessel to permit air oxidation.

∏   What is the rationale for this last procedure?

As a result of open air oxidation reactions, the A and B chains are permitted to form the proper disulphide bonds. Under these conditions, the yield of insulin is approximately 60 percent of theory relative to the amount of B chain used. Isolation and purification are by column chromatography and crystallisation. By-products are recycled.

An extraordinary variety of products may result from the chemical combination of the two insulin polypeptide chains, particularly if the disulphide bonds are formed randomly.

SAQ 4.4.   Give some examples of potential products that could be formed during the combination of A and B chains by the procedure described above (Use Figure 4.15. to help you).

In fact, a reversed-phase HPLC analysis of a typical combination mixture displays considerable heterogeneity (see Figure 4.16.). However, the high yield of insulin shows clearly that the combination reaction does not proceed in a totally random fashion. In addition, purification studies indicate that most of these side products are related to monomeric A or B chains. These can be recycled.

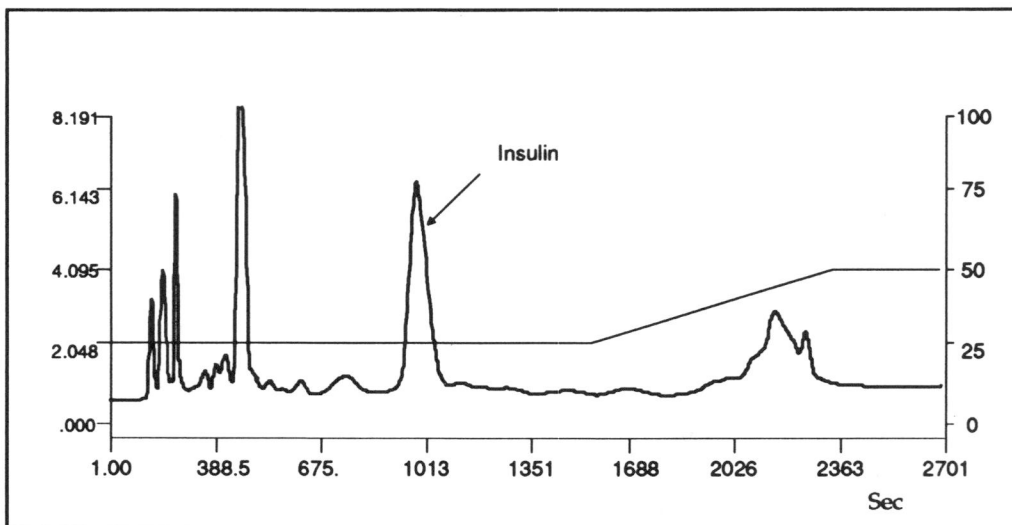

Figure 4.16 HPLC profile of the products of a completed chain combination reaction.

∏   Write down factors which you think could lead to a high yield of insulin in the correct conformation?

The arrangements of the SSO_3^- groups on the A and B chains must enable the correct combinations to react. In other words, SSO_3^- : SSO_3^- interactions are governed by the structure of the A and B chains. Also it appears that the insulin produced is chemically stable.

Several variables were studied in detail to optimise the insulin yield, including pH, protein concentration, temperature, the A:B ratio and the $SH:SSO_3^-$ ratio. Of these, the $SH:SSO_3^-$ ratio is the key factor permitting a single step, single solution combination reaction to work successfully (Figure 4.17.).

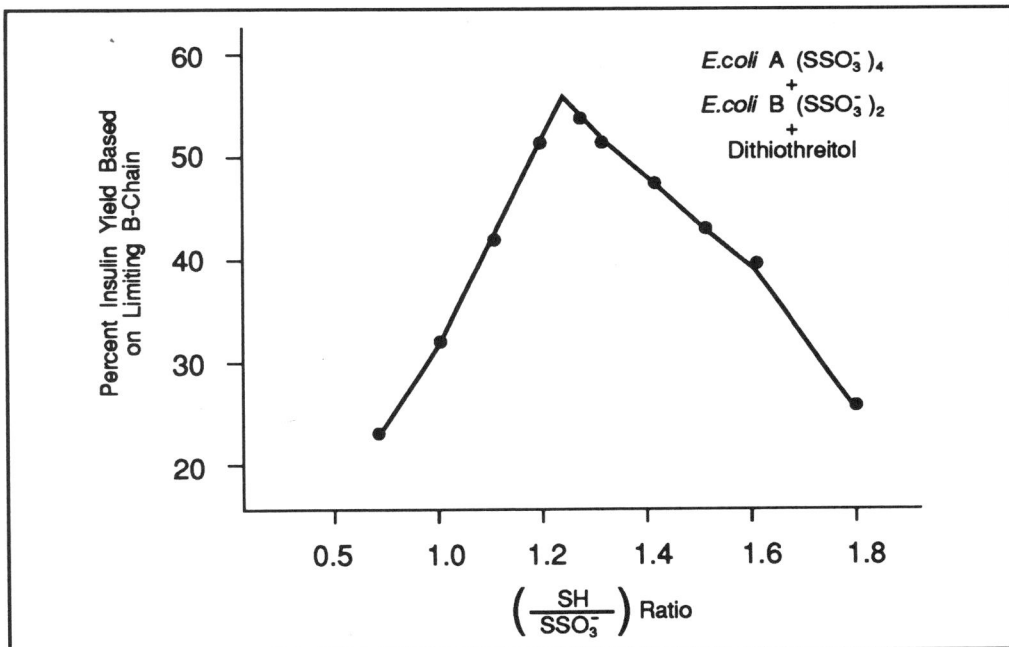

Figure 4.17. The effect of different molar ratios of $SH:SSO_3^-$ on the yield of insulin in the chain combination method. Insulin yield was based on an HPLC analysis as described in Figure 4.16.

Method II: rDNA human Insulin: "Pro-Insulin Route"

Figure 4.18. outlines the process of producing rDNA human insulin by the pro-insulin route. In this case, the chimeric protein (Tryp-E Met-Pro-insulin) is produced in a single fermentation. After cleavage with CNBr, the pro-insulin released is in the wrong configuration. To achieve the correct configuration, the -S-S-bonds are first broken by oxidative sulphitolysis (step 2) and then re-formed by reduction (step 3) to produce pro-insulin in the correct configuration. The pro-insulin is then purified (step 4).

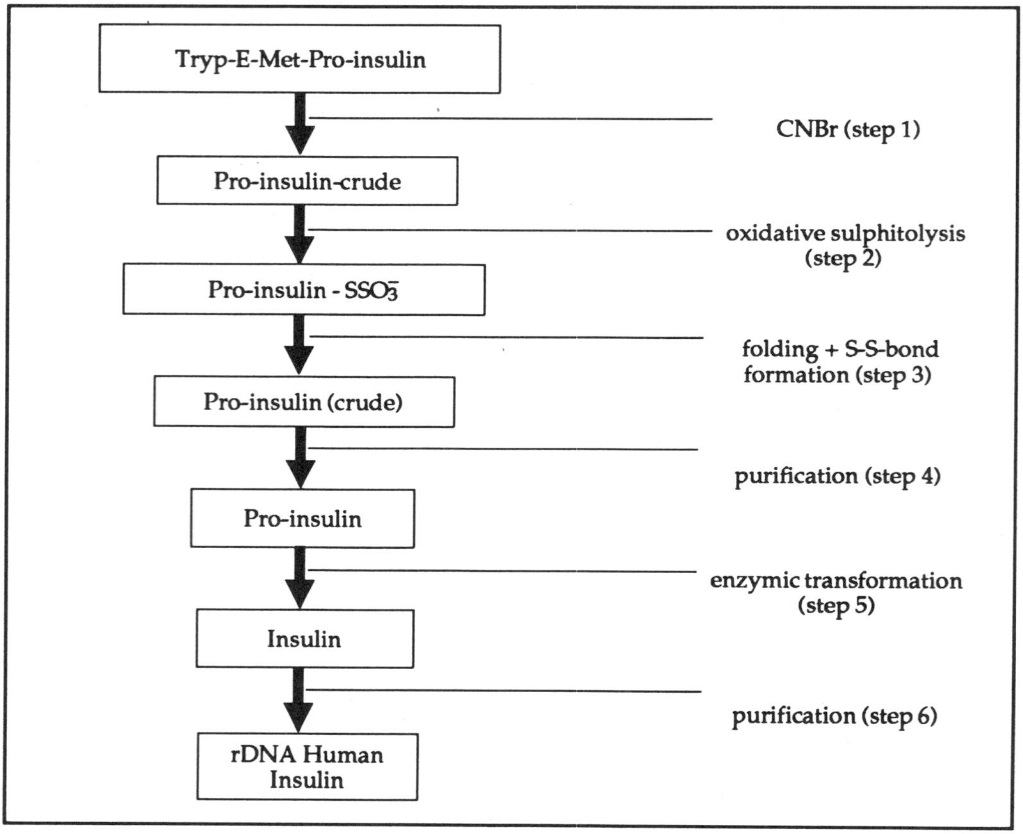

Figure 4.18. An outline of the scheme to produce rDNA human insulin by the pro-insulin route

Finally, proinsulin is converted, in greater than 95 percent yield, into insulin using a combination of the enzymes trypsin and carboxypeptidase B (step 5). This enzymic conversion of pro-insulin to insulin is very rapid. Figure 4.19. shows the progress of the reaction monitored by HPLC.

At the beginning of the reaction, only pro-insulin is present. Five minutes later, the reaction has clearly begun. After 15 minutes substantial conversion has occurred, whilst by 30 minutes the reaction is complete, and insulin has been obtained.

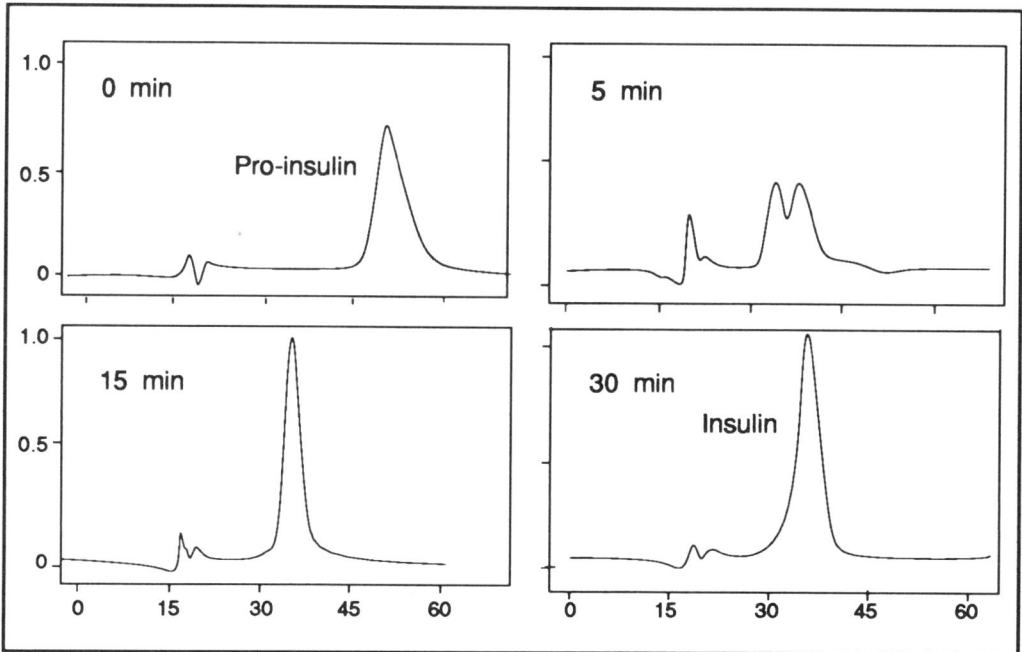

Figure 4.19. The enzymic conversion of pro-insulin to insulin monitored by HPLC.

In Figure 4.20 the sites of action of trypsin and carboxypeptidase on pro-insulin are described.

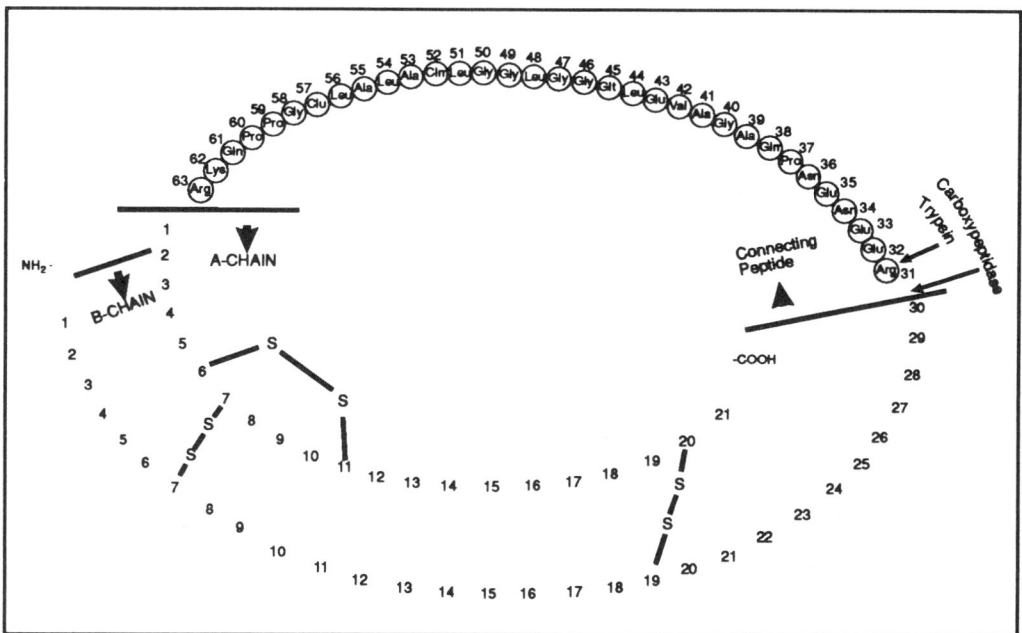

Figure 4.20. Conversion of pro-insulin to insulin

The rDNA human insulin is purified by gel filtration, ion-exchange chromatography and crystallisation (step 6).

SAQ 4.5.	Why did Eli Lilly choose:-

1) *Escherichia coli* strain K12 as the host system to produce human insulin?

2) to use a tryptophan synthetase promoter rather than a β-galactosidase promoter?

3) to use pro-insulin cDNA rather than cDNAs for separate A and B chains?

4) to include a codon for methionine adjacent to the cDNA insert?

5) to first oxidise pro-insulin to break disulphide bonds and re-reduce them to reform disulphide bonds?

6) to use enzymic rather than chemical methods to convert pro-insulin into insulin?

Now that we have examined the development of the processes by which human insulin may be produced by using recombinant DNA technology, we turn attention to aspects of product quality assurance and evaluation. It will be recalled that in order to gain a marketing licence a wide variety of criteria need to be fulfilled. These include characterising the product and making comparisons with natural (or standard) materials in order to evaluate its pharmacological activity to ensure that it is safe to use. The next section is concerned with those studies which were undertaken to fulfil the authorisation requirements.

4.4. Characterisation of rDNA human insulin

4.4.1. General

A wide variety of tests has been used to examine rDNA human insulin. The results of all of these tests can be summarised by saying that rDNA human insulin has been shown to be chemically, physically, biologically and immunologically identical to pancreatic human insulin used as a standard.

4.4.2. Chemical Analysis

HPLC

One of the most sensitive methods now available for analysis of mixtures of substances, particularly chemically similar substances, is High Performance Liquid Chromatography (HPLC). The HPLC assay offers a number of advantages over the rabbit bioassay and nitrogen assay for the determination of insulin potency. The rabbit bioassay currently in use for the certification of human insulin measures the bioactivity of insulin, i.e. the lowering of blood glucose. Compared to HPLC, this biopotency assay is substantially more variable.

bioassay relative to HPLC

The imprecision of the bioassay relative to the HPLC assay can be seen in the comparative data in Table 4.2. A series of 29 lots of human insulin were assayed by both techniques. The variability (expressed as a relative standard deviation) across lots was about 1% for HPLC and approximately 3 to 4% for the bioassay.

Lot Number	Potencies, Unit/mg by	
	Rabbit bioassay	HPLC Method
791CJ2*	29.4	28.5
989FX2	28.0	28.3
990FX2	26.9	28.6
801PP2	27.8	28.1
992FX2	26.5	28.1
283DL2	26.4	28.7
568PP2	27.1	28.8
284DL2	26.5	28.2
285DL2	29.1	28.5
286DL2*	28.8	28.6
398DC3*	28.4	28.6
339DC3	26.2	28.3
400DC3	26.5	28.5
401DC3	27.2	28.3
402DC3*	28.0	28.2
933CN3*	28.5	28.3
525PP2	28.7	28.6
934CN3	26.0	28.2
529PP3	27.6	28.7
936CN3	28.4	28.8
936CN3	27.6	28.7
937CN3*	27.5	28.1
233FR3	28.9	28.7
234FR3*	26.9	28.7
236FR3*	28.3	28.5
237FR3	27.3	28.6
135FU3*	29.2	28.8
860FU3	28.3	28.9
136FU3	30.4	29.0
Mean (X)	27.81	28.51
RSD	3.9%	0.9%
Range	26.0 - 30.4	28.1 - 29.0

Table 4.2. Bioassay and HPLC analysis of insulin

Π Now examine the data derived from 21 lots of human insulin reported in Table 4.3. Use this data to calculate the mean value of these samples by each technique and the range of values obtained by each technique. Put your answers in the appropriate column on the table.

Lot Number	Biopotency (Units/mg)	HPLC Potency (Units/mg)
779FW2	28.3	28.2
780FW2	27.0	28.4
781FW2	29.0	28.4
782FW2	29.1	27.9
783FW2	27.5	28.4
487FN2	28.4	28.3
737FU2	28.5	28.3
503PP2	27.4	28.5
488FN2	28.2	27.8
738FU2	28.4	28.7
686FW2	27.0	28.2
489FN2	27.9	28.1
490FN2	27.0	28.1
532PP2	28.6	28.6
491FN2	28.7	28.4
787CJ2	27.6	28.1
630CJ2	28.5	28.5
788CJ2	26.4	28.4
533PP2	28.8	28.0
789CJ2	29.0	29.0
790CJ2	27.8	28.7
Mean		
RSD		
Range		

Table 4.3. Data derived from biopotency and HPLC potency tests

We calculate that the mean value for these samples obtained by the rabbit biopotency was 28.06 Units/mg while the HPLC potency test gave a mean value of 28.33 Units/mg. More importantly the range from the rabbit-based assay was 26.4 - 29.12 Units/mg while the HPLC potency assay was 27.8 - 29.0 Units/mg. This set of data again supports the conclusion that the bioassay gave more variable results than the HPLC assay. Could you calculate the relative standard deviations (RSD) for these two sets of data? We calculate that the RSD values for these data are (bioassay) 2.0% and (HPLC) 1.0%. It would appear therefore that the data presented in Tables 4.2. and 4.3. support the conclusion that the HPLC based assay gives lower variability.

A second significant advantage of HPLC is its molecular specificity. HPLC may be used to distinguish insulins from different species. The bioassay is not specific, in that it will not distinguish human insulin from bovine or porcine insulin.

Additional advantages of HPLC relative to the bioassay are that it is fast, inexpensive and requires fewer resources. The bioassay procedure requires the use of 144 rabbits in each test to determine the potency of a master lot of human insulin. In addition, the bioassay test takes approximately two weeks to complete and each rabbit colony (144 rabbits) must be "rested" for a period of several days before being used in another assay for potency.

The personnel, space and analytical resources required to handle the necessary number of rabbit colonies and to assay the multitude of blood samples taken from each rabbit at different time points makes this test inefficient and costly.

Figure 4.21. is a record of an HPLC-analysis of a batch of rDNA human insulin. The main peak, the top of which is off the chart, represents rDNA human insulin and has the same Rf-value as pancreatic human insulin. In Figure 4.21. there are also some minor peaks: these are insulin-related substances, which have biological activities that are 90-100% of the biological activity of insulin.

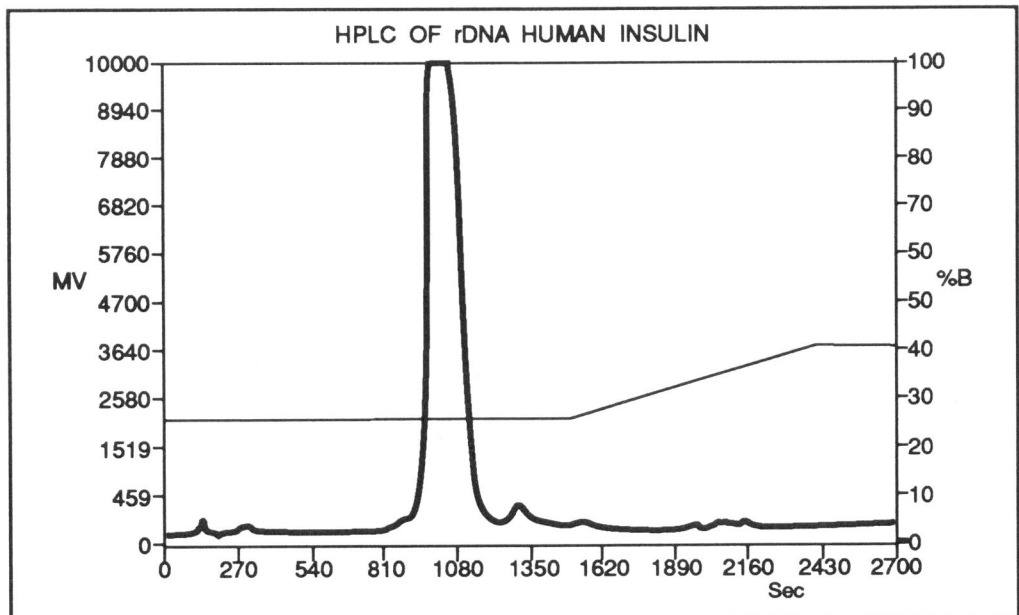

Figure 4.21. HPLC elution profile of rDNA human insulin

Further evidence of the equivalence of rDNA and pancreatic human insulins is shown in Figure 4.22., where HPLC of a mixture of the two insulins gives only one peak.

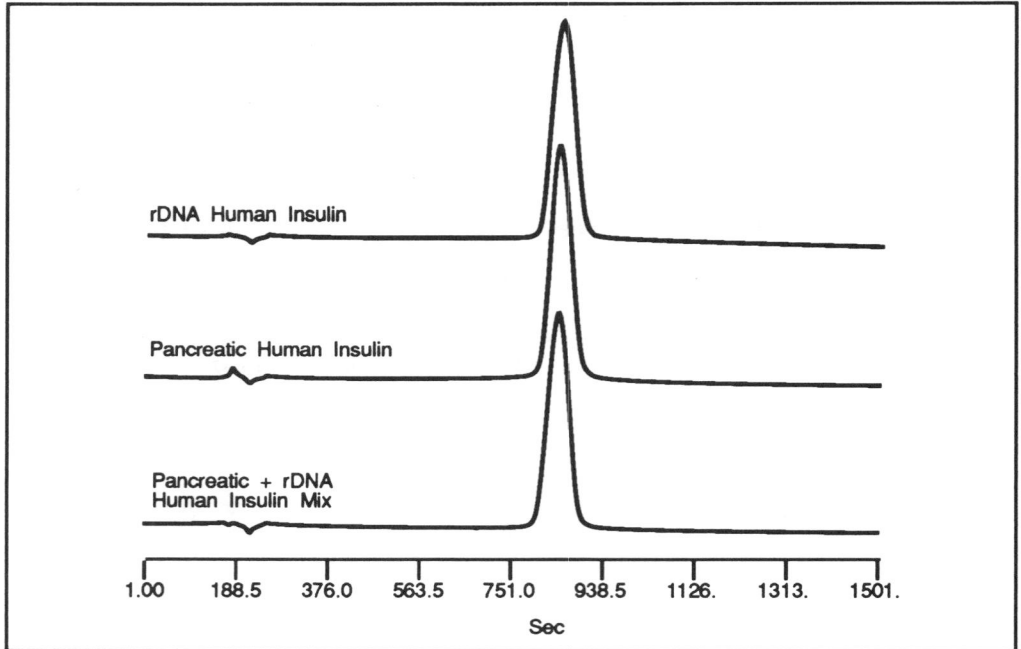

Figure 4.22. Comparison of HPLC elution profiles for rDNA human insulin, pancreatic human insulin and a mixture of the two human insulin preparations

The amino acid compositions of rDNA human insulin and true human insulin have also been found to be the same (Table 4.4.).

The data presented in Table 4.4. was generated by first separating and measuring the amounts of amino acids in a known amount of each sample. The relative molar proportions of each were then calculated by dividing the amount of each amino acid by their molecular weights. Since it is known that each molecule of insulin contains 3 aspartic acid residues, then the molecular ratios of the amino acids present in each of the samples can be calculated.

Amino acid residue	Biosynthetic human insulin	Pancreatic human insulin
Aspartic acid	3.00 (3)	3.00 (3)
Threonine	2.77 (3)	2.77 (3)
Serine	2.56 (3)	2.63 (3)
Glutamic acid	7.11 (7)	7.10 (7)
Proline	1.03 (1)	0.99 (1)
Glycine	3.98 (4)	3.98 (4)
Alanine	0.97 (1)	0.99 (1)
Half-cystine	5.31 (6)	5.43 (6)
Valine	3.76 (4)	3.71 (4)
Isoleucine	1.66 (2)	1.61 (2)
Leucine	6.16 (6)	6.14 (6)
Tyrosine	3.91 (4)	3.90 (4)
Phenylalanine	2.99 (3)	2.91 (3)
Histidine	1.97 (2)	1.99 (2)
Lysine	0.97 (1)	0.97 (1)
Arginine	1.00 (1)	1.00 (1)
Ammonia	6.89	6.95

Table 4.4. Molar amino acid ratios for biosynthetic and pancreatic human insulin calculated using aspartic acid as unity, which was 160 nmoles/mg for rDNA human insulin and 156 nmoles/mg for pancreatic human insulin. Each value is the average of three determinations. Theoretical values are listed in parenthesis.

stereochemistry

Because insulin is a large molecule, it is not enough to demonstrate that rDNA human and true human insulin have the same chemical composition and that the two chains are linked at the correct positions. The stereochemistry of the two molecules needs to be shown to be identical.

Figure 4.23. shows that the circular dichroic spectra of rDNA human insulin and porcine insulin are identical indicating identical spatial arrangement of the molecules. Porcine insulin and true human insulin are known to have identical circular dichroic spectra.

Another technique which has been used to show that rDNA human insulin and true human insulin are completely identical is enzymic cleavage of the molecules by means of a protease.

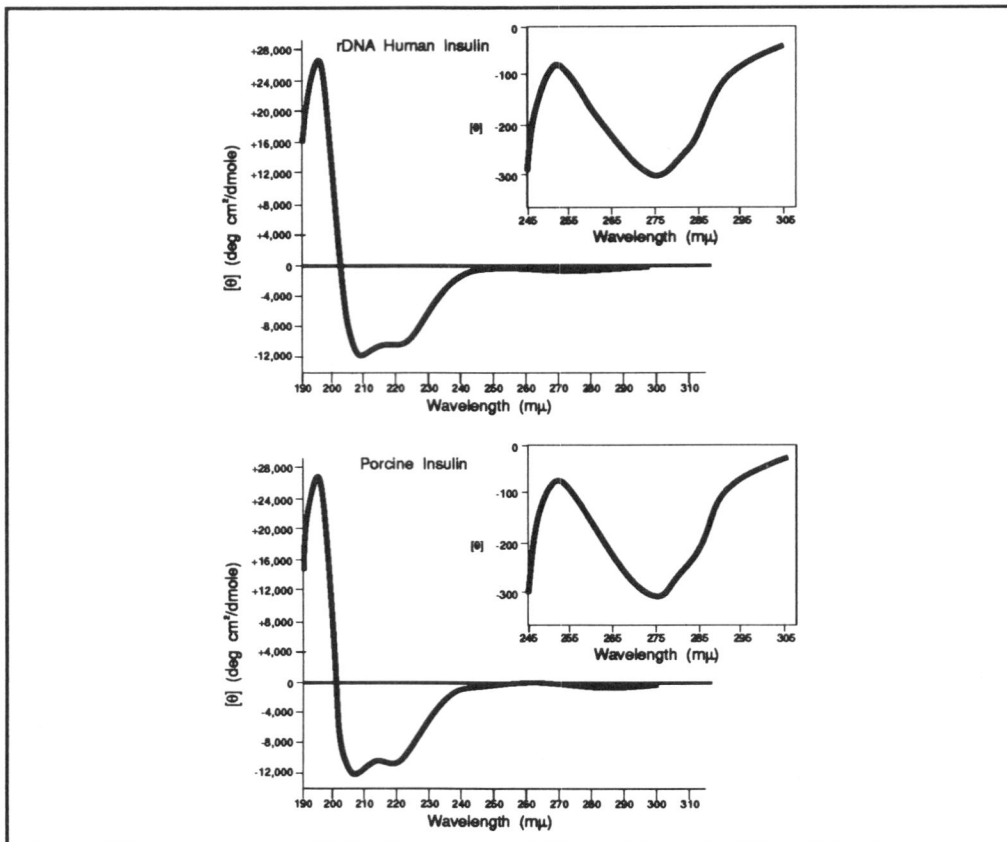

Figure 4.23. Circular dichroic spectra of rDNA human insulin (recombinantDNA) and purified porcine insulin

Staphylococcus aureus

V8 protease

Figure 4.24. shows how the enzyme V8 protease, produced by the bacterium *Staphylococcus aureus*, hydrolyses insulin. Figure 4.25. shows how the peptides produced by V8 protease action on insulin are separated by HPLC analysis. We might call this technique HPLC-finger printing. This enzyme specifically hydrolyses peptide bonds on the carboxyl side of glutamic acid residues, which gives rise to fragments containing the various disulphide bonds. In addition, it was found that an unexpected cleavage occurred at the ser (12)-leu (13) bond in the A-chain. The enzyme reactions were conducted on crystalline zinc forms of rDNA human insulin and a semisynthetic human insulin. A 2 mg sample of each human insulin preparation was dissolved in 0.2 ml of 0.01 M HCl followed by the addition of 0.8 ml of 0.05 M NH_4HCO_3 containing 100µg of enzyme. The final pH was 7.9. A control sample containing only 100µg of the enzyme was also included.

After incubation at 37°C for 24 hours, the proteolysis was terminated by lowering the pH to 2.5, and the products were then analysed by HPLC as shown in Figure 4.25. The resulting profiles were essentially indistinguishable.

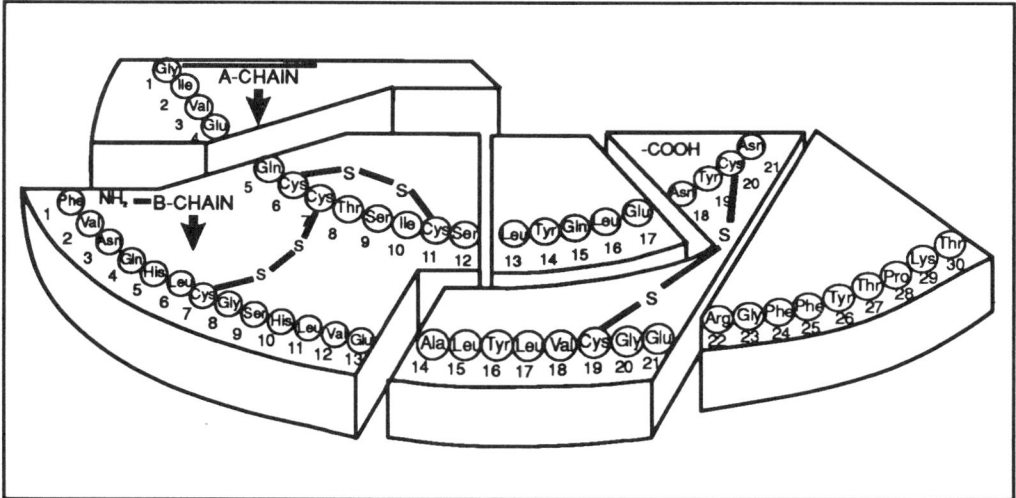

Figure 4.24. Three-dimensional diagram of the peptide fragments that may be produced by enzymic cleavage of the human insulin molecule

Π   Examine Figure 4.25. and see what you can conclude from this figure.

The evidence from these HPLC profiles is that the disulphide bridges between the A and B chains and the amino acid arrangements were identical in the insulins obtained from different sources.

Chromatographic analysis of the control sample of the protease indicated that none of the peaks observed in Figure 4.25. were due to the enzyme itself.

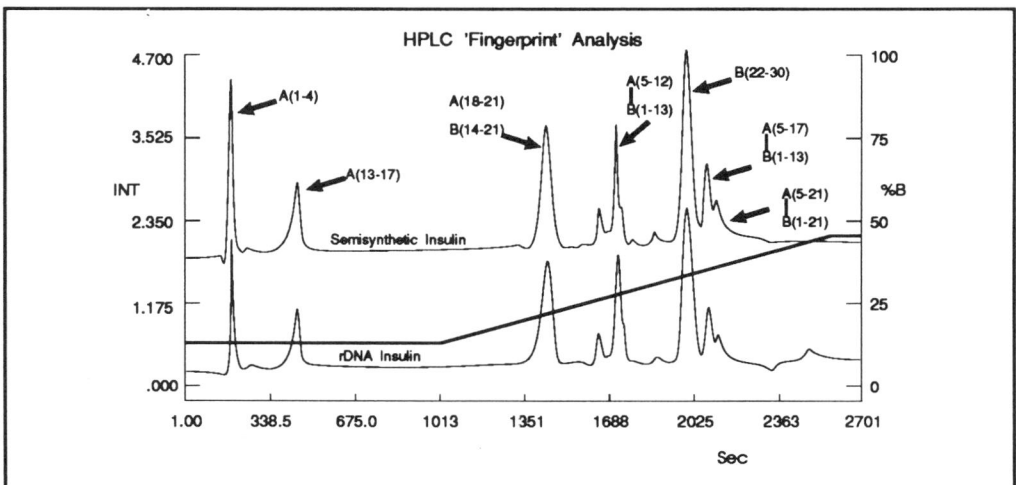

Figure 4.25. HPLC profiles for the S.aureus V8 protease digests of rDNA biosynthetic and semisynthetic human insulin. The chromatography was accomplished with a ZorbaxC-8 column using a combination of isocratic and gradient elution

4.4.3. Toxicity Tests

In the case of any product such as insulin, which is intended for continuous use over prolonged periods of time, there is a need to ensure not only that the material is pure, authentic and active, but also that it shows as few adverse effects as possible. Normally, information regarding adverse effects is first sought via studies in animals. However, in the case of rDNA human insulin, such an approach is not wholly logical.

∏   Write down the reasons why it is not entirely logical to test rDNA human insulin in animals?

Your reasons should have included the following points:

• human insulin is a foreign protein in another species of animal;

• its main effect is to reduce blood sugar levels and this makes the prolonged survival of animals to which it is administered questionable;

• one might ask whether a protein to which all normal human beings are exposed throughout their life truly requires toxicological study.

Even so, to exclude effects of possible impurities undetected in previous tests or differences in glycosylation (i.e. addition of sugars to the protein), rDNA human insulin was subjected to acute toxicity studies in mice, rats and dogs, to studies over fourteen day periods in monkeys and to studies over 30 day periods in rats and dogs. In all these studies there was evidence only of reduction in blood sugar levels, and not of any other adverse effects.

Many protein materials used in medicine are at risk of contamination with pyrogens, i.e. substances which initiate a set of inflammatory reactions made evident by a rise in body temperature.

Limulus
Amoebocyte
Test

Two types of test for pyrogens have been used in the study of rDNA human insulin. One type of test is the conventional method involving use of rabbits; the other is the so called *Limulus* amoebocyte test (Limulus Amoebocyte Lysate test, LAL test), which is widely accepted as the most sensitive test for endotoxins available. It may be recalled (see chapter 4) that addition of pyrogens to a lysate prepared from the amoebocytes of *Limulus* causes the lysate to gel. The test is regarded as extremely sensitive. On a number of batches of rDNA human insulin tested, none of the lots proved to be pyrogenic (Table 4.5.).

4.4.4. Biological characterisation

Having shown that rDNA human insulin is chemically identical to true human insulin, that it is not subject to undue contamination by impurities, and that such minor amounts of impurities as are present are similar in biological activity to true human insulin itself, it was also essential to show that rDNA human insulin is identical in biological activity to true insulin. As Figure 4.26. shows, the effects of rDNA human insulin, true human insulin and porcine insulin were not statistically different in studies in rabbits.

Lot no	LAL result (ng endotoxin/mg insulin)	Pyrogen result (mean temp rise/no of rabbits)
615 7ON 174 9	<0.6	0.19°C/N = 3(nonpyrogenic)
615 84S 30A	0.8-1.6	0.17°C/N = 3(nonpyrogenic)
989BAO	0.4-0.8	0.11°C/N = 3(nonpyrogenic)
46L 295	0.1-0.2	0.01°C/N = 3(nonpyrogenic)
44L 55	0.1-0.2	0.20°C/N = 3(nonpyrogenic)
142CY1	<0.05	0.15°C/N = 3(nonpyrogenic)
143CY1	0.2-0.4	0.06°C/N = 3(nonpyrogenic)
46L 296	0.05-0.1	0.16°C/N = 3(nonpyrogenic)
44L 79	<0.05	0.04°C/N = 3(nonpyrogenic)
46L-297	0.4-0.8	0.27°C/N = 3(nonpyrogenic)
905CY1	0.2-0.4	0.07°C/N = 3(nonpyrogenic)
544CJ1	0.05-0.7	0.07°C/N = 3(nonpyrogenic)

Table 4.5. Limulus Amoebocyte Lysate (LAL) and pyrogen data for human insulin lots

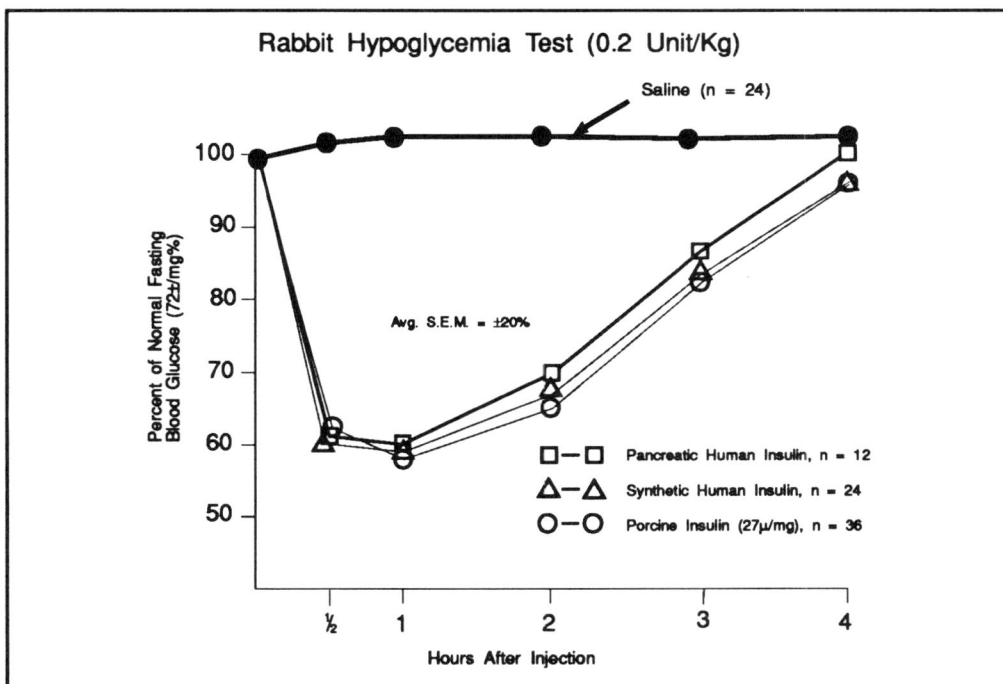

Figure 4.26. Rabbit hypoglycaemia test (0.2 U insulin / kg)

The rabbit is used in a method of measurement of insulin potency which is regarded as a standard procedure by many regulatory bodies throughout the world (European Pharmacopoeia, III-1975, page 75). However, HPLC-techniques are highly sensitive, and can reveal differences not detectable in the rabbit assay. For example, insulins from man, pig and rabbit differ in constitution by only one or two amino acids yet the three types of insulin are separable by HPLC-techniques (Figure 4.27.).

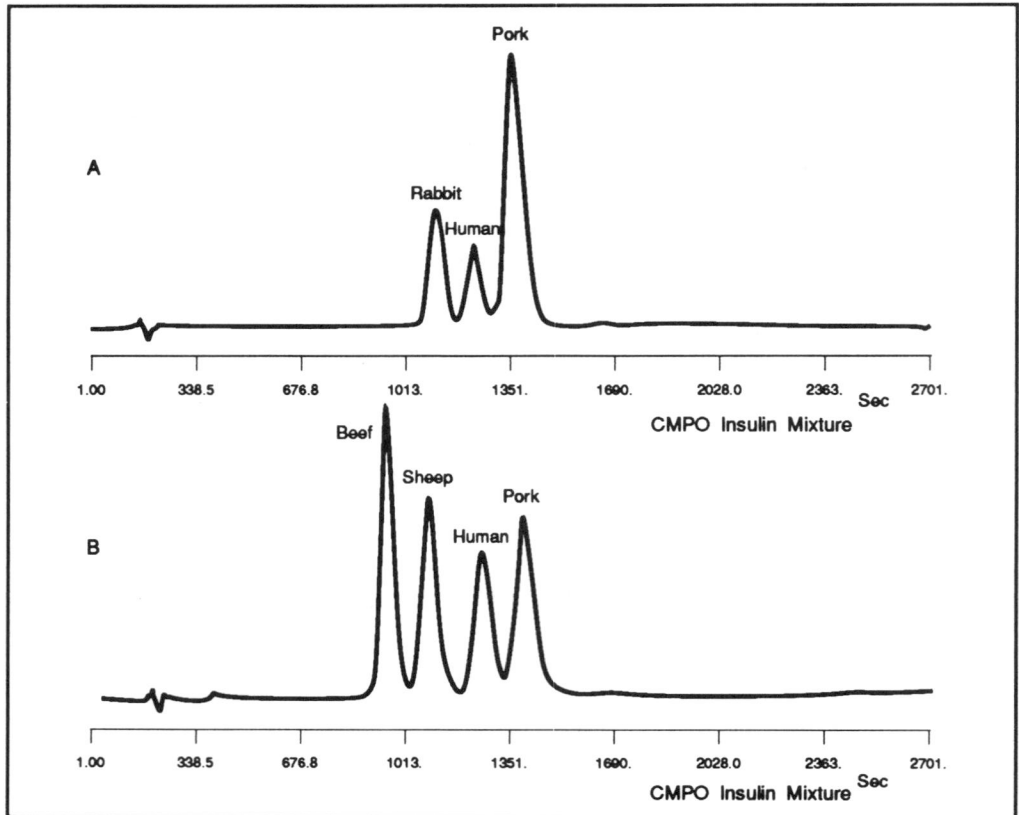

Figure 4.27. HPLC elution profiles for various mixtures of pancreatic insulins that illustrate the sensitivity of the assay to separate proteins differing by only one or two amino acids.

4.4.5. Immunological analysis

A possible source of protein contamination of rDNA human insulin is the *E.coli* used in its manufacture.

To test this, plasmids identical in all respects to those used in the production of human insulin A-chain, B-chain or pro-insulin, except that they did not contain the genetic determinants for the A-chain, the B-chain or pro-insulin, were inserted into the strain of *E.coli* used for production of these materials. Cultures were grown on a large scale and subjected to the purification schemes used to prepare rDNA human insulin.

ECPs

All that was produced were barely detectable amounts of a mixture of small peptides, which were designated ECPs (*E.coli* peptides). These peptides were studied *in vivo* and found to be almost non-immunogenic.

By using complete Freund's adjuvant (i.e. with killed mycobacteria to enhance antigenicity), it was possible to obtain antibodies and to sensitise animals to ECPs. A significant immune response to ECPs was obtained only when guinea pigs were sensitised with ECPs incorporated into complete Freund's adjuvant. No significant Arthus-type reaction was noted when human insulin or porcine insulin was administered to ECP-sensitised rats and guinea pigs. (The Arthus-type reaction is an allergic reaction at the site of injection of the antigen. If an animal has been previously sensitised to an antigen, in the Arthus type of reaction, a localised immune reaction occurs 6-8 hours after injection of the antigen).

The negative Arthus type reaction implies that there are no ECPs antigens present in the insulin preparations.

An assay method for detection of immunoreactive ECPs in human insulin was also developed. An outline scheme for these assays is given in Figure 4.28.

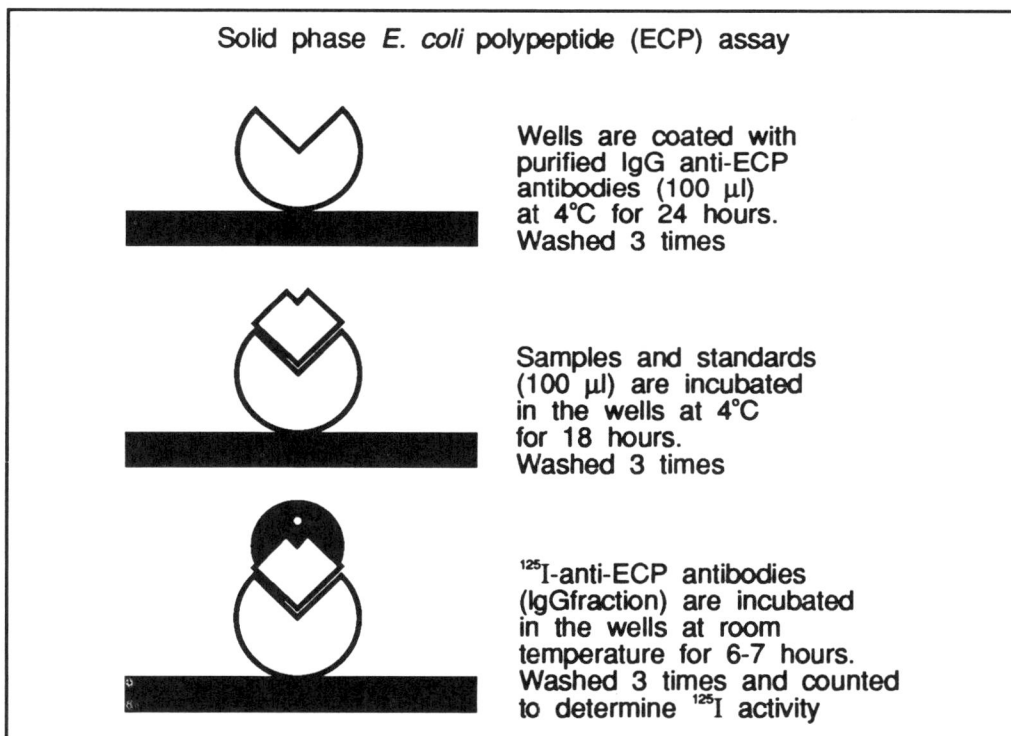

Figure 4.28. Procedure for detection of immunoreactive *E.coli* polypeptides.

A 96-well plate is coated with purified IgG against the ECPs. The ECP-standards and human insulin samples are added and incubated, as shown. After washing, a purified IgG for ECPs containing a radioactively labelled antibody is added to react with any ECP bound in the wells. By measuring the residual radioactivity after washing, ECP levels in the sample can be determined by comparison to known standards. There is no cross-reactivity with insulin. This assay is carried out on all batches of human insulin and it is sensitive to 1-3 ppm of immunoreactive ECPs. The human insulin preparations used in clinical trials contained less than 4 ppm by this assay.

The tests described in this section fulfilled many of the criteria needed for licensing. The chemical analysis included amino acid composition, HPLC profiles, circular dichroic patterns, "finger prints" derived from enzymic degradation. These tests established that, structurally, the rDNA-insulin was identical to normal insulin. The acute toxicity testing, the LAL test for pyrogens and the failure to demonstrate *E. coli* peptides in the final product indicates that it was safe to use. It is however important to proceed with caution and the product underwent a series of clinical studies.

Before we examine these, attempt the SAQs which tests your understanding of the criteria used to characterise rDNA-human insulin.

SAQ 4.6.	Insulin can be measured by bioassay using rabbits or by high performance liquid chromatography (HPLC). By recording R (for Rabbit bioassay) or H (for HPLC), assign each of the following features to one of the assay procedures.

1) Takes less space — R or H

2) Capable of quantitatively measuring insulin more precisely — R or H

3) Is specific for each insulin — R or H

4) Cheaper to operate — R or H

5) Measures biological activity of the insulin — R or H

6) Is quickest to give a result — R or H

SAQ 4.7.	A preparation of rDNA-insulin has been analysed and compared with authentic human insulin. From the data presented below can the insulin preparation be regarded as being of identical structure to human insulin.

a) HPLC profiles

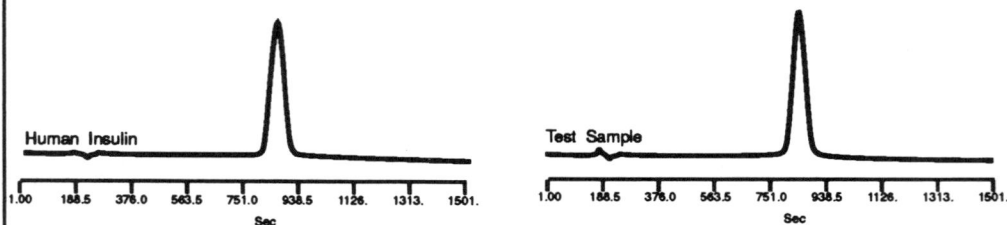

SAQ 4.7.	

b) Amino acid analysis (The ratios of the amino acids present are based on aspartic acid being present in 3 moieties per molecule).

Amino acid	Human	Test Sample
Aspartic acid	3.00	3.00
Threonine	2.77	2.69
Serine	2.56	2.60
Glutamic acid	7.11	7.12
Proline	1.03	1.00
Glycine	3.98	3.89
Alanine	0.99	0.96
Half-cystine	5.43	5.33
Valine	3.71	3.66
Isoleu	6.14	6.04
Tyrosine	3.90	3.86
Phenylalanine	2.99	2.00
Histidine	1.99	2.01
Lysine	0.97	0.95
Arginine	1.00	1.00
Ammonia	6.95	6.79

c) Circular dichroic spectra

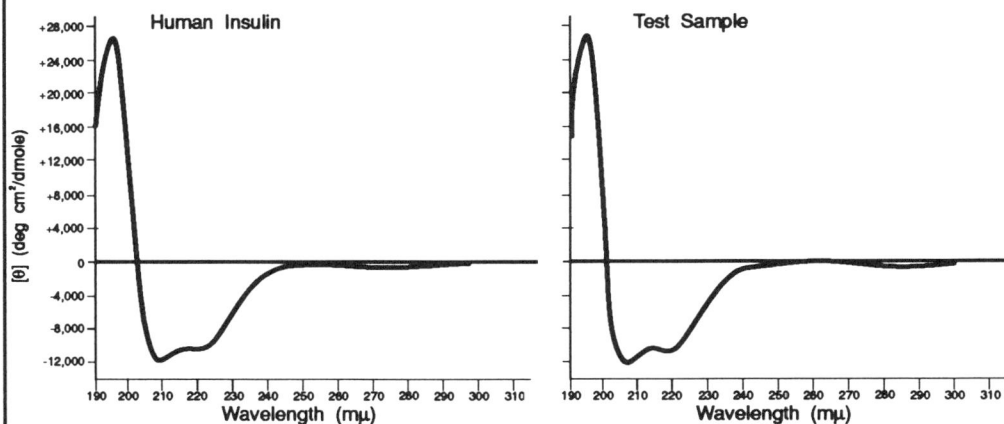

SAQ 4.7.

d) HPLC profiles of a *S.aureus* V8 protease digest of rDNA and authentic human insulin.

HPLC 'Fingerprint' Analysis

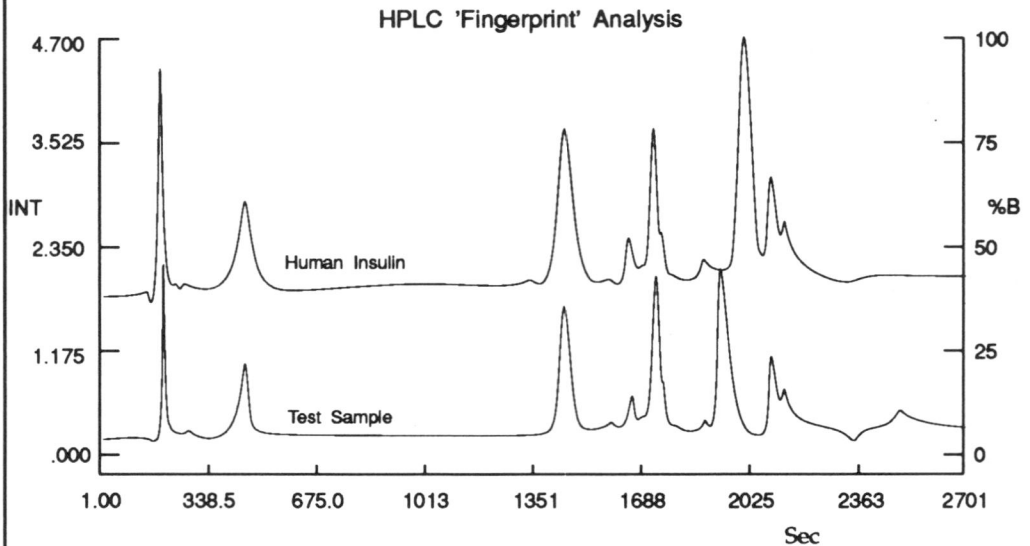

4.5. Clinical studies

4.5.1. Pharmacological studies

absorption characteristics

The pharmacology of rDNA-human insulin has been examined in a number of studies using several formulations. Almost all of these studies have found rDNA human insulin and purified porcine insulin to be equivalent. There are a few exceptions to this. For example, it has been shown that regular human insulin is absorbed slightly more rapidly than porcine insulin; several studies have suggested more rapid absorption of NPH human insulin than NPH porcine insulin. [NPH insulin stands for Neutral Protamine Hagedorn insulin, an extended-action insulin, which contains little or no excess protamine to bind regular (=normal) insulin]. In addition, some studies suggest that NPH human insulin may have a shorter duration of action than NPH porcine insulin. Interestingly, there appears to be very little difference in the time course of action of several of the regular:NPH mixtures. For example, in one study, no differences were detected when 20:80, 25:75 and 30:70 mixtures were compared, or when 10:90, 15:85 and 20:80 mixtures were compared. (It must be appreciated that only 20 units of insulin were used in the test, thus making these mixtures differ by only 1-2 unit(s) of insulin). The titles of some pharmacologic studies that have been performed illustrate the diversity of experimental work that is required:

• Clinical Pharmacologic Studies with Human Insulin (rDNA);

• Absorption Kinetics of Semisynthetic Human Insulin and Biosynthetic (rDNA) Human Insulin;

- Plasma Insulin and C-Peptide after Subcutaneous and Intravenous Administration of Human Insulin (rDNA) and Purified Porcine Insulin in Healthy Men;

- Clinical Pharmacology of Human Insulin of rDNA Origin in Healthy Volunteers;

- Crossover Study with Human Insulin (rDNA) in Type I Diabetic Subjects;

- Insulin Concentration and Time Action Profiles of Three Different Intermediate-acting Insulin Preparations in Non-diabetic Volunteers Under Glucose-controlled Glucose Infusion Technique;

- Search for the Most Practical Regular/NPH Mixtures for Type I Diabetic Patients

- Comparative Study of NPH Human Insulin (rDNA) and NPH Bovine Insulin in Diabetic Subjects;

- Comparative Study of NPH Human Insulin (rDNA) and Porcine Insulin in Diabetic Subjects;

- Time Action Profile of Zinc Human Insulin (rDNA) in Young Volunteers as Compared with Zinc Porcine Insulin.

Let us examine why these studies have been conducted. Insulin must be absorbed the subcutaneous tissues in order to be effective in the plasma. In the plasma, it passes around the body and subsequently binds with insulin receptors on the surface of a wide variety of cells. Thus a particular feature of insulins (human insulin, rDNA human insulin, porcine insulin etc) is the kinetics of their distribution and absorption. Thus many of the studies cited relate to measuring the rate of absorption and the kinetics of the insulin pool in the plasma after administration. The other important feature is the pharmacological action of insulin (i.e. producing a reduction in blood glucose levels). Thus several studies were directed towards measuring this effect with the rDNA insulin. Note that many of the studies are carried out using other sources of insulin as comparisons.

4.5.2. Biological effects

It has been demonstrated that the dose response curves for insulin action, in terms of both suppression of hepatic glucose output and stimulation of peripheral glucose utilisation, are identical in diabetic subjects using either purified porcine insulin or rDNA human insulin.

counter-regulatory hormone

Other studies which have been performed regarding the biological effects relate to counter-regulatory hormone (hormones that have the opposite effects to insulin) responsiveness to insulin-induced hypoglycaemia (glucagon, growth hormone, catecholamines). *In vitro* experiments have demonstrated that human insulin (rDNA) and porcine insulin are equipotent in stimulating lipogenesis in human adipocytes. A list of some biological studies that have been carried out include:

- The Therapeutic Efficacy of Human Insulin (rDNA) in Patients with Insulin-dependent Diabetes Mellitus: A Comparative Study with Purified Porcine Insulin;

- Different Counter-regulatory Responses to Human Insulin (rDNA) and Purified Porcine Insulin;

- Comparative Studies on Intermediary Metabolism and Hormonal Counter-regulation Following Human Insulin (rDNA) and Purified Porcine Insulin in Man;

- Less Pronounced Changes in Serum Potassium and Epinephrine During Hypoglycaemia Induced by Human Insulin (rDNA);

- Inhibition of Pancreatic Glucagon Responses to Arginine by Human Insulin (rDNA) and Purified Porcine Insulin in Normal and Diabetic Subjects;

- Comparison of Biological Potency of Human Insulin (rDNA) and Purified Porcine Insulin (PPI) in the Rat and Human Adipocyte Lipogenesis Model;

- Receptor Binding Properties of Human Insulin (rDNA) and Human Pro-insulin and Their Interaction at the Receptor Site.

In many ways these studies extend the pharmacological studies by analysing the fundamental biological properties of the insulin.

4.5.3. Immunological effects

circulating antibodies

The immunological studies with human insulin (rDNA) are particularly interesting. A number of parameters related to circulating antibodies have been studied in patients transferred either from mixed bovine-porcine insulin or purified porcine insulin to rDNA human insulin. It has been found that in patients using either mixed bovine-porcine or purified porcine insulin, the hormone because it was bound to circulatory anti-insulin antibodies, decreased after transferring to human insulin. Those patients transferred from mixed bovine-porcine (but not from purified porcine) insulin to human insulin also showed decreases in species-specific binding of bovine and porcine insulin, and decreases in both the high- and low-affinity binding capacities of circulating anti-insulin antibodies. There was also an increase in the affinity constant of the high-affinity binding antibodies. These observations indicate that circulating antibody levels are likely to decrease on transfer of patients from mixed bovine-porcine to human insulin, with only minimal changes occurring on transfer from purified porcine insulin to human insulin. In other words there appears to be the expected immunological advantages expected of transfer to human insulin.

The affinities of IgG anti-insulin antibodies for porcine and human insulin have also been studied. In 29 patients, the affinities were similar. In 4 patients with immunological insulin resistance and high levels of circulating anti-insulin antibodies, the affinity for human insulin was significantly less than that for porcine insulin.

Other studies have made it clear that rDNA human insulin, despite its sequence homology to native insulin, is immunogenic in human beings if administered subcutaneously. This may not be surprising since it is possible for many homologous substances to be immunogenic if given subcutaneously in the proper adjuvant. One could speculate that such immunogenicity might be obviated if newly diagnosed patients were treated from the onset of their disease only with unmodified (i.e. regular) human insulin either by continuous subcutaneous insulin infusion (CSII), which has but a minimal subcutaneous depot and no real repository, or ideally by continuous intravenous infusion, thus avoiding the subcutaneous route and any repository. The inherent difficulties with the latter approach preclude its being tested, but such use of CSII might be possible. The contribution to antibody formation of other additives in commercial insulin preparations is not yet known.

It is important to note that although much attention is paid to the immunogenicity of insulin preparations, there is no convincing evidence of adverse effects of circulating insulin antibodies, except in the rare cases of immunological insulin resistance. Some authors have argued that since circulating antibodies can serve as a reservoir for insulin, this may serve as a stabilising factor facilitating glycaemic control. Nevertheless, it would seem desirable to avoid the influence of circulating insulin antibodies, since release of the hormone from such a reservoir may create unwanted hypoglycaemia.

A list of immunological investigations is provided below:

- Immunologic Improvement Resulting from the Transfer of Animal Insulin-treated Diabetic Subjects to Human Insulin (rDNA);

- Affinity of IgG-Insulin Antibodies to Human (rDNA) Insulin and Porcine Insulin in Insulin-treated Diabetic Individuals With and Without Insulin Resistance;

- Development of IgE Antibodies to Human (rDNA), Porcine, and Bovine Insulins in Diabetic Subjects;

- Insulin-specific IgG and IgE Antibody Response in Type I Diabetic Subjects Exclusively Treated with Human Insulin (rDNA).

4.5.4. Clinical trials

One extensive trial involved the administration of rDNA human insulin to patients who had never previously been treated with insulin. In these patients, human insulin was found to be relatively effective and apparently safe. They remained free of local or systemic allergy and free of lipo-atrophy (at the injection-site), although 2 out of 101 patients developed lipo-hypertrophy during the first six months of use of human insulin. When compared to non-human insulins, 5-7% give rise to lipo-atrophy and approximately 10% give rise to lipo-hypertrophy. Patients did not show intradermal sensitivity either to human insulin or *E.coli* peptides, and they remained free of antibodies to *E.coli* peptides. Anti-insulin-antibodies were detected in 50% of subjects after 6 months of treatment with human insulin.

There have been three major clinical trials in which subjects previously treated with mixed bovine-porcine, bovine or porcine insulins were transferred to rDNA human insulin. These were multicentre double blind studies in which half of the patients were maintained on their "old insulin" and the other half were transferred to rDNA human insulin. All three studies found rDNA human insulin to be safe and relatively effective. In some patients, there was some evidence of deterioration of glycaemic control, particularly in the fasting state, on transfer to human insulin. This is consistent with the pharmacologic studies, which suggested a shorter duration of action of NPH human insulin due to a more rapid absorption of NPH human insulin than NPH porcine insulin.

A list of Clinical Trials performed:

- A Double-Blind Crossover Trial Comparing Human Insulin (rDNA) with Animal Insulins in the Treatment of Previously Insulin-treated Diabetic Patients;

- The U.S. "New Patient" and "Transfer" Studies;

- Treatment with Human Insulin (rDNA) in Diabetic Subjects Pretreated with Porcine or Bovine Insulin: First Results of a Multicentre Study.

4.5.5. Case reports and clinical experience

Human insulin has been used in the treatment of diabetic ketoacidosis and hyperosmolar coma, in infusion pumps and in a variety of special types of patients such as patients with receptor abnormalities and a vegetarian patient who refused animal insulin.

A list of Case Reports and Clinical Experience studies performed:

- Human Insulin (rDNA) in the Treatment of Patients with Newly Diagnosed Insulin-dependent Diabetes Mellitus;

- Receptor Binding Studies and Clinical Effects of Human Insulin (rDNA): Studies in Patients with Newly Diagnosed Type I Diabetes, Type II Diabetes, Insulin Resistance (Type A and Type B), Insulin Antibodies, Insulin Allergy, and "Brittle" Diabetes;

- Efficiency of Human Insulin (rDNA) in the Treatment of Diabetic Ketoacidosis and Severe Non-ketoacidotic Hyperglycaemia;

- Intradermal Desensitisation with Human Insulin (rDNA) in a Patient with Severe Allergic Skin Reaction due to Insulin;

- Insulin Allergy Treated with Human Insulin (rDNA);

- Decrease of Circulating Insulin Antibodies in Two Patients Treated with Continuous Subcutaneous Infusion of Human Insulin (rDNA);

- Successful Treatment of Immunemediated Insulin Resistance by Human Insulin (rDNA);

- Treatment of a Diabetic Vegetarian with Human Insulin (rDNA): A Case Report.

4.5.6. Closing remarks

Human insulin produced by recombinant DNA technology has been extensively evaluated in diabetic patients in a series of published studies. With few exceptions, human insulin has been found to be equivalent to porcine insulin. Regular human insulin may be more rapidly absorbed than porcine insulin and NPH human insulin may have a shorter duration of action than animal insulins, differences which may have clinical significance.

Human insulin is not entirely non-immunogenic. On the other hand, anti-insulin-antibody levels appear to decrease when patients are transferred from animal insulins to human insulin.

Thus, human insulin produced by recombinant DNA technology appears to be safe and effective in the management of diabetes mellitus. In comparison with purified porcine insulin, it may offer no unique advantages.

On the other hand, in comparison with bovine insulin or mixed bovine-porcine insulin, human insulin is less immunogenic and less antigenic in that levels of preformed anti-insulin-antibodies decrease on transfer to human insulin. The real advantage of human insulin production by recombinant DNA technology is that it provides an infinite resource of a highly purified product. This technology also offers the possibility of intentionally altering the amino acid sequence of human insulin to create a series of analogues with different biologic characteristics, as well as for the production and clinical evaluation of human pro-insulin and human C-peptide.

SAQ 4.8.

Answer true or false to the following statements regarding the clinical testing of rDNA insulin.

1) Patients being transferred from porcine insulin onto rDNA human insulin sometimes suffered some loss in glycaemic control probably because they more rapidly adsorb the human insulin

2) The sera from patients treated with rDNA human insulin were examined for antibodies against *E.coli* peptides to ensure that the patients were not receiving pyrogens

3) Transfer of patients from bovine-porcine insulin to human insulin led to some immunological benefits

4) There is little point carrying out extensive trials with rDNA human insulin in animals to determine its biological activity

5) There is little point in carrying out extensive clinical trials with rDNA-human insulin as we all know it will be safe to use human insulin in humans

6) The fact that anti-insulin antibodies are produced in 50% of the cases receiving rDNA insulin is a good thing because the antibody:insulin complex forms a reservoir for insulin which can be slowly released into the blood stream as the free-insulin becomes degraded

Summary and objectives

This case study began by explaining the two types of diabetes mellitus and their relationship with glucose metabolism and the function of insulin. It was argued that the anticipated growth in the number of sufferers from diabetes and the limited availability of animal insulins made insulin an attractive target for the application of recombinant technology.

Now that you have completed this case study you should be able to:-

- describe the symptoms of the two types of diabetes mellitus;

- describe two strategies for producing human insulin by genetic engineering including the sequence of manipulations needed to produce and identify clones of host cells producing insulin polypeptides;

- calculate which are the most effective promoters to use if the desired protein is produced as a chimeric protein;

- explain why chemical and physical analysis of insulin is more desirable than using bioassays;

- explain why a single set of criteria is often insufficient to offer as quality assurance with many biological products;

- explain why it was important to conduct immunological evaluation of human proteins produced by recombinant DNA technology;

- explain why patients being treated with rDNA-human insulin should be monitored despite market authorization being gained.

5

Case Study: Erythropoietin - a growth factor produced by rDNA technology

5.1. Introduction   102

5.2. Anaemia and blood transfusions.   102

5.3. Red blood cells: physiology   103

5.4. Anaemia   109

5.5. Erythropoietin and Eprex®   111

5.6. Production of Eprex®   112

5.7. Pre-clinical and clinical studies   117

5.8. Clinical studies on the anaemia of chronic renal failure   128

5.9. Therapeutic interest of Eprex®   138

Summary and objectives   140

Case Study: Erythropoietin - a growth factor produced by rDNA technology.

5.1. Introduction

In many ways erythropoietin production using recombinant DNA technology has its parallels with the production of human insulin by recombinant DNA technology. Both involve the isolation of a human gene coding for a hormonal polypeptide and its cloning and expression in a suitable host system. Although this case study follows the same format as that used in the insulin case study, the emphasis is different. Thus we will first explore the clinical conditions in which rDNA human erythropoietin might be used and the technology which led to the production of this hormone. Emphasis has, however, been placed on the pre-clinical and clinical trials. The reasons for this are two fold. Firstly, it will broaden your experience of the demands made by the licensing authorities on the introduction of new pharmaceuticals. Secondly, erythropoietin offers a completely novel approach to the treatment of particular medical conditions. Its introduction is not based upon long medical experience in using an analogous treatment. This, of course, markedly contrasts with the long term experience of using insulin (albeit of different origin) for the treatment of diabetes mellitus.

In examining the pre-clinical and clinical trials, you should not attempt to remember all of the finer details of the study. You should however attempt to gain a clear idea of the sequence and objectives of each set of trials. In this way you will extend your knowledge of the ways in which biotechnology can offer novel solutions to medical problems and how to respond to the conditions imposed by the licensing authorities.

5.2. Anaemia and blood transfusions.

anaemia

The red blood cell, or erythrocyte, is a small, non-nucleated cell, responsible for oxygen transportation and delivery in many living organisms. Since oxygen is a basic requirement for most metabolic pathways in the human body, it can easily be understood that a shortage of red blood cells, a condition called anaemia, will result in severe and even life threatening situations.

transfusion

Anaemia is a fairly common disorder which can be on an occasional (e.g. severe bleeding) or on a chronic basis (e.g. end-stage renal disease). Until recently, the only available treatment was blood replacement or transfusion. Transfusions, however, require an extensive network of donors, transfusion facilities, physicians, laboratory personnel and clinical staff and blood can only be stored for a limited time. Furthermore, there is a substantial risk to the patient receiving the blood. Cautious matching of blood groups between donors and recipients has resulted in fewer transfusion reactions, although they still occur. Repeated blood administration, as occurs in transfusion dependent renal patients may cause iron storage disease (haemochromatosis).

∏   Why can repeated blood transfusions cause an iron overload?

Erythrocytes store oxygen by means of a protein called haemoglobin. This protein has a haem group which consists of a tetrapyrrole ring with four binding sites for oxygen. The binding site for oxygen is iron. In a disease with high turnover of red blood cells, considerable amounts of iron are released into the body. Although iron can be stored in several different locations (liver, spleen, bone marrow, skin), it eventually causes slow destruction of these tissues and organs.

transfusion
HIV

The introduction of viruses via blood transfusion also poses problems. The acquired immune deficiency syndrome (AIDS) has lead to important considerations about transfusion policies. Blood can only be tested for known diseases. AIDS, however, existed for some time before the virus was isolated and adequate testing methods developed. Meanwhile some transfusion patients were infected by HIV (the causative agent of AIDS). Likewise hepatitis B has also presented difficulties. Testing for hepatitis B causes no difficulties nowadays, although it still remains difficult for the non-A non-B categories of this disease, which still cause accidental infections by transfusion.

A drug which could avoid these risks in a simple and efficient way would open new and beneficial perspectives for chronic anaemic patients, saving blood components for more specific conditions.

SAQ 5.1.

Which of the following could be cited as benefits to society and patients as consequences of a reduced need for blood transfusions?

1) Lower cost

2) Lowered infection risk

3) Fewer transfusion reactions

4) Lower incidence of iron intoxication

5) More patient comfort

The isolation and cloning of the gene for erythropoietin has provided a potent and safe source of the hormone. Its stimulating effect on red blood cell production has opened new perspectives in reducing the need for blood transfusions in a variety of indications. Before documenting these properties, it will useful to review some of the characteristics of anaemia and the role of erythropoietin in this condition. This should provide insight into the indications and restrictions of erythropoietin treatment.

5.3. Red blood cells: physiology

The definition of anaemia is based on a reduced haemoglobin level in the peripheral blood. Normal levels are in the range of 12 to 15 g/dl blood. (N.B dl = decilitre = 100 ml). This means that in a normal adult haemoglobin accounts for approximately 1% of total body weight! Haemoglobin is entirely contained in the red blood cells (RBCs). Disruption of the cell membrane releases it into the plasma where it is inactivated by peroxidation and metabolized in the liver. Disruption of RBCs is called haemolysis and is a feature of some clinical conditions.

5.3.1. Red blood cell production (erythropoiesis).

erythropoiesis
CFUs

Red blood cells originate from precursor cells present in the bone marrow. After passing through different stages of maturation they synthesise haemoglobin. On microscopic examination of the bone marrow, it is possible to recognise the sites of RBC production. There are distinct foci, called colony forming units (CFUs), which, on appropriate stimulation, form islets of immature red blood cells. Each cell of such a CFU originates from the same precursor cell.

reticulocytes

In its early stages an immature RBC contains a large nucleus which, after further division and differentiation steps, gradually becomes smaller and finally disappears. During this process the RBC also reduces in size and takes its typical biconcave shape. Throughout all phases of maturation, haemoglobin synthesis occurs at a high rate. After expulsion of the nucleus, the RBCs are finally released into circulation. At this stage they still contain some mitochondria and ribosomes. These cells are called reticulocytes and represent young RBCs indicating a functioning bone marrow. They represent approximately 1% of the circulating red cell mass. A red blood cell is incapable of repairing itself; the average lifetime of such cells is about 120 days.

SAQ 5.2.

From the list below, select the reasons why mature RBCs are unable to perform reparative and regenerative processes. (This question tests your understanding of what you have read as well as previous knowledge you may have had about red blood cells).

1) They have no nucleus

2) They do not have any metabolism

3) They have no apparatus for making new protein

4) They cannot produce ATP

5) They contain a lot of iron which inhibits protein synthesis.

∏   Write down two possible mechanisms by which a shortage of RBCs or anaemia could develop.

Anaemia can develop by two major mechanisms:

• insufficient production of red blood cells

• increased turnover or loss of red blood cells

Some disease-related or nutritional deficiencies can rapidly induce a state of anaemia. These deficiencies usually result in lower and/or incomplete erythropoiesis (red blood cell development) with specific features, each related to the deficient component.

A high rate of production implies a high rate of cell division with a corresponding demand for precursors of the special components of the red cells. Haemoglobin is the most important substance produced by these cells. Each molecule of haemoglobin consists of two pairs of polypeptide chains and four porphyrin structures known as haem. Each haem group consists of four cyclic pyrrole groups each of which binds one iron atom (Figure 5.1.). This iron is the binding site for oxygen. Haem is bound to globulin to form haemoglobin.

The globin polypeptides group consists of 2 paired α and β chains in each molecule. The α and β polypeptide chains are encoded by separate genes. This globin cluster binds to 4 haem groups forming the haemoglobin molecule.

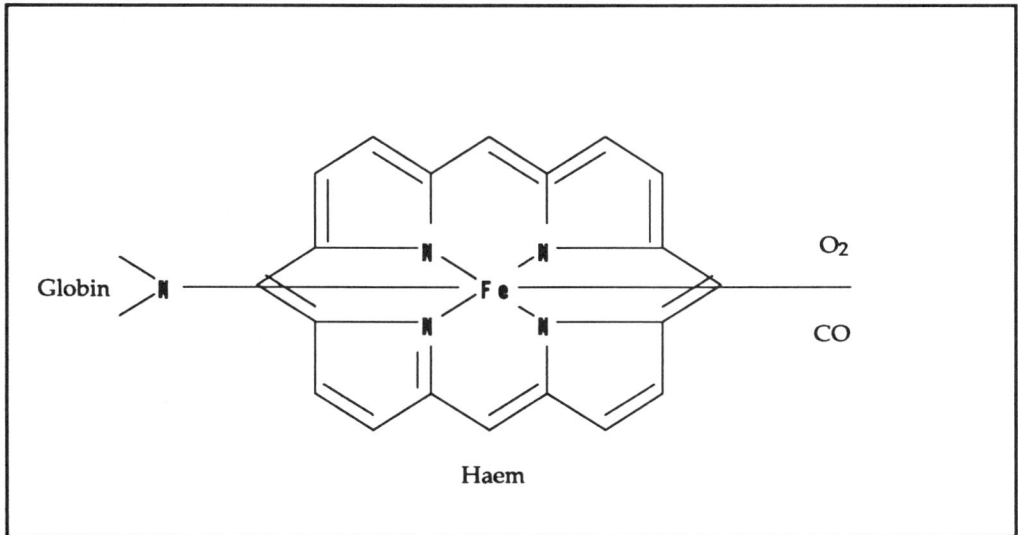

Figure 5.1. The schematic structure of haem and haemoglobin

Ⅱ By examining the structure of haemoglobin given in Figure 5.1., see if you can think of the most likely cause of anaemia.

hypochromia
microcytosis

The most likely answer for you to give is iron deficiency.

Iron is indeed one of the requirements for efficient haemoglobin synthesis. Iron, however, is largely reutilised from the destruction of old erythrocytes. Furthermore, the body contains iron reserves bound to different transportation and storage proteins (e.g. transferrin, ferritin). The only way to lose large amounts of iron is by blood loss. Severe blood loss by trauma, menstruation or urinary bleeding during infection can cause iron deficiencies. In addition, certain diseases such as chronic inflammatory disorders or infections can cause a relative iron deficiency due to disturbances in reutilisation. The special features of RBCs related to iron deficiency are microcytosis, (or RBCs of smaller volume), and hypochromia (or reduced haemoglobin content of the cell).

intrinsic factor

pernicious
anaemia

macrocytosis

There are however alternative causes of anaemia. Vitamin B_{12} is an essential co-enzyme for DNA synthesis. This vitamin is resorbed at a special localisation in the gut. It requires binding to a glycoprotein to facilitate absorption. The glycoprotein is produced by the stomach and is called intrinsic factor (IF). Hence certain surgical procedures in which the site of resorption in the gut or the stomach are removed, will result in a deficiency of vitamin B_{12}. Disease of the stomach with loss of IF production will give rise to the same problem. This type of anaemia is called pernicious anaemia and is characterised by large spheroid RBCs with a normal haemoglobin content (macrocytosis). Gas exchange, however, is severely disturbed. These aberrant RBCs also have much shorter life in circulation.

∏ Why would macrocytotic RBCs have a shorter life-span in circulation? (Think about the vessels RSCs have to pass through).

The reason is that to deliver oxygen to the tissues RBCs have to deform to pass through the smallest vessels. Structural or morphological abnormalities compromise this ability, resulting in less efficient oxygenation of tissue and greater mechanical damage to the RBCs.

A comparable anaemia with large RBCs occurs in folic acid deficiency. Folic acid is required for synthesis of thymidylate, a DNA precursor.

hypochromia Pyridoxine (vitamin B₆) is a necessary co-factor in the synthesis δ-amino levulinic acid, a precursor of haem. The RBCs will have a reduced haemoglobin content in vitamin B₆ deficiency (hypochromia). There is a resemblance to iron deficiency, although iron in the patient's serum remains high in this case.

Other necessary components for normal erythropoiesis are: pantothenic acid, riboflavin, copper and cobalt.

5.3.2. Breakdown of red blood cells

bilirubin RBCs have a finite life span of about 120 days. During this time there is progressive reduction in the activity of the enzymes of glycolysis which provides the ATP required to maintain cell shape and volume. A reduction in enzymic activity by aging produces deformed RBCs, which thereby lose their capacity for efficient oxygen delivery. Phagocytic cells, especially in the spleen, remove and digest deformed RBCs. There is an almost complete recycling of all the digested elements except for the haem group which is metabolized and excreted in the bile as a yellow pigment called bilirubin.

∏ What is the distinct clinical feature of severe haemolysis?

A striking feature of severe haemolysis is jaundice. This jaundice develops when large amounts of bilirubin are released into circulation. On the other hand, if bilirubin is prevented from excretion, as for instance by gall-stones, this will also result in jaundice.

5.3.3 Regulations of erythropoiesis

oxygen sensor As we already mentioned earlier, the most important role of the RBC is oxygen delivery. It seems logical that somewhere must be continuous a "measuring sensor" to detect the oxygen content of the blood. If oxygen decreases this sensor will trigger an alarm system which will pass a signal to the bone marrow, resulting in an increased RBC and haemoglobin synthesis.

∏ See if you can write down some features that such a sensor might have? (Think about where you might find such a sensor and whether or not the site would usually have a low or high oxygen tension).

You may have written the following requirements.

- The sensor should be situated at a site receiving a considerable amount of blood. This blood delivery must be continuous at the same rate and not dependent on activity of the tissue. In muscles, for instance, blood flow is largely dependent on activity.

- The sensor must be placed in or after the smallest vessels. If you, place this detector in large vessels it will not detect impaired oxygen delivery in the tissues due to abnormal RBCs (see earlier).

- The situation of the sensor must allow measurement of oxygen levels in situations where haemoglobin normally contains substantial amounts of oxygen. In such a situation the sensor could detect depression in oxygen levels. If the tissue or organ requires large amounts of oxygen (e.g. muscle) the oxygen levels will always be low in or after the smallest vessels (or capillaries).

The sensor function has in fact, been located in the kidney. Renal blood flow accounts for about 25% of the cardiac output and it is maintained at a constant value by fine regulatory mechanisms. The renal cortex has a widespread capillary network for filtering the blood. In this process, which is mainly pressure dependent, little oxygen is required. Thus the "oxygen sensor" is situated in the renal cortex.

EPO
hypoxia

Signal transduction to the bone marrow is performed by a hormone-like substance: erythropoietin (EPO). This EPO is produced in the renal cortex. The exact mechanism of action on bone marrow is still not fully understood. It seems to act as a growth and/or differentiation factor for stem cells, causing them to differentiate into erythroblasts, the key cells for the colony forming units (CFU). The production of EPO is stimulated by decreased oxygen tension (hypoxia) and also by androgens (male sex hormones). This relationship with androgens explains why the mean haemoglobin level is higher in males. The feedback loop between oxygen tension in the kidney and the stimulation of red blood cell production in the bone marrow is shown in Figure 5.2.

∏ The opposite disorder to anaemia is polycythaemia. Using what you have learnt about oxygen tension in the kidneys and its effect on erythropoietin secretion, can you suggest situations in which polycythaemia might occur?

polycythaemia

Since erythropoietin secretion is mainly dependent on oxygen tension, any situation with low circulating oxygen will result in increased RBC production. Persons living at high altitudes, where oxygen is scarcer than at sea level, show polycythaemia. In some situations, the oxygen binding site on haemoglobin can be occupied and irreversibly bound to carbon monoxide. This can also cause polycythaemia. This situation can likewise occur in accidental intoxication or in smoking. Some lung diseases compromise normal oxygen exchange in the pulmonary circulation which also results in increased RBC production.

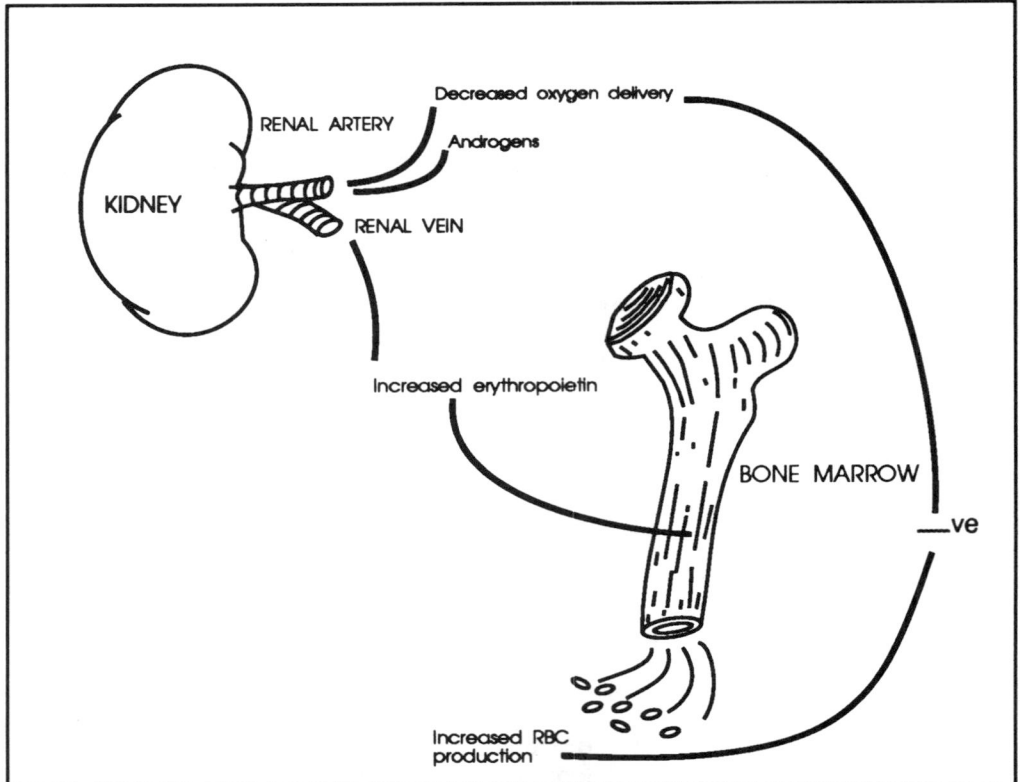

Figure 5.2. Mechanism of erythropoietin secretion-regulation

SAQ 5.3a	Indicate which of the following reasons might lead to anaemia.

1) Haemorrhage

2) Iron deficiency

3) Low oxygen tension in the kidneys

4) Folic acid deficiency

5) Kidney infection

6) Living at high altitude

7) Vitamin B_{12} deficiency

8) Pyridoxine deficiency

9) Failure to produce the intrinsic factor (IF) in the stomach

10) An over production of androgens

SAQ 5.3b   Which of the reasons you have indicated in SAQ 5.3a could be treated by administering erythropoietin?

5.4. Anaemia

Let us summarise what we have learnt about anaemia and extend our knowledge of the conditions which may be treated by erythropoietin.

We have learnt that decreased production of RBCs can be caused by an insufficient supply of the required basic materials such as iron, vitamin B_{12}, vitamin B_6, folic acid etc. Erythropoietin levels in those patients will usually be high. The treatment in such cases is simply by repletion of the missing materials.

Myelofibrosis   A second cause of decreased RBC production can be a primary (originating in the bone marrow) or secondary (from external non-marrow related source) marrow disease, which induces a decrease or complete destruction of the bone marrow. Myelofibrosis or replacement of the marrow by "scar tissue" is an example of primary disease.

tumour in bone marrow   The bone marrow is often a preferred site for tumour spread. A rapidly growing tumour can replace the normal marrow by tumour cells, which thereby provide an external source of bone marrow destruction. In such situations natural EPO secretion will increase. Treatment with exogenous EPO could ensure maximal stimulus for production of RBCs.

The third mechanisms by which RBC production can be impaired is by a decreased output of erythropoietin.

∏   In which disease will decreased erythropoietin output occur? (Think about the site of erythropoietin synthesis).

renal failure   The answer is in renal disease since the kidneys are the sites of production. Indeed in severe renal failure we often see the development of anaemia. EPO production is impaired due to loss of healthy kidney tissue. Such conditions would be relieved by the addition of exogenous EPO.

Other causes of decreased production of RBCs are often related to drug use or inborn errors of metabolism.

These are all conditions which are produced by reduced RBC production.

increased RBC turnover   The alternative cause of anaemia is as a result of increased turnover.

The most frequent cause of anaemia in this setting is acute or chronic bleeding. EPO can be beneficial in such cases since it can give a maximal stimulation of production of RBCs. However, sudden acute bleedings cannot be treated by EPO since the production of RBCs process takes too long to compensate for the loss.

⊓ Are there situations in which the administration of EPO in relation to acute bleeding could be beneficial?

autologous
transfusion

The answer is yes. There are some surgical procedures, such as open heart surgery, in which considerable amounts of blood can be lost. These patients are dependent on donor blood for restoring the loss. However, EPO can be administered a few weeks before surgery. The increased RBC production allows regular donation of blood, which is then re-administered to the patient when needed during surgery, thus avoiding complications with donor blood. The patient is his own blood donor (autologous transfusion).

chronic
inflammation

A second condition which gives rise to increased RBC turnover is chronic inflammatory disease. In such diseases, there is a continuous damage of tissues, resulting in many minor vascular leaks with loss of RBCs. Similar chronic losses can occur in certain drug treatments (eg. aspirin). It is clear that in these situations an increased production and replacement of the RBCs can be obtained with EPO.

spleen
hyperactivity
malaria

Apart from bleeding, the RBCs can be lost by a hyperactivity of the spleen, retaining an excessive proportion of the RBCs which pass through it. Malaria is a disease caused by an intra-cellular parasite (Plasmodium). The preferred location of this parasite is the RBC. After each reproductive cycle, the infected RBC is destroyed and the released organisms will infect new RBCs. Neither malaria nor hyperactivity of the spleen would be cured by treatment with EPO.

Table 5.1 summarises the different causes of anaemia and shows possible applications of EPO. It should be noted that until now the only approved indication for EPO is in the treatment of anaemia in chronic renal failure. Other applications are under current investigation.

Mechanism	Examples	EPO
Decreased output	Deficiencies: Iron, Vitamin B_{12}, Vitamin B_6	No
	Bone Marrow Disease: Cancer	Yes
	Myelofibrosis	Yes
	Decreased EPO: Renal disease	Yes*
	Bleeding: Acute	Yes/No
	Chronic	Yes/No
Increased turnover	Hyperactivity of spleen	No
	Infection: Malaria	No
	Haemolysis	No

Table 5.1. Summary of some common causes of anaemia related to the possible indications for EPO administration. * Renal disease related anemia however is still the only approved indication for EPO traded under the name Eprex®.

∏ What is the common feature of the anaemic conditions which could be treated by EPO?

The next section will mainly deal with the investigations required for official registration. It is preceded by a short description of the development and production of EPO by recombinant DNA techniques; the trade name for EPO made in this way is Eprex®.

5.5. Erythropoietin and Eprex®

In Chapter 3, we spent some time considering the regulatory issues involved in producing and marketing pharmaceutically active materials for use in the treatment of clinical conditions. You may perhaps take this opportunity to remind yourself of the contents of a licensing file.

You will recall that there is:

- an administrative part (who is the manufacturer, where is it to be manufactured);

- a technical part relating to the method of manufacture, characterisation of the product;

- pre-clinical testing (pharmacology, toxicology) leading to clinical trials with volunteers and patients.

Here we will discuss the steps taken to prove that the erythropoietin (EPO) produced by recombinant DNA techniques is a safe and active drug. It is not intended that you memorise all of the results. It is however important that you understand and can explain the logical sequence of the steps taken to prove that the new product has the desired properties and is safe to use.

5.5.1. History of the development of Eprex®

The story of erythropoietin started more than 80 years ago, when Carnot and Deflandre postulated in 1906 that the production of red blood cells is controlled by a humoral factor which they called "haemopoietin". It took more that 40 years before more proof for the existence of this hormone, which was more appropriately named "erythropoietin" by Bonsdorff and Jalevisto, was provided. Reissmann found that erythropoiesis in normal rats could be stimulated with plasma obtained from hypoxic (oxygen limited) rats. Erslev demonstrated the presence of an erythropoietic factor in plasma of anaemic rabbits. In 1957, Goldwasser suggested the kidney as the site of erythropoietin production and the detection of lack of oxygen. Then in the 1970s Goldwasser and his colleagues isolated EPO from plasma of anaemic sheep and from human urine. The quantities of pure EPO obtained by isolation from plasma or urine were however too small to carry out extensive pharmacologic and therapeutic studies. These became possible in the 1980s, when genetic engineering was applied to the production of recombinant human erythropoietin (r-HuEPO; Eprex®) in quantities which were sufficient to start a joint Cilag, Amgen, Ortho and Kirin research programme.

5.6. Production of Eprex®

5.6.1. Isolation and expression of the erythropoietin gene

This is quite a long and complicated process. To help you we have first divided the process into a number of stages which we now outline in turn.

a) Development of probes to identify the EPO gene

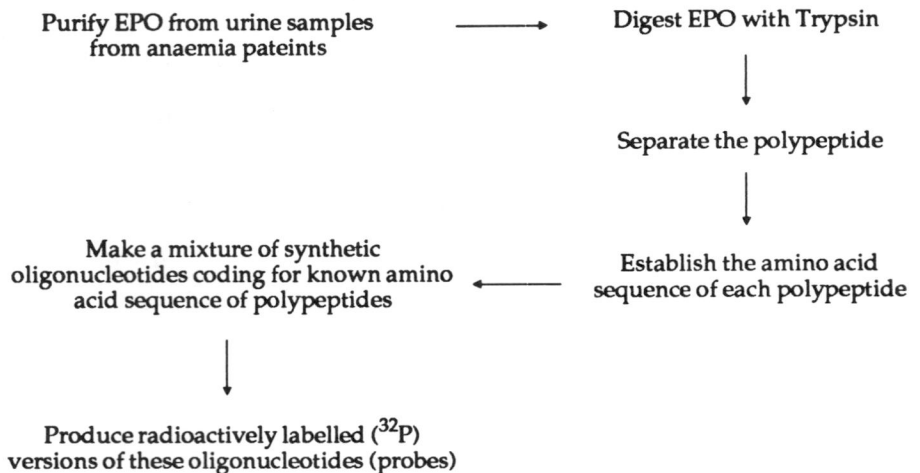

Purify EPO from urine samples from anaemia pateints $\longrightarrow$ Digest EPO with Trypsin

$\downarrow$

Separate the polypeptide

$\downarrow$

Make a mixture of synthetic oligonucleotides coding for known amino acid sequence of polypeptides $\longleftarrow$ Establish the amino acid sequence of each polypeptide

$\downarrow$

Produce radioactively labelled (^{32}P) versions of these oligonucleotides (probes)

b) Production of a genomic library of human DNA and the identification of clones containing the EPO gene.

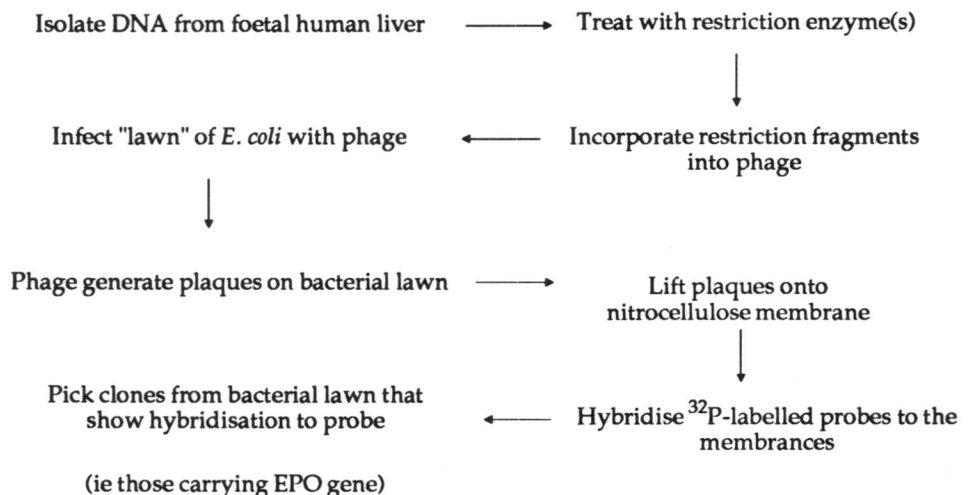

Isolate DNA from foetal human liver $\longrightarrow$ Treat with restriction enzyme(s)

$\downarrow$

Infect "lawn" of *E. coli* with phage $\longleftarrow$ Incorporate restriction fragments into phage

$\downarrow$

Phage generate plaques on bacterial lawn $\longrightarrow$ Lift plaques onto nitrocellulose membrane

$\downarrow$

Pick clones from bacterial lawn that show hybridisation to probe $\longleftarrow$ Hybridise ^{32}P-labelled probes to the membrances

(ie those carrying EPO gene)

c) Transfer of the EPO gene into a production system

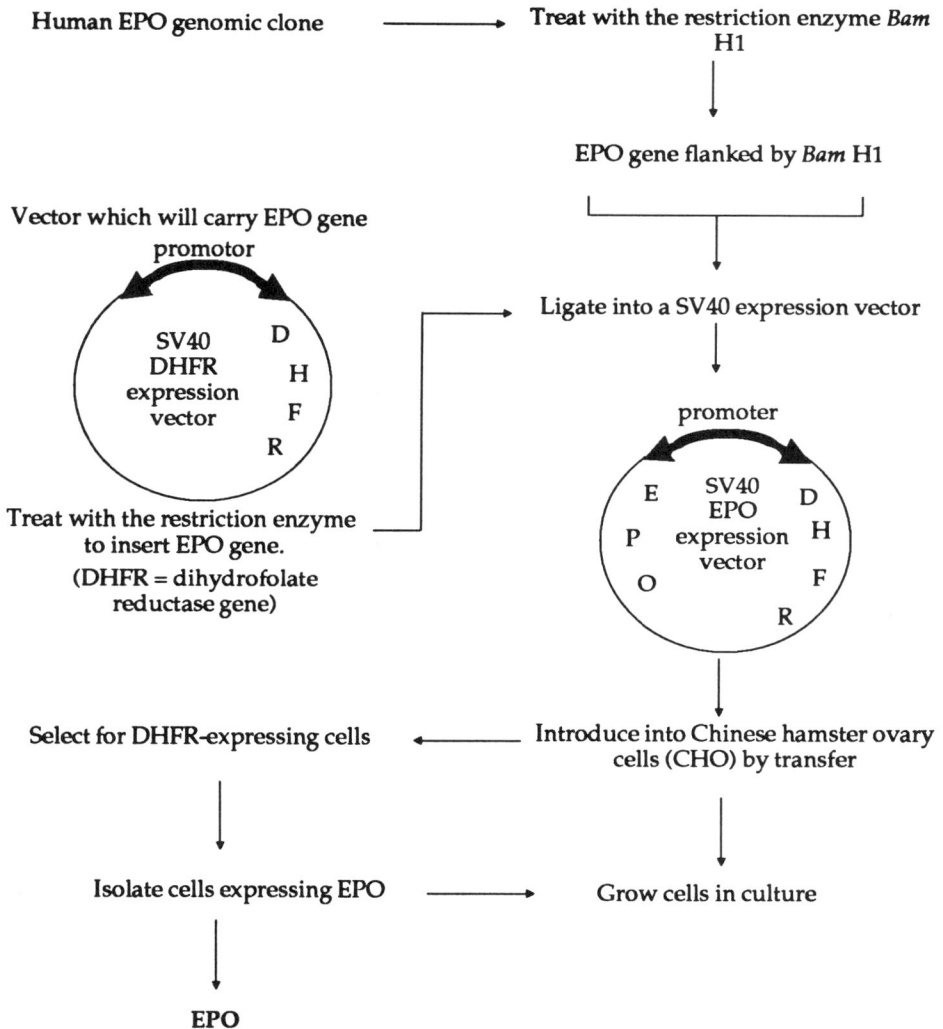

Human EPO genomic clone   ⟶   Treat with the restriction enzyme *Bam* H1

⟱

EPO gene flanked by *Bam* H1

Vector which will carry EPO gene

SV40 DHFR expression vector (D H F R)

⟶ Ligate into a SV40 expression vector

Treat with the restriction enzyme to insert EPO gene.
(DHFR = dihydrofolate reductase gene)

promoter

SV40 EPO expression vector (E P O D H F R)

⟱

Introduce into Chinese hamster ovary cells (CHO) by transfer

Select for DHFR-expressing cells ⟵

⟱

Isolate cells expressing EPO ⟶ Grow cells in culture

⟱

EPO

Thus the isolation of the human EPO gene was accomplished by cloning human DNA, as outlined below:

- isolation and restriction of DNA from liver;

- insertion of the DNA fragments into a bacteriophage;

- radioactive labelling of short, synthetic DNAs which enable identification of the EPO gene;

- identification of the bacteriophage that contained the DNA segment coding for EPO by hybridisation with the radioactive DNA probe;

- isolation of the bacteriophage carrying the EPO gene.

∏ Using the description we have given, draw out a flow diagram of the cloning procedure for the EPO gene.

synthetic oligonucleotide probes

The key to the successful isolation of the erythropoietin gene was to devise a system for identifying the gene in the large number of gene fragments incorporated into the bacteriophage (ie the genomic library). Note that the identification tools used were synthetic oligonucleotides made on the basis of the amino and sequence of erythropoietin. How does this work? Since the amino acid sequence of erythropoietin is known (Figure 5.3.) and we know which nucleotide triplets code for each amino acid, oligonucleotide sequences can be predicted that would code for segments of the erythropoietin molecule. Such nucleotide sequences should be capable of forming hybrids with portions of the erythropoietin gene. They can thus be used to check which of the bacteriophage plaques contain the erythropoietin gene sequence. We call such oligonucleotide sequences "probes".

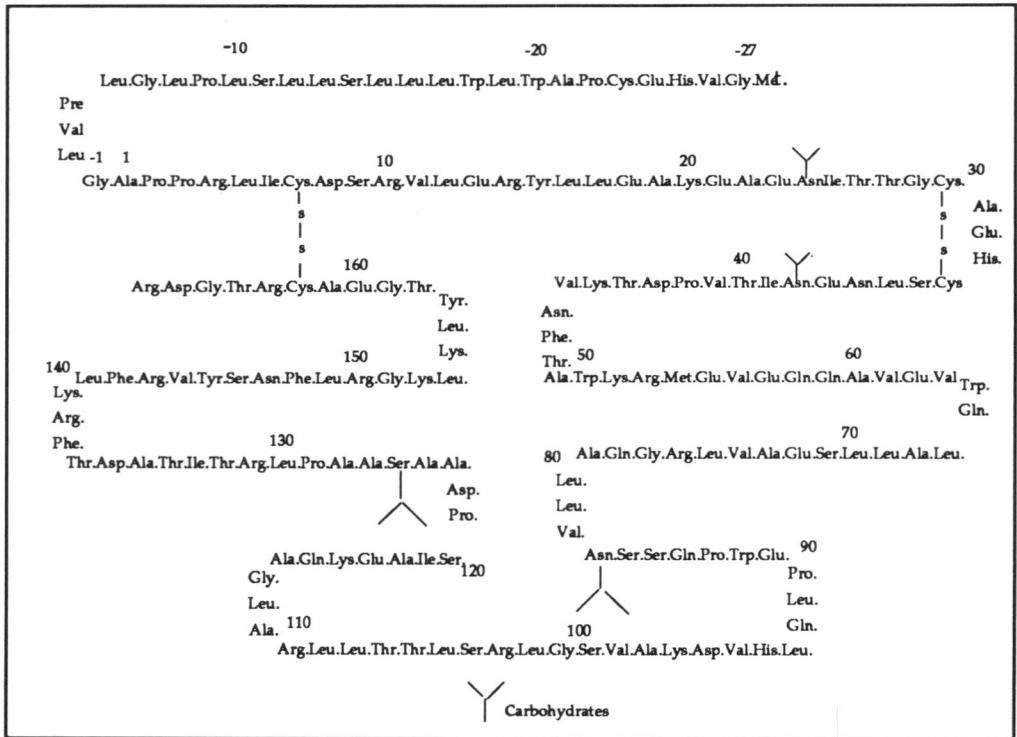

Figure 5.3. Structure of human erythropoietin

Once such bacteriophage plaques had been identified, the appropriate clones were picked off and cultivated and the bacteriophage DNA isolated. The DNA was treated with a restriction enzyme (Bam H1) which cuts the bacteriophage DNA on either side of the erythropoietin gene.

SV40   vector   The isolated erythropoietin gene was then inserted into an expression vector. Look back to the vector used. You will see that we have put a double headed arrow between the EPO gene and a gene labelled DHFR. This double headed arrow represents a promotor which controls RNA synthesis on either side. Thus this promoter controls both the expression of the EPO gene and the DHFR gene. DHFR stands for the enzyme dihydrofolate reductase.

After the gene had been inserted into the expression vector, the vector was used to transform Chinese hamster ovary (CHO) cells. Cultures of these cells are used to produce human EPO.

SAQ 5.4.

A company is attempting to isolate a human gene coding for a protein labelled X. The amino acid sequence X is not fully known, but does include the sequence

- Methionine - Methionine - Tryptophan - Isoleucine - Tyrosine -

The genetic codes for these amino acids in mRNA are:-

AUG codes for Methionine

UUG codes for Tryptophan

AUU or AUC or AUA codes for Isoleucine

UAU or UAC codes for Tyrosine

U = Uracil; G = Guanine; C = Cytidine; T = Thymine.

a) Which of the following oligonucleotides would be a suitable probe to detect the gene coding for protein X in a gene library? (We remind you that mRNA is translated from the 5' end to 3' end).

1) 3'GTAGTATTATAT5'

2) 5'GTAGTATTATAT3'

3) 5'ATGATGTTGATTTAT3'

4) 3'AAAAAAAAAAAAAAA5'

5) 5'ATGATGTTGATATAC3'

b) Give reasons why the amino acid sequence shown above is a good choice to produce an oligonucleotide sequence for probing.

5.6.2. Production of large quantities of r-HuEPO

The production of large quantities of Eprex®was accomplished by modifying the gene through addition of a regulatory piece of DNA (a promoter), linking it to the gene for dihydrofolate reductase (DHFR), increasing the number of copies of the gene within each cell and, finally, with a selection technique, isolating cells that contain many copies of the erythropoietin gene. A master working cell bank was then constructed containing vials of erythropoietin positive cells used to seed growth medium for large-scale culture.

Both the master cell bank and growth culture are subjected to stringent testing to prevent contamination from pathogenic viruses and other proteins. Again, these have their parallels in the insulin story where *Escherichia coli* peptides were looked for. It is important not to allow SV40 viruses and hamster proteins to contaminate the final product. These can be detected by appropriate immunoassay procedures. The r-HuEPO produced is tested extensively for purity and for biological and immunological equivalence with human erythropoietin from urine. The final step in the production process is formulation of r-HuEPO in ampoules of buffered solution, stabilised with addition of human serum albumin.

∏ Why is a vector used which has a separate gene coding for DHFR linked via a promoter to EPO? (To answer this question you will need to know that the enzyme DHFR is essential for the synthesis of certain amino acids and the the Chinese hamster ovary cells used were deficient in DHFR).

The EPO gene linked via the promoter to the DHFR gene are inserted into a Chinese hamster ovary cell line by transfection. However, this cell line is deficient in DHFR so that it cannot survive in media lacking certain amino acids. Since these amino acids are not necessary for DHFR - producing cells, their depletion in the culture medium will allow survival only of those cells which were efficiently transfected and thus contain the DHFR gene. Since this DHFR gene is linked to the EPO gene, these cells will also express EPO.

This procedure resulted in recombinant human erythropoietin being produced from the cloned gene for human erythropoietin. It is biologically and immunologically indistinguishable from human erythropoietin. Recombinant human erythropoietin (Eprex®) is now available as a pure pharmaceutical preparation for clinical use.

5.6.3. Characterisation and comparison of r-HuEPO with native EPO

Think back to the insulin case study. You will recall that it is essential for licensing purposes to characterise a product thoroughly and, where possible, compare it with authentic material. You will also recall that it is insufficient to rely on a single type of criterion (eg amino acid composition). In the case of insulin, chromatographic behaviour, amino acid composition, circular dichroic spectra, enzymic hydrolysis products and biological activity were used. In the case of EPO the protein was characterised by radioimmunoassay, Western blot, (a technique for characterising proteins separated by electrophoresis), structural analyses, and *in vitro* and *in vivo* biological assays to show that the product had biological and immunological properties identical to human erythropoietin. The structure of r-HuEPO is shown in Figure 5.3. It is a 165 amino acid, extensively glycosylated protein.

Biological activity was also determined. The dose-response curve of the gene product assayed in mice was identical to that of standard erythropoietin isolated from urine; recombinant human erythropoietin raised the haematocrit (volume of RBCs as % of total blood volume) in normal mice from 47 percent to 73 percent.

To produce sufficient natural EPO exhypoxic polycythaemic mice were used, exhypoxic mice are mice that have been kept under an atmosphere of reduced oxygen content. Such mice produce an excess of RBCs (ie are polycythaemic). This conditioning of mice gives maximal stimulation of EPO secretion and is necessary to obtain sufficient amount of EPO for the comparative study with the recombinant EPO.

| SAQ 5.5. | Which of the following criteria were used to select Chinese hamster ovary (CHO) cells which might be carrying EPO genes after transfection with the SV40 expression vector? |

1) Ampicillin resistance of CHO cells

2) Immunoassay for SV40 coat protein on the surface of CHO cells

3) Ability of the CHO cells to grow in medium with a restricted amino acid content

4) By the demonstration of the enzyme for dihydrofolate reductase.

5) By immunoassay for EPO production

Π Before we leave the production of EPO, can you think of reasons which may have influenced the choice of Chinese hamster ovary cells for EPO production. Do you think they are the cheapest cells to produce human proteins in? (Think back to the insulin story). Do you think that if EPO genes were placed in *E. coli*, the protein would be produced in the correct conformation? Do you think *E. coli* would have the machinery for glycosylating the protein?

It perhaps is obvious that Chinese hampster cells are more expensive to cultivate than *E.coli*. They are slower growing and more fragile. Thus the choice of using Chinese hampster cells must have been made for reasons other than economic ones. It was important that the product (EPO) had the correct structural conformation in order to display the right activity and to produce minimal immunological response. Particularly important is the glycosylation of the protein. Hence cells with the appropriate glycosylation mechanisms had to be used. In this case Chinese hampster cells proved to be satisfactory.

5.7. Pre-clinical and clinical studies

We can divide these studies into the following types: pharmacological studies in animals; toxicological studies in animals; pharmacokinetic studies; human pharmacological studies pilot studies and large scale mulitcentre studies.

In the diagram below we have arranged these studies into two stages, pre-clinical and clinical. The types and objectives of the studies are also given.

Π Examine the information provided in the diagram, and see if you can work out the rationale for the sequence of studies that are undertaken.

Stage	Type of study	Objectives
Pre-clinical	Pharmacological	-tests carried out in animals to evaluate pharmacological activity over a wide dose range
	Toxicological	-tests in animals to examine possible toxicological consequences of the drug-often including excessively high drug loads
	Primary pharmacokinetic studies	- tests in animals to measure the adsorption, distribution metabolism and excretion of the drug
	Human pharmacological studies	- tests in healthy volunteers to determine the pharmacological activity over a dose range, extended into chronic patients
	Human pharmacokinetic studies	- to measure the adsorption, metabolism, excretion of the drug in healthy volunteers, then in chronic patients
Clinical Studies	Pilot studies	- to test the efficacy/risks of the drug in small controlled groups of patients
	Multicentre studies	- larger group analysis of risk/benefits of the drug.

It is important to remember this sequence. It has a logical basis.

First it is important to establish that the new drug has some pharmacological activity (if it has no activity, it is pointless to continue) and to establish that it has no, or acceptable, toxic effects. It is also important to begin to establish how quickly the drug is accumulated, metabolised and excreted since this will influence the size and frequency of dose. There is also a need to establish if the route of administering the drug is important and, if so, in what ways. These studies are, if appropriate, first conducted in animals and then, cautiously, with humans.

If at this stage, the drug performs satisfactorily it may enter clinical trials, firstly on a small (pilot) scale and, subsequently on a larger (multicentre) scale. If the signs are satisfactory, it may then progress to more general use.

Using this framework we will now explore the data relating to these tests for r-HuEPO.

5.7.1. Pharmacology

Pharmacological studies have shown that recombinant human erythropoietin (r-HuEPO) is essentially identical both biologically and immunologically to human urinary erythropoietin. The ability of r-HuEPO to stimulate red blood cell production has been demonstrated both in normal animals and in animal models used to simulate end-stage renal disease (ESRD). These data suggest a role for r-HuEPO in the treatment of anaemia associated with renal disease, as well as anaemias of various other aetiologies. In general pharmacological testing, r-HuEPO was devoid of significant pharmacological activity, (other than its expected effect on erythrocyte production), at intravenous doses up to 2000 Units/kg (or up to 1000 U/ml for *in vitro* studies) on the central nervous system, the cardiopulmonary system, isolated smooth muscle, renal function and the blood clotting system. The only unexpected pharmacological effect was a decrease in body temperature in mice after 2000 U/kg intravenous doses of r-HuEPO. This dosage is greatly in excess of the clinically effective dose (50-500 U/kg/3x/week i.v. in renal disease) and therefore this finding should form no objection to the clinical use of r-HuEPO.

5.7.2 . Toxicology

route of administration

In assessing its toxicological potential, r-HuEPO was administered by intravenously (directly into the blood), subcutaneously (under the skin), orally (through the mouth), intraperitoneally (in the abdominal cavity) and intramuscularly (in the muscle), to mice, rats, guinea pigs, rabbits, dogs and monkeys. We will use the following abbreviations in much of the discussion:

i.v.: intravenous, i.m.: intramuscular, s.c.: subcutaneous, p.o.: oral, i.p.: intraperitoneal

acute toxicology myelofibrosis

In acute toxicological studies, intravenous, oral and intramuscular doses up to 20,000 U/kg r-HuEPO (40-400 times the highest proposed human dose) were well tolerated and there was no mortality . Most of the changes associated with the i.v. and s.c. multidose studies represent the expected pharmacological action of r-HuEPO including increases in red blood cell (RBC) count, haematocrit (Hct), haemoglobin (Hgb) levels and reticulocyte number. Major clinical signs included reddening of the limbs in rats and dilations of the vessels of the eye in monkeys and rats at higher doses. Bone marrow fibrosis occurred in multidose studies in dogs. Whether this is related to antibodies against human EPO, which develop in some dogs, is not known. However, high doses (1000 U/kg/day) of r-HuEPO for 90 days failed to cause myelofibrosis in Cynomolgus monkeys. In studies of dogs in which drug withdrawal followed the treatment periods, biochemical and histopathologic examinations indicated that most of these finding, including myelofibrosis, lessened in severity, or returned to normal. Despite the fibrosis, the bone marrow remained functional throughout the entire period. Some degree of transient anaphylactic (immune) reaction to foreign protein occurred in dogs which was indicated by reddening and swelling of the areas around the mouth and eyelids.

☐ Why would such a reaction not be expected in humans?

Anaphylactic (or allergic) reactions occur when a protein, which is recognised as a foreign protein, is administered. This protein will act as an antigen sometimes inducing severe immunological reaction. Since the r-HuEPO is indistinguishable from the natural EPO in humans, this reaction will probably not occur. Remember however that we learnt while discussing insulin that sometimes the artificial introduction of human proteins into humans may stimulate the immune system.

foetal weight teratogenicity mutagenicity

No impairment of fertility occurred when male and female rats were administered r-HuEPO at doses up to 500 U/kg. Maternal toxicity such as decreased food consumption and gain in body weight, gastric erosions and mild redness of the limbs have been observed at 100 and 500 U/kg. Foetal weight gain was decreased and ossification was delayed after treatment with 500 U/kg r-HuEPO in rats. Since the foetal changes were associated with maternal toxicity, they may be secondary rather than due to a direct foetal effect. There was no indication of teratogenicity, (capability of a drug to induce foetal abnormalities) in rats or rabbits and no toxicity in two subsequent generations at doses up to 500 U/kg. In a series of mutagenicity (capability of a drug to induce mutations) studies, r-HuEPO failed to induce bacterial gene mutations at the HGPRT (hypoxanthine: guanine phosphoribosyl transferase) locus or micronuclei in mice, and was therefore not considered to have mutagenic potential.

pyrogen test

antigenicity test

Recombinant human erythropoietin was also non-pyrogenic when tested in rabbits and in the LAL assays. We remind you that the Limulus Amoebocyte Lysate test detects pyrogenic or fever inducing substances by causing clot formation when the LAL reagent is added. The LAL reagent is isolated from the horseshoe crab (Limulus). In antigenicity tests, r-HuEPO was tested for passive haemagglutation (PHA) and 4-hour passive cutaneous anaphylaxis (PCA) in rabbits, for active systemic anaphylaxis and cutaneous reaction in guinea pigs and mice. Results of these studies indicated weak to strong positive antibody formation in rabbits and guinea pigs, but not in mice. The reason for the difference in the mouse is not clear. However, since r-HuEPO is of human origin, anaphylaxis is not anticipated when administered clinically to humans. Studies of the seizure threshold in mice and water content of the brain tissue and electrolyte distribution in rats by administrating intraperitoneal doses of 1500 U/kg r-HuEPO showed these had no effects.

5.7.3. Absorption, distribution, metabolism and excretion

In the primary studies of the pharmacokinetics and metabolism of exogenously administered r-HuEPO in animals, the i.v. route was the mode of drug administration. Consequently, the rate and extent of absorption was not a parameter under investigation.

In studies in which radiotracer techniques were utilised, concentrations of radioactivity following administration of ^{125}I-r-HuEPO to rats were highest in the bone marrow, spleen and kidneys.

Studies of the kinetics of elimination of i.v. administered r-HuEPO were conducted in rats, dogs and Rhesus and Cynomolgus monkeys. The distribution and elimination of r-HuEPO observed in most studies conformed to a two-compartment model. The initial volume of distribution approximated to that of the plasma.

Although only limited data are available, the metabolic fate of erythropoietin appears similar to that of other sialylglycoprotein hormones. The metabolism of both endogenous EPO and exogenously administered r-HuEPO appears to occur primarily in the liver.

SAQ 5.6.

Although r-HuEPO showed no evidence of teratogenicity or mutagenicity and the LAL test detected no pyrogenic factor, some deleterious effects were detected when r-HuEPO was administered to dogs, rats, rabbits and guinea pigs. For example in dogs, bone marrow fibrosis was detected; in pregnant rats there was some loss in foetal weight gain and some systemic anaphylaxis. Cutaneous immune reaction was detected in rabbits and guinea pigs.

Explain why these observations were not sufficient to abandon the use of r-HuEPO for the treatment of certain types of human anaemias.

5.7.4. Human pharmacology (pharmacodynamics)

Human EPO from urine and r-HuEPO appear to be identical on the basis of all comparative analyses conducted to date, and it is appropriate to review the action of the natural hormone. Erythropoietin serves as one limb in an oxygen-mediated feedback loop that controls the rate of red cell production (see Figure 5.2). It is produced in response to hypoxia in the kidney. More specifically, EPO m-RNA appears within 30 minutes and peaks within 6-10 hours following an hypoxic stimulus. Peritubular interstitial cells (presumably endothelial cells) are implicated as the EPO generation site. This was demonstrated only recently using the technique of *in situ* hybridisation and more than thirty years after the kidney was postulated to be the most important site of EPO production. After release, the hormone interacts with specific receptors on the surface of committed erythroid cells in the bone marrow, with colony forming units (CFU) being the most sensitive population. The effects of r-HuEPO are believed to be entirely upon red cell production and the natural hormone has not been found to have any observable effect on other cells.

In the anaemia of renal disease, radioimmunoassays have confirmed data provided by bioassays in showing that this anaemia is caused primarily by a decreased erythropoietin production. Figure 5.4 shows the causal relationship between a decline in renal function and the degree of anaemia. EPO titres have also been shown to be low in patients with rheumatoid arthritis and very high in patients with aplastic anaemia.

In conclusion, r-Hu-EPO can be considered as a potential therapeutic agent in patients with renal disease and possibly also in patients with rheumatoid arthritis.

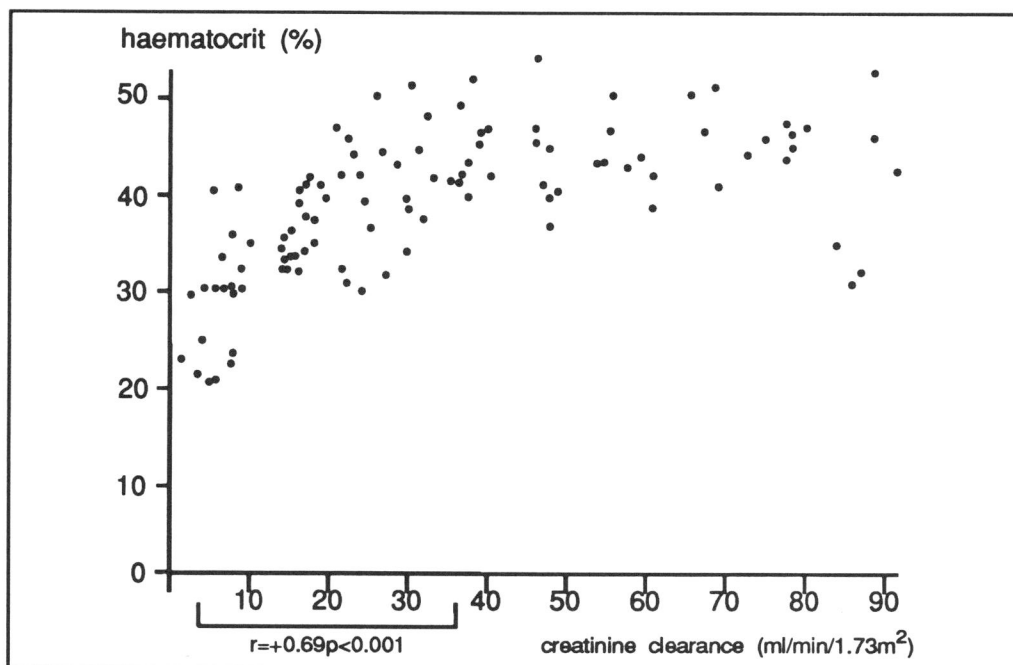

Figure 5.4. Relation between anaemia and renal function. NB creatinine clearance is taken as a measure of renal function, haematocrit is the % of the blood volume taken up by RBCs.

Pharmacological studies in healthy volunteers

First study

The earliest study was a double-blind, placebo-controlled trial of the safety and haematological effects in 48 normal male volunteers. Eprex® or placebo was administered via the subcutaneous (s.c.) or intravenous (i.v.) route at a wide variety of doses (1, 10, 50 or 100 U/kg), as the therapeutic dose was still undetermined. No pharmacological responses were measurable in reticulocyte count, erythrocyte count or haemoglobin with the single doses tested.

Second study

Another placebo-controlled, double-blind study did show an effect (ie increased RBC production) when multiple doses of 10 and 20 U/kg were administered, either i.v. or s.c. Eprex® was given once a day for 14 consecutive days. The mean reticulocyte and haematocrit response was statistically significant and dose-dependent. The placebo-treated subjects also showed a moderate increase in reticulocytes, probably due to the frequent blood sampling.

Ⅱ Why would reticulocytes increase during frequent blood sampling in such placebo treated subjects?

Frequent blood sampling mimics acute bleeding resulting in an endogenous EPO stimulation and release of new RBC (reticulocytes).

Third study

In a double-blind, placebo-controlled study of safety and efficacy, 32 healthy male volunteers received Eprex® 150 U/kg or 300 U/kg or placebo i.v. or s.c. on days 1, 4, 6, 8 and 10. There was a statistically significant relationship between the dose of Eprex® and the haematocrit and haemoglobin responses. Both routes of administration were efficacious. There was a trend toward higher reticulocyte count with increased dose; however, the overall number of reticulocytes did not increase in a significant manner. The drug was well tolerated by all subjects.

Fourth study

Eprex and Iron deficiency

A further double-blind, placebo-controlled study was performed to determine the effect of multiple doses of Eprex®i.v. (100, 200 and 300 U/kg). Subjects were dosed on day 1, then rested for two days, and then dosed once daily on days 4 through 17. Increases were seen in haematocrit for all dosage groups but the responses were not statistically different from the placebo-group. There was, however, a substantial drop in ferritin and serum iron in all the Eprex® subjects. It was therefore hypothesised that the minimal effect on erythropoiesis was due to iron depletion. After rechallenge to identical doses as in the original study and oral iron supplementation, all subjects on Eprex®showed a large increase in reticulocyte count and haematocrit while there was no response in the placebo-group. This study illustrates that Eprex® can only be effective if all of the other pre-requisites (iron, vitamins) for erythrocyte production are available.

Pharmacological studies in patients with chronic renal failure

Eschbach study

ferrokinetic studies

erythron transferrin uptake

In this study, ferrokinetic studies were performed before the first Eprex® injection and 24 hours after the fourth injection. At doses of 50 U/kg i.v., thrice weekly, consistent dose-dependent increases in the reticulocyte count and erythron transferrin uptake were observed. Erythron transferrin uptake is a synonym for iron uptake by the red blood cell producing system (erythron). Iron does not circulate freely in the blood but is transported by a protein called transferrin. A rise in haematocrit was also observed. Also a statistically significant correlation, for doses up to 150 U/kg, between the change in corrected reticulocyte count and the change in erythron transferrin uptake, was observed. This indicated that the reticulocyte count is an adequate monitor of changes in erythropoiesis, as reflected by changes in iron kinetics. A further important observation was the development of functional iron deficiency at doses of 15 U/kg or more.

Winearls study

In the pilot study of the Winearls-Cotes group, ferrokinetic studies were performed. Erythron transferrin uptake increased to within or above the normal range following successful treatment in 9/10 of the patients. Red cell life span, subnormal before treatment, was increased in the four patients who had previously been given blood transfusions. In the same study, the responses of erythroid progenitor (colony forming units) cells were examined. Following treatment, a significant rise in CFU mitotic rate was observed.

5.7.5. Human Pharmacokinetics

Healthy volunteers

Three studies in normal volunteers, designed to address safety and haematological considerations arising from the administration of Eprex® to humans, also included a pharmacokinetic component. A fourth investigation in healthy volunteers aimed to define the pharmacokinetic profiles of Eprex® after both i.v. and s.c. administration. In these volunteers, Eprex® was administered over the dose range of 1 to 300 U/kg for both i.v. and s.c. routes. A radioimmunoassay was used to determine EPO in serum; this procedure does not distinguish between endogenous and recombinant hormone. The baseline serum EPO in a young, healthy population has been estimated to be 10 to 30 mU/ml.

Intravenous administration

systemic clearance

first order decline in serum EPO

At 1 U/kg i.v., serum EPO levels were similar to these from placebo-treated subjects. Following single i.v. doses of 50 and 100 U/kg, Eprex® was eliminated by a single, first-order process yielding an harmonic mean half-life of 4.8 hrs (range 3.8 to 5.6 hrs). A minimum i.v. dose of 100 U/kg was required to maintain serum EPO levels above the endogenous baseline for 24 hrs. The time for the EPO concentration to decline to baseline was dependent upon the administered dose. The mean estimates of systemic clearance and volume of distribution were 7.1 ± 1.2 ml/hr/kg and 49.3 ml/kg respectively. The distributional space for Eprex® showed good correlation with plasma volume. At 100, 200 and 300 U/kg serum EPO concentrations were found to increase in a dose-proportional manner. After multiple i.v. dosing (on day 1 and then for 14 consecutive days from day 4 to day 17), Eprex® was eliminated more rapidly; the harmonic mean half-life being 4.7 hrs (range 3.7 - 6.2 hrs) on day 1 and 3.3 hrs (range 2.5 - 4.2 hrs) on day 17.

In the major kinetic study, Eprex® was administered i.v at 150 and 300 U/kg on days 1, 4, 6, 8 and 10. Mean serum EPO concentration time profiles following i.v. administration on day 1 and day 10 are presented in Figure 5.5 for both the 150 U/kg and the 300 U/kg dose group.

∏ Examine Figure 5.5, carefully. What can you conclude about the decline in serum EPO concentrations after dosing? Does multiple dosing have an effect?

Your main conclusion should have been that EPO concentrations decline **exponentially** (Note the scales on the axis). You may also have detected a slight increase in the rate of decline in serum EPO concentrations after multiple dosing.

This is indicated by the changes from day 1 to day 10 in the area under the curve, the systemic clearance, the serum EPO concentrations 24 hrs post-dose, the elimination rate constant and the resultant harmonic mean half-life. The apparent volume of distribution appeared to be a constant parameter. There was a trend to a slower elimination of Eprex® as the dose was increased from 150 to 300 U/kg.

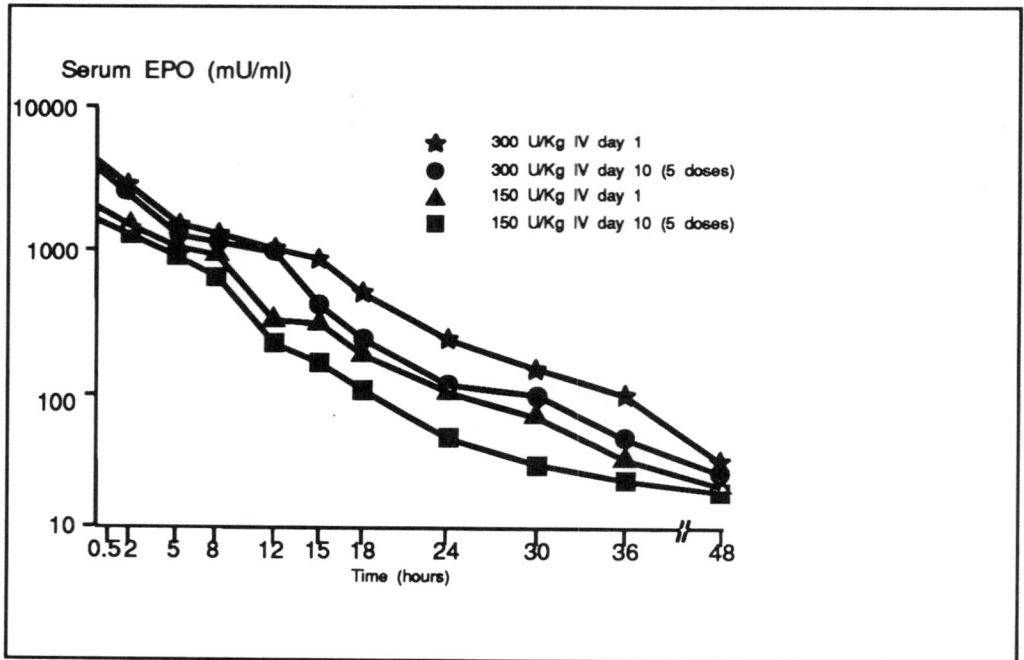

Figure 5.5. Decline in the serum levels of EPO after intravenous administration

It would appear therefore that both the duration of treatment and, to a lesser extent, the dose, may influence the disposition of Eprex®. Despite these findings, clinical data confirm the long-term therapeutic efficacy of Eprex® (over 2000 patients, more than 1 year treatment). The variables affecting the shortening of the serum EPO half-life over time have yet to be characterised. Anti-EPO antibodies could not be found in any healthy volunteer or any patient. Monitoring the serum EPO concentrations indicated that the greatest change in the half life $t_{1/2}$ occurred from the first to the second dose.

This finding suggests that the decrease of this parameter is not a time-dependent process and the possibility exists that the kinetic profile after a single (the first) dose may be regarded as "unique".

∏ Could you make any reasonable assumption about these observed changes in the half life ($t_{1/2}$) described above?

Perhaps the most likely is that Eprex® administration might induce more receptor formation. This may be reflected in a short half life after the second dose since more EPO will then be bound to the receptors.

Subcutaneous administration

Following subcutaneous (SC) administration, Eprex® is slowly absorbed from the injection site giving rise to a low concentration in serum which is maintained for several hours throughout the dosing interval. Consequently, dosing every 48 hours s.c. results in higher pre-dose serum EPO levels than from i.v. administration at comparable doses. Figure 5.6 represents mean serum EPO concentration-time profiles following a single s.c. or i.v. administration of 300 U/kg Eprex®.

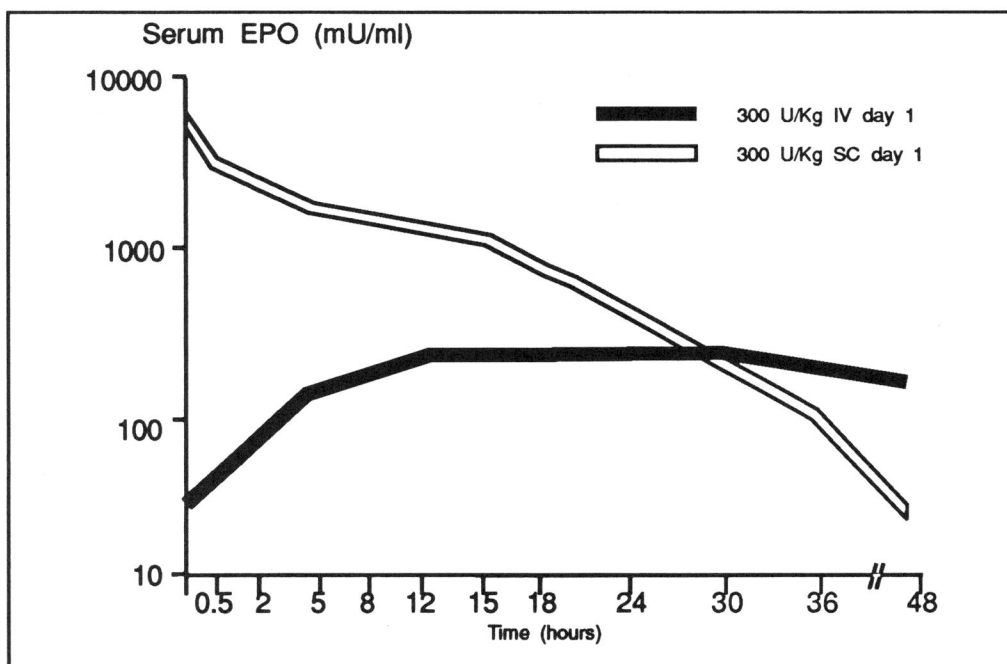

Figure 5.6. Serum EPO levels after intravenous and subcutaneous adminstration of 300U/Kg EPO

∏ Does the route of administration have much effect on the serum concentration of EPO?

Figure 5.7. compares the pharmacokinetic profiles of either a 150 U/kg or a 300 U/kg s.c. dose, given on day 1 and after multiple doses on day 10. On day 1 there appeared to be a reasonable dose proportionality in the maximum serum EPO concentrations achieved, even though this did not occur for several hours post-dose. This relationship did not hold at day 10, after 5 doses, when serum EPO levels were very similar after both 150 and 300 U/kg. (Note for example the serium level over the first 12 hours)

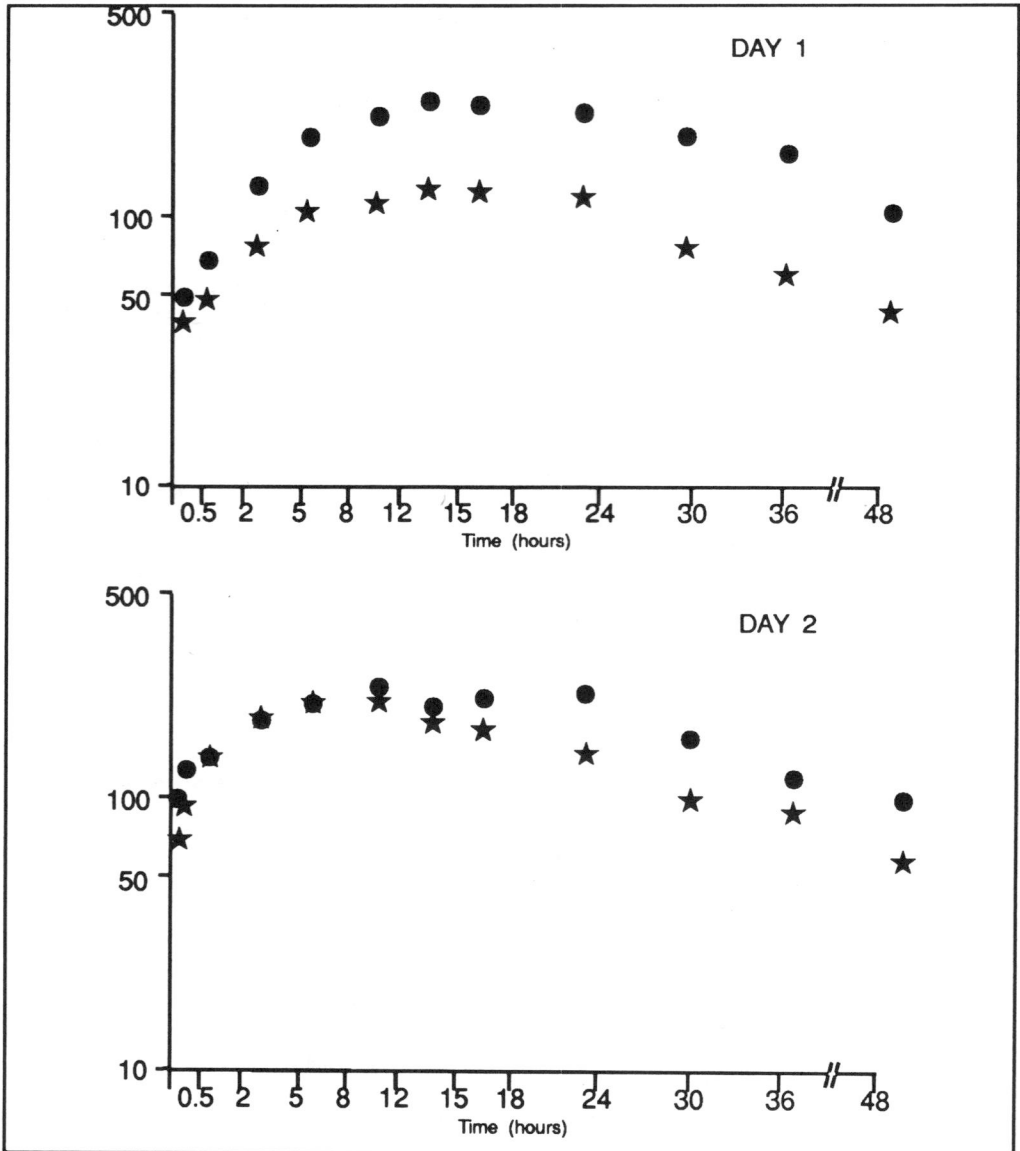

Figure 5.7. Mean serum EPO concentrations-time profiles following subcutaneous administration on days 1 and 10 of either 150 U/kg (★) or 300 U/kg (●)

These findings were confirmed by a Swedish study. Danielson and collaborators found that the maximum concentration after s.c. administration is reached after 12 hours with a peak which is about 1/20 of the concentrations reached by i.v. administration. The half life following s.c. injection is 24 hours and the bioavailability 30% but with a wide variation between individuals.

independance
of response
on route of
administration

Despite the differences in the pharmacokinetic profiles of Eprex® when administered by the i.v. or s.c. routes, it appears to produce a comparable erythropoietic response. Thus the s.c. route of administration may provide a useful alternative to i.v. administration.

Chronic renal failure patients

Pre-dialysis patients

Results from a pharmacokinetic study conducted in "pre-dialysis" patients, i.e. patients with end-stage renal disease but not yet on (haemo)dialysis, indicate a decrease in the ability of chronic renal failure (CRF) patients to eliminate Eprex®. This was indicated by a significantly lower K_{el} (elimination rate constant) and a higher serum EPO concentration 24 hours post-dosing in the CRF patients, as compared to healthy volunteers. The degree to which the rate of elimination of Eprex® is decreased is, however, substantially less than that observed with compounds cleared predominantly via the renal route. Indeed, animal data suggest that the majority of circulating EPO is catabolised by non-renal tissues, possibly the liver, prior to excretion by the kidney.

Haemodialysis patients

As part of the first open phase clinical study in 10 patients with chronic renal failure maintained by haemodialysis, pharmacokinetic studies were carried out at the time of administration of the first Eprex® dose and on a 2nd occasion after a period of some 14 - 54 weeks treatment (still administered thrice weekly as an i.v. bolus after dialysis).

Π Why would Eprex® be administered after and not before haemodialysis?

Haemodialysis means cleaning a patient's blood from toxic substances with filters. These filters retain cellular components and large proteins but smaller peptides or proteins up to MW 50,000 pass through the pores of this membrane (ie erythropoietin would be lost).

In a second multiple dose study, Eprex® was administered intravenously or subcutaneously to 11 haemodialysis patients. When administered i.v. (6 patients at doses of 15, 150 and 500 U/kg), the following values were found:

	after dose 1	after dose 7	after 3 months
$t_{1/2}$(hrs) of EPO	9.3 ± 3.2	6.2 ± 1.8	stable
V_D (% of body weight)	5.5%	stable	stable

When administered s.c. (5 patients at doses of 15 or 150 U/kg), peak serum concentrations were only 10% of those obtained by i.v. injection of the same dose. They were achieved within 8-12 hours and maintained at the peak level for at least the next 12 - 16 hours.

CAPD patients

CAPD
peritoneal
dialysis

In a pilot study conducted by Boelaert et al., Eprex® was administered by the i.v., s.c. and i.p. (intraperitoneal) route to patients with chronic renal failure maintained by continuous ambulatory peritoneal dialysis (CAPD). The results were similar to those obtained in haemodialysis patients. After s.c. administration, the serum concentration rose slowly to a plateau of 20% of that achieved after i.v. administration. After i.p. administration, the serum concentration was even lower and 80% of the dose was recovered in the first peritoneal effluent. NB: In peritoneal dialysis, the peritoneum (in the abdominal cavity) is used as an osmotic exchange membrane. Fluid is administered in the abdomen with certain concentrations of electrolytes etc. After osmotic equilibration (2-3 hrs) this fluid is removed and replaced by fresh fluid. The toxic components in the patient's circulation are thus eliminated by osmosis since the administered fluid does not contain these.

SAQ 5.7.

Answer yes or no to each of the following questions unless you feel that the answer is not clear cut.

1) In the first study with healthy volunteers, was a wide range of doses used because the researchers wanted to find out what levels of the drug were toxic?

2) Is a single dose of erythropoietin unlikely to have an observable pharmacological effect because its mode of action is fairly slow and it is quite quickly removed from circulation?

3) Is a raised reticulocyte count after administration of erythropoietin indicative of a stimulation of erythropoiesis?

4) Are healthy volunteers good candidates for measuring the pharmacological effects of erythropoietin?

5) Are the diets of volunteers used in testing erythropoietin important?

6) Is the method of administering erythropoietin critical to its erythropoietic effect?

5.8. Clinical studies on the anaemia of chronic renal failure

5.8.1. Introduction: the anaemia of chronic renal failure

anephric patients

normochromic, normocytic and hypoproliferative cells

Chronic renal failure (CRF) is almost always accompanied by anaemia. The anaemia is usually most severe in patients maintained by long-term haemodialysis and this is particularly the case in anephric patients (patients with no kidneys). The anaemia is characteristically normochromic, (red blood cells contain normal amounts of haemoglobin), normocytic (red blood cells are normal size) and hypoproliferative (red blood cells not produced in sufficient quantities). The debilitating effects of the anaemia associated with chronic renal failure, such as chronic fatigue, lack of energy, cold intolerance, sleep disturbance, chest pain etc, obviously deprive patients from leading a normal life. Given the fact, that, in Europe alone, about 84,000 patients were treated by dialysis in 1985, it appears that the anaemia associated with CRF is an important medical problem. Until recently, the anaemia associated with CRF could only be treated by blood transfusion. However, apart from the fact that transfusion brings about only temporary relief, regular blood transfusions should as far as possible be avoided because of risks of infection (particularly non - A and non - B hepatitis), iron overload and sensitisation to histocompatibility antigens which reduces the chances of successful kidney transplantation (we will deal with problems of histocompatibility in the case study on OKT3).

cause of
anaemia is
CRF

Although many factors such as shortened red-cell survival, blood loss associated with dialysis, aluminium intoxication and bone marrow suppression by "uraemic toxins" may contribute to the anaemia of CRF, the main cause of the anaemia of renal patients is thought to be inadequate erythropoietin production.

Taking into account the urgent need for an effective treatment for anaemia associated with CRF and the large number of patients involved, it is obvious that the primary indication for Eprex® treatment should be the anaemia associated with CRF. Very shortly after the production of r-HuEPO in 1985, two pilot studies were started with this in mind, in Seattle and in London.

5.8.2 .The Seattle pilot study

Twenty five patients undergoing regular haemodialysis for chronic end-stage renal disease were included in this study. The patients' ages ranged from 21 to 69 years, and seven were anephric. At the onset, the mean haematocrit of the patients was 19.3% (range 15 - 25%) and the mean reticulocyte count, corrected for haematocrit, was 1.06% (range 0.2-2.7%). Eighteen patients required transfusions in the four months preceding entrance into the study, and 12 were transfusion-dependent, in that they required one or more transfusions at least twice a month. It was an open study, with six dosage groups containing 4 or 5 patients each; the dose levels were 1.5, 5, 15, 50, 150 and 500 U/kg.

R-HuEPO was administered as.an intravenous injection three times a week, within 12 hours of the termination of dialysis, for three weeks. The trial developed in the following way. Two patients had to complete the initial treatment period without evidence of toxic effects before additional patients could be entered into the next dose level. All patients who had an acute erythroid response to r-HuEPO continued to receive the drug as part of a maintenance protocol.

Three of 4 patients treated with 15 U/kg responded and were transferred to maintenance therapy. At the time of preparation of this article, they had been treated for 22-29 weeks and their haematocrits had increased from baseline mean (SD) 20.2% to peak at 24.4%. The response was more obvious in the higher dosage groups: in 2 of 5 patients treated with 50 U/kg per dose and in 9 of 10 patients treated with 150 or 500 U/kg per dose, the haematocrit increased to 35%. The haematocrit in the other patients continued to rise (observation period at the time of reporting 14-27 weeks). If haematocrit exceeded 35%, the dose of r-HuEPO was reduced to maintain a stable value between 35-40%. The effect of different doses of r-HuEPO in the range of 50-500 U/kg on the haematocrit was dose-dependent (see Figure 5.8).

At the time of reporting, 17 patients had been treated with r-HuEPO for 3-7 months; most were receiving doses between 25-100 U/kg per injection to maintain haematocrit values between 35-40%. At these doses, no evidence of cardiac, pulmonary, hepatic, or other organ dysfunction that could be attributed to the treatment have been seen.

hypertension
grand mal

When the haematocrit exceeded 30%, hypertension developed or worsened in 4 patients, two of whom were receiving antihypertensive therapy. In one of the patients, the exacerbated hypertension was associated with a grand mal seizure (this patient had had a similar seizure three years earlier, but was not receiving anticonvulsant therapy). Grand mal is a certain type of epilepsy. Hypertension can cause cerebral edema which in its turn can induce these seizures. With 500 U/kg, the patient's haematocrit had risen from 16% to 37% in five weeks. A phlebotomy (phlebotomy: blood letting) to reduce the haematocrit and an increase in antihypertensive medication controlled the blood

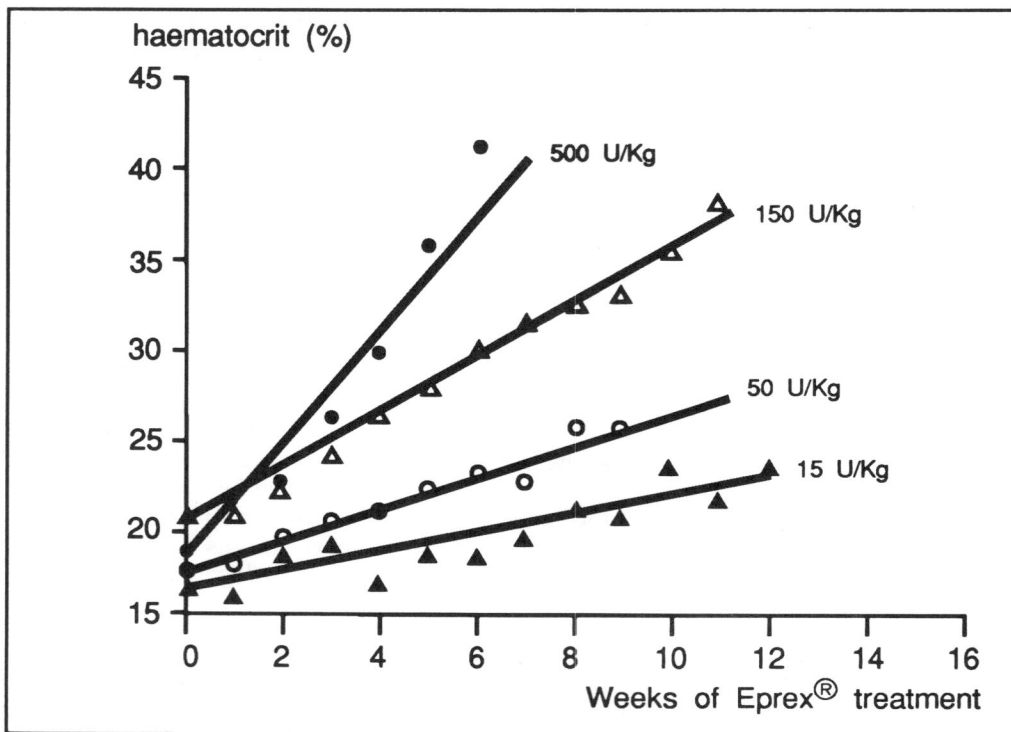

Figure 5.8. Seattle pilot study
Increase in haematocrit as a result of treatment with Eprex® (Note dose dependancy)

pressure, and r-HuEPO treatment was reinstituted later at 150 U/kg. Predialysis creatinine levels increased slightly in patients whose haematocrits exceeded 30%. Predialysis serum levels of urea nitrogen and potassium also increased, and this was presumed to be secondary to decreased dialyser clearances and increased food intake. Creatinine, urea and potassium are excreted by the kidney and form a parameter of renal function. In terminal CRF, the elimination of these substances is reduced to zero since there is no kidney function left.

It should be emphasised, that, before r-HuEPO therapy, blood transfusions had been required by nearly three quarters of the patients, and this need for transfusion was eliminated in all the patients who responded to r-HuEPO. Three patients who received low doses of r-HuEPO, underwent successful renal transplantation.

5.8.3 . The London pilot study

efficacy defined

This study was designed as an open, non-randomised escalating dose study in 10 patients (8 males, 2 females) with chronic end-stage renal disease, dependent on dialysis. The mean age was 45 years (range 26-65) and mean duration of haemodialysis 6.6 years (range 3-18 years). The mean haematocrit at entry was 18% (range 14-26).

There was an initial two-week control period of administration of normal saline to collect baseline efficacy and safety parameters. After the control period, each patient was treated three times a week by the i.v. route, starting with 12 U/kg per administration. Following a one-week course with this dose, in the absence of toxic effects and no therapeutic effect, the dose was increased at weekly intervals by doubling the previous dose up to 48 U/kg. For the purpose of this study, efficacy was defined as an increase in Hb of 2 g/dl (decilitre = 100 ml) above the mean value of the six observations obtained during the control period. Further increases in dose to achieve that target were performed by doubling the dose at two-week intervals, with a maximum allowed dose of 768 U/kg per administration. When therapeutic efficacy, as defined above was seen, the patient was placed on maintenance therapy, with the objective of titrating the thrice weekly dose to achieve a haematocrit of 35%.

Safety was evaluated by standard physical examinations, objective clinical evaluations, and laboratory tests done at regular intervals. At the end of the trial all 10 patients had been treated for at least 48 weeks. All met the criteria for response (ie an increase of 2g/dl Hb); the median time to reach target was 7 weeks (range 5-8). Nine patients achieved a haematocrit of 35%. The median dose to reach that target was 96 U/kg thrice weekly (range 96-192). Estimates of maintenance doses required to maintain 35% were between 48-96 U/kg thrice weekly.

Only one side effect appeared related to r-HuEPO per se, namely bone pain and chills following injection. Other effects were attributed to change in haematocrit consequent upon efficacy of r-HuEPO. These were an increase in blood pressure in 4 patients (three requiring additional therapy), thrombosis in av-fistula requiring intervention in one patient, a significant increase in pre-dialysis creatinine levels in one patient and increased heparin requirements for dialysis in 6. No neutralising antibodies to r-HuEPO were detected.

The results of this study with a follow-up of 12 weeks have been published in the Lancet. Since publication, a follow-up report has been prepared which describes the results obtained in these same 10 patients who have now received Eprex® treatment for more than 2 years. The response obtained in all 10 patients within three months has been maintained during the 2 years follow-up period. The dose required to maintain the Hb concentration between 10 and 12 g/dl has remained relatively constant at a mean of 50 U/kg thrice weekly. IgG antibodies to r-HuEPO have not been found.

5.8.4. Conclusions from the pilot studies

Both the Seattle and the London study showed Eprex® to be effective for the correction of anaemia associated with CRF. It was shown that the response was dose-dependent and that side effects were manageable. Consequently the European Cilag-companies decided to start two multicentre studies including larger patient populations in order to confirm the preliminary data. One study included 141 haemodialysis patients which were anaemic but not transfusion-dependent, while the second study enroled 137 transfusion-dependent patients.

5.8.5. First European multicentre study

The first European multicentre study enroled 141 non-transfusion dependent patients from 13 European dialysis centres. Participation in the trial required a haematocrit lower than 30%. Eprex® was started at 24 U/kg intravenous administration, 3 times weekly for at least two weeks. If the mean Hb level increased by less than 10% as compared to basal values, the dose was doubled.

All 141 patients achieved an individual Hb increase of at least 2 g/dl. In the large majority of the patients (93%) this was obtained with a dose of 192 U/kg thrice weekly. Four patients did not respond within 16 weeks but were found to be iron-deficient. After iron supplementation, these 4 patients eventually responded.

There were 12 drop-outs during the study because of protocol violation or early cut-offs leading to 129 patients who were eligible for "full response", arbitrarily defined as an Hb range of 10 to 12 g/dl. All 129 patients could be dose-titrated to reach this range. As could be expected, patients with a low Hb baseline seemed to need higher doses and needed longer treatment to achieve that goal. In order to maintain the target Hb-range of between 10 and 12 g/dl, a weekly dose of about 200 U/kg in 2 or 3 applications per week was generally adequate. As to adverse reactions, the cardiovascular system deserves special attention. In spite of the statistical fact that mean pre-dialysis blood pressure remained fairly normal and stable throughout the study, 31 patients were reported to have hypertensive problems; all but one (drop out) were adequately controlled by adjustment of antihypertensive medication. Fourteen out of these 31 patients had no hypertension in their medical history. There was no indication that r-HuEPO had any acute effects of blood pressure or other effects on vital signs. Fourteen patients had clotting problems with their vascular access. The most frequently reported adverse experience was "pain"; a precise incidence cannot be stated, because the same patients appear in several subcategories. Bone pain occurred repeatedly in two patients; six others reported less frequent or single occurrences of bone pain. Only 3 of those patients reported concomitant "flu-like symptoms".

In summary, the risk/benefit-ratio seems to be low when r-HuEPO is used appropriately. The need for blood transfusion is eliminated, because almost every patient can be "titrated" into the Hb range of choice.

A later report of this same study, on 96 patients who had been treated by Eprex® for a full year, confirmed the conclusions of the shorter term observation period. In addition, it appeared that there is no loss of Eprex® efficacy over the treatment period of one year.

5.8.6. Second European multicentre study

This second multicentre study was undertaken particularly in order to investigate whether Eprex® administration could eliminate the need for blood transfusions in patients who were transfusion-dependent before r-HuEPO therapy. In this study, 137 transfusion-dependent patients maintained by chronic haemodialysis, were enroled.

The dose regimen called for a starting dose of 100 U/kg of Eprex®, given i.v. 3 times weekly for 3 weeks. Further increases in dose were performed in steps of 25 U/kg/dose until the target Hb-range of 10 g/dl was reached. The frequency and size of the dose was then optimised in order to maintain Hb between 10-12 g/dl.

At data cut-off, 120 patients were eligible for a 3 month treatment evaluation. Out of the total of 120 patients eligible for 12 weeks analysis only 8 patients failed to reach the target level of 10 g/dl. However, 6 of these 8 patients were maintained below 10 g/dl for medical reasons (n = 4) or were slow responders and had not yet reached 10 g/dl at the time of data cut-off (n = 2).

hypersplenism Accordingly, there were only two true non-responders; one of those patients had hypersplenism (hyperactivity of the spleen) and the other had folate deficiency associated with macrocytosis.

The median Hb increase over time shows that it requires about 8 weeks therapy with r-HuEPO 100 U/kg x 3 weekly to reach the target level of 10 g/dl when the starting point is the Hb value measured after the last blood transfusion.

At data cut-off, 50 patients were eligible for a one year Eprex® treatment evaluation. After the arbitrarily defined optimal Hb level of 10 g/dl had been achieved, the median weekly dose could be reduced to 200 U/kg in the 50 patients treated for one year. For practical treatment the weekly dose appears to be somewhere between 100 - 300 U/kg, and has to be defined individually. When the Hb has stabilised within the chosen optimal range the required weekly dose can be apportioned as two injections in approximately half of all patients.

As to adverse reactions, these were similar to those observed in the first multicentre study. Forty six patients experienced a total of 84 hypertensive events during the treatment with r-HuEPO. Ten patients in this study presented with symptoms suggestive of hypertensive encephalopathy. Most patients had a history of hypertension and 3 had experienced seizures before embarking on r-HuEPO therapy. It is noteworthy that the absolute levels of Hb were sometimes only in the same range as those achieved after a blood transfusion. The seizure events with repeated grand-mal convulsions were followed by complete neurologic recovery after appropriate therapy. There is no evidence that r-HuEPO acts as a direct CNS irritant.

Because the occurrence of hypertensive encephalopathy is not directly related either to the absolute Hb level at the time of event or to the rate of rise in Hb from baseline levels, the only precaution which can be taken is to monitor blood pressure carefully in r-HuEPO treated patients, especially in those developing headache during the therapy.

fistula thromboses

Fistula or av-fistula is the name given to the vascular access which is created by connecting a vein with an artery to assure sufficient blood flow to the dialysis machine. Fistula thromboses occurred in 13 patients and the clotting of the extra component (blood circulating from the patient through the dialysis machine) occurred in 5 patients. These conditions had however, no clear-cut relationship with the level of Hb and haematocrit. There was a statistically significant rise in the platelet count throughout the study but the levels were within the normal range. The apparent increased tendency to clotting in the dialysis circuit and thrombosis of av-fistulas may reflect improved platelet adherence to the endothelial wall of the vessel. At data cut-off there was no evidence of development of antibodies to r-HuEPO.

5.8.7. Conclusions from the two European multicentre studies

Intravenous administration of Eprex® in both non-transfusion-dependent and transfusion-dependent patients with end-stage renal disease is a very effective treatment to correct anaemia and to abolish the need for blood transfusions. Because the majority of the adverse effects during the study were attributed to the therapeutic effect of Eprex®, i.e. increasing the haematocrit, it was decided to recommend lower starting regimens, i.e. 50 U/kg thrice weekly and to adopt a more prudent dose escalation i.e. in steps of 25 U/kg every 4 weeks if necessary. Consequently a third European multicentre study with this "low and slow" dose regimen was set up.

5.8.8. Third European multicentre study

The third European multicentre trial is being conducted in over a thousand patients with end-stage renal disease maintained by chronic haemodialysis.

The "low and slow" dose regimen used was as follows:

The initial dose was 50 U/kg body weight intravenously three times weekly (post dialysis) at an injection speed of at least 1 to 2 minutes. This dose was to be maintained for 4 weeks waiting for response. An increase of 1 g/dl in 4 weeks was considered to be a good response.

Depending on the rate of response, the dose was increased by adding 25 U/kg/dose and then maintained for 4 more weeks. The dose was titrated at intervals of 4 weeks by adding 25 U/kg/dose until the target haemoglobin between 10 and 12 g/dl (6.2 - 7.5 mmol/l) was achieved. A maximum dose of 200 U/kg given three times per week was not to be exceeded. If the haemoglobin value exceeded 12 g/dl r-HuEPO treatment was to be discontinued until the haemoglobin returned to values between 10 and 12 g/dl and then restarted with r-HuEPO at a dose reduced by 25 U/kg.

Reduction of dose frequency to twice a week was considered after a patient's haemoglobin concentration was stable (10-12 g/dl) and the study continued beyond 3 months. The study is still continuing, but an interim report has been prepared. At data cut-off, 355 patients had been evaluated for a period of 12 weeks therapy. Their increase in haemoglobin is presented in Figure 5.9 and the evolution of Eprex® doses used to achieve this, related to the % of patients was as follows:

	End of week 4	End of week 8	End of week 12
3 x 50 U/kg/week	94%	30%	27%
3 x 75 U/kg/week		64%	26%
3 x 100 U/kg/week			35%

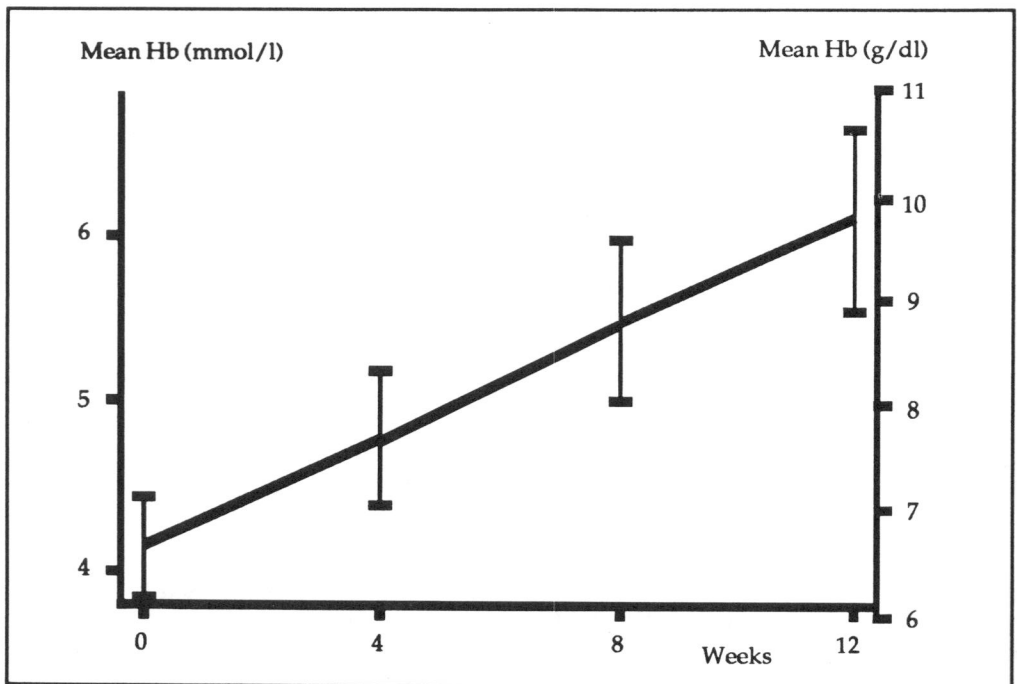

Figure 5.9 Monthly haemoglobin mean +/- SD

At data cut-off, 71 patients have been evaluated for a total of 24 weeks of Eprex® therapy. The evolution of their Hb levels is presented in Figure 5.10, and the evolution of Eprex® doses related to the percentage of patients was as follows:

	End wk 4	End wk 8	End wk 12	End wk 16	End wk 20	End wk 24
150 U/kg/week or lower	100%	28%	27%	30%	42%	49%
225 U/kg/week		72%	25%	28%	29%	30%
300 U/kg/week			48%	32%	22%	13%
375-400 U/kg/week				8%	3%	3%

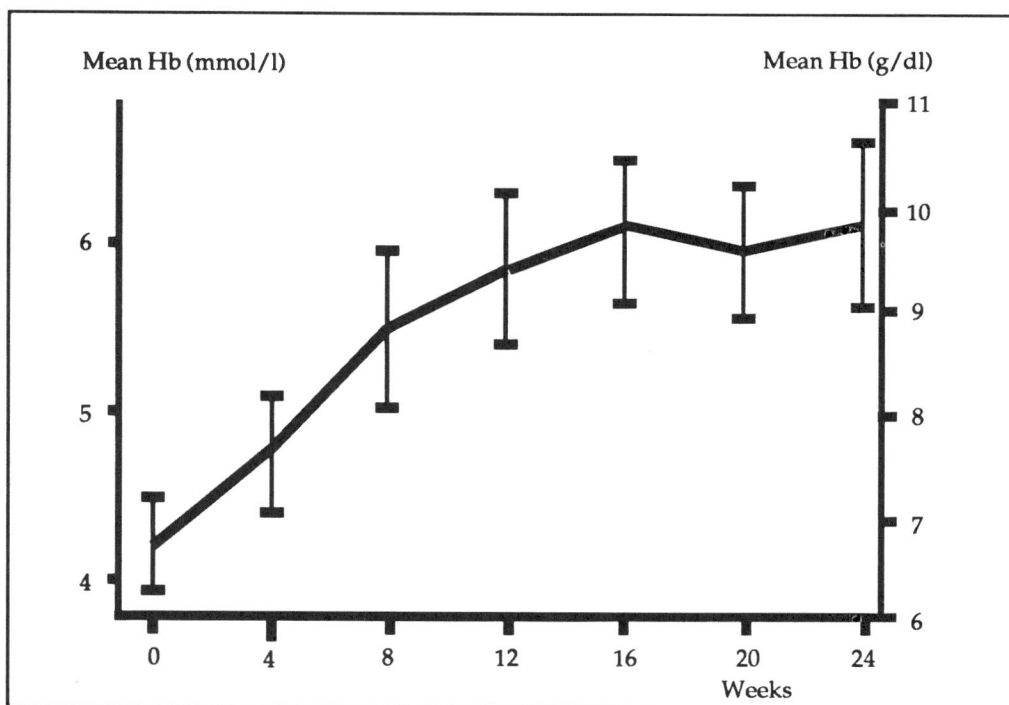

Figure 5.10. Monthly haemoglobin mean +/- S.D.: 24 week evaluable patients

∏ What are the changes in dose rate between the patients after 3 months of treatment and after 6 months treatment?

It can be concluded that after 3 months of therapy half of the patients need a weekly dose of 150 U/kg or 225 U/kg and the remainder a weekly dose of 300 U/kg. By that time the anaemia is usually corrected and the Eprex® doses can be decreased so that at

6 months of therapy, half of the patients are back at the initial starting dose of 150 U/kg/week, while 30% of the patients receive 225 U/kg/week.

As to adverse experiences, it should be emphasised that hypertensive problems were seen in only 9.4% of the patients while this was the case in 33 to 34% of the patients in the first and second European multicentre studies. Epilepsy and grand mal convulsions occurred in only 1.7% of the patients which compares favourably, particularly with the second multicentre study (7.3%).

In conclusion, the "low and slow" dose regimen leads to a somewhat slower achievement of anaemia correction but appears to be safe and very well tolerated, giving rise to a very favourable benefit/risk ratio.

5.8.9. Eprex® in pre-dialysis patients

The anaemia associated with renal disease is more pronounced in patients maintained by chronic haemodialysis. Consequently, all foregoing studies were performed in such a population of patients. However, patients who are progressing towards end-stage renal disease but with a residual renal function which is still sufficient for dialysis not to be needed, so-called pre-dialysis patients, can also become anaemic. Consequently, studies were undertaken to evaluate the efficacy and safety of Eprex® in the treatment of the anaemia of pre-dialysis patients.

Up till now three reports are available, one from Europe and two from the U.S.A. The protocols of the three studies were very similar in that the patients were randomly assigned to fixed dose levels of 50 U/kg, 100 U/kg and 150 U/kg administered thrice weekly by the i.v. route. In the U.S.A. studies, a placebo group was also introduced, but this was not the case in the European study. At the completion of an 8 week observation period or whenever a patient's haematocrit exceeded the target value (37% males, 39% females) by two percentage points, the patient was entered into the maintenance phase, in which Eprex® was injected only once a week. The dose size used at the beginning of this maintenance phase was three times the size of each dose given per injection during the correction phase. The haemocrit was carefully monitored and the dose size adjusted if the haemocrit began to rise or full. Dose adjustments were made in steps of 75 U/kg per week, at least every 4 weeks. The three studies led essentially to the same conclusions.

5.8.10. Eprex® in CAPD patients

One group of patients with chronic renal failure which has not yet been discussed are those who are maintained by continuous ambulatory peritoneal dialysis (CAPD). As these patients are ambulatory and do not come to the hospital very often, self administration of Eprex® via the subcutaneous or the intra-peritoneal route seems preferable over i.v.- administration. A significant amount of EPO is not absorbed when administered via the i.p. route. Consequently, a number of studies were started in which CAPD patients administered Eprex® to themselves via the s.c. route. So far it appears that in CAPD, s.c. administered Eprex® is acceptable, safe and efficient for the correction of anaemia in these patients. It is even suggested that the efficiency of CAPD is increased by augmenting peritoneal ultrafiltration.

5.8.11. Eprex® and quality of life in patients with CRF

quality of life   In all the studies discussed up till now, most patients reported an improvement in their feeling of well-being after correction of their anaemia with Eprex®. This sort of statement is, however, quite subjective. Consequently, a number of studies which would allow objective measurement of a number of aspects pertaining to the quality of life of patients with end-stage renal failure were set up. Most studies are still continuing; the evaluation of the results will be important in giving a full clinical picture of the use of Eprex®. Preliminary results indicate that there is a real improvement in the quality of life of patients with CRF.

5.8.12. Conclusions

From the studies with Eprex® in patients with anaemia associated with end-stage renal disease, conducted and reported so far, a number of conclusions can be drawn, which are probably best divided into two categories: what has been learned and what do we still have to learn? See if you can demonstrate that you have understood what the studies on EPO have revealed by answering the following SAQ.

| SAQ 5.8. | Which of the following statements are true?

1) The anaemia of chronic renal failure can be completely reversed with Eprex® therapy in many patients.

2) Statement 1 indicates that the uncomplicated anaemia of CRF is largely due to a hormone deficiency. If in CRF, inhibitors of EPO exist, they are easily overcome by Eprex®.

3) Exprex® treatment can largely abolish the need for blood transfusions with their inherent complications and risks. It has been shown that iron overload is corrected and that formation of cytotoxic antibodies is reduced, the latter giving rise to improved chances of successful renal transplantation.

4) Stimulation of erythropoiesis by Eprex® is dose-dependent. The "low and slow" dose regimen i.e. a starting dose of 50 U/kg administered thrice weekly by the i.v. route and dose adjustments in steps of 25 U/kg every four weeks (if necessary), allows adequate correction of anaemia with minimal adverse effects.

5) When the "low and slow" dose regimen is applied, Eprex® appears to be a safe drug. However, hypertensive episodes are seen in about 10% of the treated patients, making close monitoring of blood pressure a necessity during Eprex® therapy.

6) Iron deficiency may develop during Eprex® therapy, even in previously iron overloaded patients. Eprex® treatment will be less efficient in iron depleted patients and oral or even i.v. iron supplementation might be necessary for an adequate increase in red cells.

7) Many clinical symptoms and signs in patients with CRF have been attributed to uraemia (poisioning by urea). The reduction or elimination of many of these symptoms upon correction of anaemia, suggest that at least part of the "uraemic" syndrome appears to be due to anaemia and not uraemia.

8) Eprex® treatment allows a dramatic improvement of the patients' quality of life, rather than merely survive, the patients can resume an active life.

If you have successfully answered this question you have understood what knowledge has been gained from the clinical studies on Eprex®.

5.8.13. What do we still have to learn?

There are many questions, we raise a few here for you to speculate about.

1) Why do a significant percentage of patients who increase their haematocrit develop hypertension? Many speculations including increased blood viscosity and increased peripheral vascular resistance have been presented but it remains an open question.

   If we consider the blood vessel as a hollow tube, the pressure in this tube will be dependent on several mechanical conditions.

   First its diameter, second the flow rate, which in this case is depending on cardiac output, third the viscosity of the fluid. Adding red blood cells to the circulating blood, considerably increases viscosity.

2) How and to what extent do other conditions (eg as aluminium excess, inflammation) interfere with the effectiveness of Eprex®?

3) What will be the impact of a healthier population of dialysis patients upon the results of renal transplantation? What will be the impact of a more or less normal haematocrit on early graft function?

5.9. Therapeutic interest of Eprex®

The availability of Eprex® for therapeutic use represents a real breakthrough in pharmacotherapy and will provide real benefit to a large number of patients. We cite the following circumstances.

1) Patients with end-stage renal disease can be kept alive via dialysis. The "artificial kidney" takes over the major part of the normal kidney functions (excretion of unwanted products, regulation of electrolyte and acid-base balance etc.) but unfortunately, it is not able to take over the hormonal function of the normal kidney. Consequently, for more than two decades dialysis patients, their families and the medical staff caring for them have had to cope with the inevitable erythropoietin deficiency, the resulting anaemia and its debilitating consequences: lack of energy, fatigue, cold intolerance, anorexia, impaired sexual function, sleep disturbance, brain dysfunction, angina etc. are the results of this anaemia and prevent these patients from leading a normal life. Instead of really living, a lot of patients with chronic renal failure were merely surviving.

   The introduction of Eprex® allows the correction of the anaemia and its consequences, and offers to the patients a second chance for life. Eprex®-treated patients restart physical activities, they regain interest in various activities and take up social and even professional activities. The dramatic impact of Eprex® on the quality of life of patients with end-stage renal disease has convinced many nephrologists to describe Eprex® as the only true breakthrough in clinical nephrology in over 20 years.

2) The extremely rapid development of Eprex®from a concept (1983) to a drug (1988) merits special attention. A lot of modern drugs are quite often the result of the synthesis of a new molecule and the characterisation of its pharmacological properties. In the case of Eprex® it was the other way round. It was postulated

that anaemia, particularly the anaemia associated with chronic renal failure, was due to a deficiency of erythropoietin. Consequently efforts were made to produce the hormone in large quantities sufficient for therapeutic use. In addition all this was done in an extremely short period of time. Only five years lay between the first experiments leading to the isolation of the human EPO gene and the registration for its large scale therapeutic use. It is obvious that such a rapid development was not only due to a joint Cilag/Amgen/Ortho/Kirin extensive research program but also to the enthusiasm of both the clinicians and the patients involved in the clinical development of Eprex®.

3)   The availability of Eprex® has led and will certainly continue to lead to a better understanding of erythropoiesis and anaemia. Only two examples to illustrate this will be given. The hypothesis that the anaemia associated with CRF is due to an EPO deficiency has now been proven. In addition, it has been demonstrated that at least part of the uraemic syndrome is not due to uraemia but to anaemia.

4)   In Europe alone, at least 84,000 patients are suffering from end stage renal disease. Consequently the population of patients who can benefit from Eprex® treatment is substantial. However, from the preliminary results of the studies of anaemias of various aetiologies, it appears that patients with the anaemia of prematurity, anticancer chemotherapy, AZT therapy in AIDS, chronic disorders such as rheumatoid arthritis etc. could potentially also benefit from Eprex®.

Summary and objectives

In this case study, we have examined the causes of anaemia and the role of erythropoietin in the regulation of red blood cell production. This enables us to recognise anaemia as an inevitable adjunct to the other changes that accompany loss of kidney function due to disease. We then explored how the erythropoietin gene was cloned and how substantial quantities of erythropoietin can be synthesised by transfected animal cells. We also explored how pre-clinical and clinical studies confirmed that erythropoietin produced by recombinant DNA technology can alleviate the anaemia of late-stage renal disease. We have also described that the time between inception and the licensing of r-HuEPO was quite short.

Now that you have completed this study you should, therefore, be able to:

- describe a wide variety of causes of anaemia including the importance of anaemia as an inevitable addition to the other changes that accompany loss of kidney function due to disease;

- describe the role of the hormone, erythropoietin produced by the kidney in regulating the formation of red blood cells;

- identify which of the conditions resulting in anaemia can be potentially treated by erythropoietin;

- explain how olignucleotide probes may be devised from known amino acid sequences to identify DNA segments carrying a specific gene;

- describe the principles behind selecting clones of cells containing an expression vector carrying the erythropoietin gene;

- explain why animal studies are not always appropriate for the evaluation of rDNA products;

- describe the conclusions that can be drawn from pre-clinical and clinical studies about the use and limitations of erythropoietin in the treatment of anaemia;

- explain how rHuEPO has confirmed the role of erythropoietin in the anaemia which developes in renal disease;

- explain the benefits of erythropoietin in terms of alleviating anaemia and in the reduction of the need to carry out blood transfusion;

- Finally, we re-emphasis the point that despite the impression that may be gained about the complexity of the genetic manipulation and the extent of the pre-clinical and clinical studies, the time lapse from inception to product for recombinant DNA products, might be quite short. This is especially so if the need is great.

6

Case Study: Hepatitis B vaccines: a product of rDNA techniques

6.1. Introduction   142

6.2. Development of the recombinant vaccine   144

6.3. Pre-clinical studies   153

6.4. Clinical studies   158

Summary and objectives   164

Case Study: Hepatitis B vaccines: a product of rDNA techniques

6.1. Introduction

In contrast to the two previous case studies, in this chapter we will explore the use of recombinant DNA technology to produce a vaccine (i.e. a protection against infection) rather than a treatment of a clinical condition. Although the same general principles apply, there are some differences in emphasis in the pre-clinical and clinical studies that need to be undertaken. In this study particularly note the criteria used to select the host and vector systems used.

6.1.1. Hepatitis B Infection

Hepatitis B virus (HBV) is the major infective agent of liver and it can cause either acute or chronic viral hepatitis. In acute hepatitis, the virus can be cleared, leading to recovery. In chronic hepatitis B, the virus persists, and may lead to progressive liver disease, cirrhosis and primary hepatocellular carcinoma. It has been estimated that there are 250 million chronic carriers, which is about five percent of the world's population.

Π   How is the hepatitis virus transmitted?

transmission
symptoms

The virus is maintained in the population primarily via maternal transmission. In the developed countries, infection is also transmitted by exposure to contaminated needles or instruments, by blood transfusion or by sexual contact. The incubation period for hepatitis B varies from six weeks to three months. Clinical symptoms usually begin with fatigue, anorexia and myalgia. Later, jaundice, dark urine, light stools and tender hepatomegaly may occur. However, the onset of symptoms may be rapid, with early jaundice in association with fever, chills and leucocytosis. Sometimes jaundice is never recognised and the patient may be aware only of an influenza-like illness. Such people may however become chronic carriers, just like those who show more extensive clinical symptoms. As there is no effective treatment for hepatitis B, there was an urgent need for a vaccine that prevents both acute illness and its dire consequences.

HBsAg

In 1982, the first hepatitis B vaccine derived from plasma became available. This vaccine was prepared from hepatitis B surface antigen (HBsAg) that had been purified from the plasma of human carriers of hepatitis B virus infection. The hepatitis B viral or Dane particle has a diameter of 42 nm and contains a partially double-stranded DNA genome, which is enclosed within a core structure and surrounded by a lipid-containing envelope in which numerous copies of HBsAg are embedded (Figure 6.1.).

During infection, an excess of HBsAg is synthesised in the liver and assembled into particles with a diameter of 22 nm. These non-infectious particles are released in large quantities into the circulatory system of chronic carriers, from whom it can be recovered. These HBsAg particles were used as a vaccine. Although this vaccine is safe, well tolerated and highly effective, some disadvantages are attached to it:

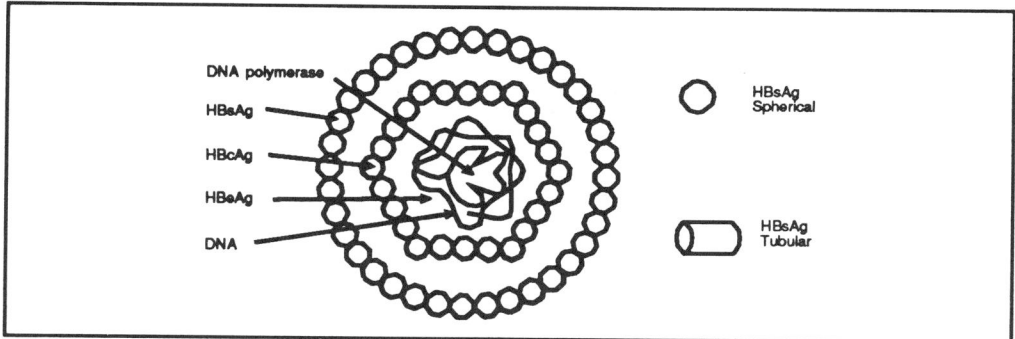

Figure 6.1. Diagram of the hepatitis B virus. The complete virion is known as the Dane particle. Spherical and tubular bodies of surface antigen circulate freely in the serum. The core of the virus contains the core antigen (HBcAg), the e antigen (HBeAg) which is secreted from the cells, the circular DNA and a DNA polymerase. HBsAg is the surface antigen.

- the supply of vaccine is limited by the availability of plasma;

- extensive processing and safety testing are necessary to ensure that the vaccine antigen is pure and free of any extraneous living agent;

- the acceptance of a vaccine derived from human plasma is less than might be expected given its safety and efficacy.

These concerns have led to alternative production of antigen, namely by cloning of the hepatitis B surface antigen gene into yeast.

SAQ 6.1.	Why was there a need for a hepatitis B vaccine?

SAQ 6.2.	Give three reasons why an alternative to the plasma vaccine is desirable.

6.1.2. Hepatitis B virus

HBsAg serotypes

The genome of HBV is a small, circular DNA molecule, which is partly double-stranded, partly single-stranded in a variable ratio. A drawing of the genome of the virus is provided in Figure 6.2. The long or minus (-) strand has a fixed length of 3200 nucleotides whilst the length of the short or plus (+) strand ranges from 50 to 75 percent of that of the minus strand. The minus strand contains four major reading frames, which encode for virus-specific proteins, including surface antigen (S-region), core protein (C-region), virus-specific polymerase (P-region) and X protein (X-region). We have included only the S-region genes on Figure 6.2.

The S region is divided into the S gene, pre S1 and pre S2 regions and codes for the viral envelope protein. The major S gene codes for 226 amino acids, the pre S2 for 55 and the pre S1 for 128. The major S gene codes for surface antigen and is labelled HBsAg in Figure 6.2. (HBsAg = Hepatitis B surface antigen). Four major subtypes of HBsAg have been described (adw, ayw, adr and ayr). These subtypes have the determinant 'a' in common. There is evidence that cross-immunity among various sub-specificities exists.

Chimpanzees vaccinated with the adw subtype, yeast-derived vaccine were protected when challenged intravenously with the adr or ayw subtype virus. To produce such a vaccine, the piece of the HBV genome coding for 226 residues of HBsAg of the subtype adw, was introduced into cells of bakers' yeast (*Saccharomyces cerevisiae*) via a specific plasmid and placed under the control of a yeast promoter.

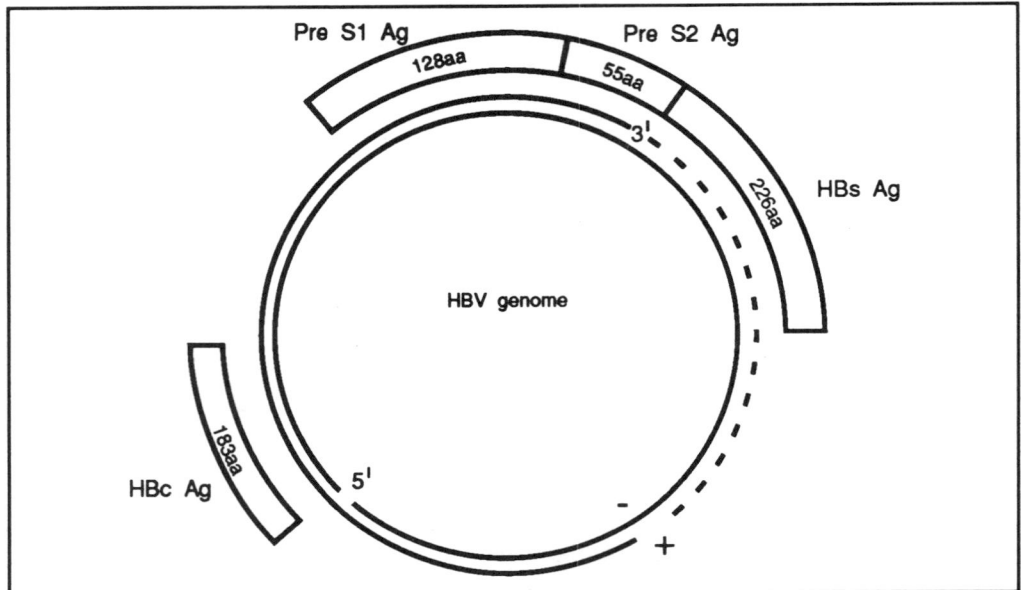

Figure 6.2. Hepatitis B virus DNA and defined antigens that are produced. aa= amino acids.

SAQ 6.3.

What surface antigen determinant have all hepatitis B subtypes in common?

SAQ 6.4.

How can cross-immunity between subtypes be accounted for?

6.2. Development of the recombinant vaccine

6.2.1. Host cell systems

Π We can perhaps think of four main types of host systems that might be used to produce recombinant DNA-derived HBsAg, these include genetically modified bacteria, viruses, yeast and mammalian cells. Which one should we use?

bacterial hosts   The first attempts to use recombinant *Escherichia coli* as a source of HBsAg were unsuccessful, despite the fact that this bacterium expresses the HBsAg gene.

∏ Can you think of a reason why *Escherichia coli* was unsuccessful as a source of HBsAg? (Think back to the production of pro-insulin by bacteria).

Although this bacterium containing the HBV genome is capable of expressing the viral antigens, the HBsAg produced does not assume the spherical or filamentous forms as found in plasma, and is considered essentially non-antigenic for anti-HBs from plasma, i.e. the antigen takes up the wrong structural configuration. Remember that in the case of pro-insulin, the molecule had to be unfolded and refolded because it had been made in the wrong configuration.

recombinant viruses

Recently, the HBsAg gene has been introduced into the vaccinia virus. (ie a recombinant virus was used - the idea is to use an attenuated recombinant virus to act as a multiple vaccine. In this case the recombinant vaccinia virus was to produce immunity against both small pox and hepatitis B. We will explore this strategy again in our discussion of producing a vaccine for Aujeszky's disease in chapter 8. The recombinant virus expressed HBsAg and, upon replication in rabbits, induces high titres of anti-HBs. Vaccinated chimpanzees, however, are not protected consistently against challenge with live HBV. If the situation proved to be similar in humans, it would be extremely difficult to evaluate the response to such a vaccine. Booster doses would probably not be effective. The feasibility of this approach is questionable for other reasons, such as safety considerations and immunity to vaccinia as a result of previous exposure. It is also known that the use of vaccinia for smallpox prevention carried a substantial hazard for patients with dermatological conditions and immune deficiencies. Among the adverse effects that have been reported are postvaccinia encephalitis, eczema vaccinatum, and vaccinia necrosum. Using a virus host is not therefore the method of choice.

mammalian and yeast hosts

The HBsAg gene has also been expressed in recombinant mammalian cells in culture and in baker's yeast, *Saccharomyces cerevisiae*.

Both host cell systems produce HBsAg, which is assembled into 22 nm lipoprotein particles and is highly immunogenic. In yeast all of the HBsAg is contained intracellularly whereas the mammalian cells secretes the antigen. All yeast-derived HBsAg is non-glycosylated whereas approximately 75% of the mammalian-derived HBsAg is glycosylated. Nevertheless, this biochemical difference is reflected neither in the antigenicity of the various particles nor in their ability to elicit virus-neutralising antibodies.

Since both *S. cerevisiae* and mammalian cells produce HBsAg, there was a need to select the most suitable system. For this selection, three general considerations were of interest:

- the biological activity of the expressed protein;

- the scale of the production process and;

- the concern for the safety of the final product relative to its host cell for production.

∏ Which of these considerations do you think proved to be most important?

With respect to the biological activity (i.e. the immunogenicity) of the yeast-derived and of the mammalian-derived HBsAg are comparable. So this criterion provided no reason to prefer one system to the other.

Yeast cultures can readily be scaled up but attempts to scale up mammalian cell cultures, although recently achieved, prove to be more difficult. In addition, the molecular stability of recombinant yeast cells during fermentation can be controlled much more precisely, and the recombinant proteins, produced in yeast cells, have higher yields at correspondingly lower costs. On these grounds, yeast is the preferred expression system on the scale needed for production.

Finally mammalian cells used for expressing foreign genes are generally cells which show good, continuous growth characteristics. Such cells have one or more properties associated with *in vivo* tumorogenicity or *in vitro* transformation, and may harbour endogenous proto-oncogenes (genetic material which can cause cancer and which originates from the mammalian cell) and retroviruses. These properties may be associated with any residual DNA in a product. Such properties are not associated with yeast cell DNA. It is virtually impossible to ensure the total removal of host cell DNA from the product. So for reasons of perceived safety, yeast is the preferred host.

| SAQ 6.5. | Give two differences between the yeast cells and mammalian cells with regard to synthesis of HBsAg. |

| SAQ 6.6. | Give two reasons why the yeast cell is preferred to the mammalian cell as a host for the production of HBsAg. |

6.2.2. Cloning of DNA from HBV Dane particles

Double-stranded DNA isolated from Dane particles was cleaved by the restriction endonuclease *EcoR1* into a single fragment of approximately 3,200 base pairs. This HBV DNA sequence was ligated into the *EcoR1* site of the cloning vector pBR325 and was used to transform *Escherichia coli* HB101. The plasmid used had a chloramphenicol resistance gene and an ampicillin resistance gene. The restriction endonuclease cuts this plasmid in the chloramphenic gene. Inserting foreign DNA at this site, caused the plasmid to lose its chloramphenicol resistance. Thus if a chloramphenicol and ampicillin sensitive host (eg *E.coli* HB101) receives pBR325 with an insert in the chloramphenicol resistance gene, the host will become ampicillin resistant but remain chloramphenicol sensitive. If, on the other hand, the host receives unmodified plasmid, it will become resistant to both ampicillin and chloramphenicol (see Figure 6.3.).

In such experiments three chloramphenicol-sensitive, ampicillin-resistant colonies were obtained, and plasmids from these colonies were isolated and examined by gel electrophoresis after digestion with *EcoR1*.

One clone, named pHBV3200, was found to contain an inserted DNA fragment of approximately 3,200 bp, the reported size of linearized viral DNA. The identity of this insert within the pHBV3200 was verified by cleavage of the plasmid with *EcoR1* and hybridisation with ^{32}P-labelled HBV particle cDNA. The entire 3,200 bp HBV genome was sequenced.

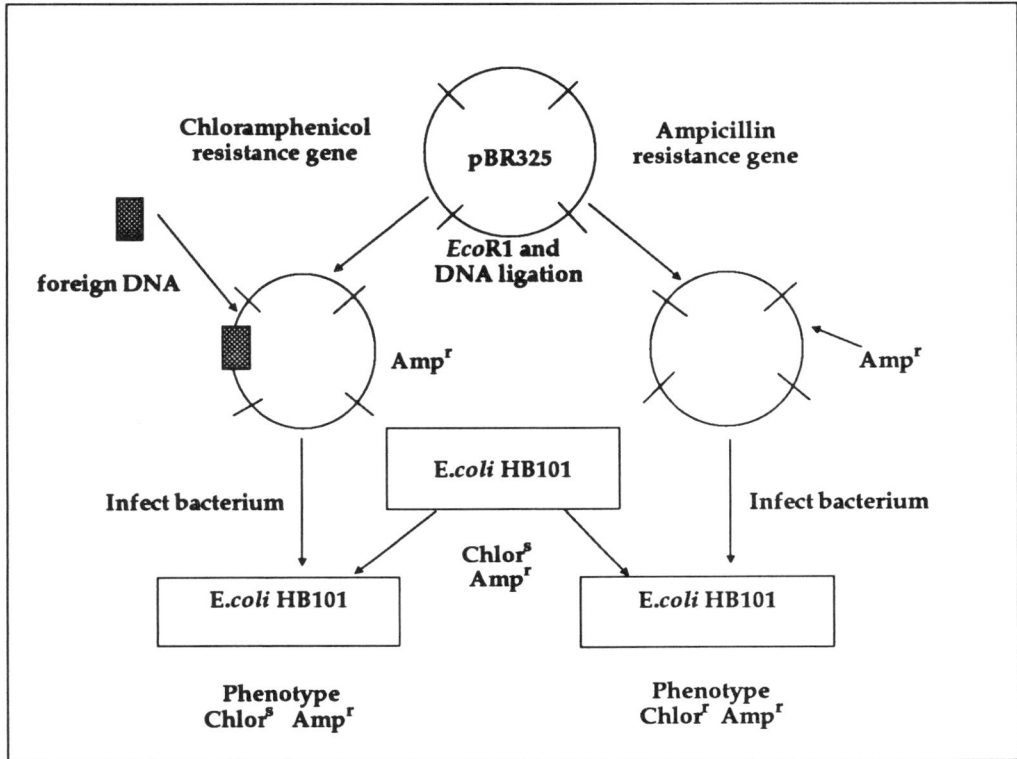

Figure 6.3. The principles behind selecting *E.coli* HB101 receiving plasmid pBR325 carrying a foreign piece of DNA. Amp^r = ampicillin resistance, Amp^s = ampicillin sensitivity, Chlor^r = chloramphenicol resistance, Chlor^s = Chloramphenicol sensitivity.

The HBsAg coding sequence was isolated from pHBV3200 by digestion with restriction enzymes and agarose gel electrophoresis. The isolated fragment was then inserted into the plasmid pBR322, which in turn was inserted into the yeast expression vector pC1/1. We will not examine in detail the construction of pC1/1 here except to indicate that the pC1/1 plasmid contains the entire yeast 2μ DNA and the yeast LEU-2 gene (leucine-2 gene). Because this plasmid has the entire 2μDNA it can be replicated within yeast cells. Inclusion of the leucine-2 gene (LEU-2) in this vector allows that cell carrying the vector to synthesis leucine. In other words, cells carrying the vector will grow in a leucine free medium. This vector (pC1/1 provides a suitable vehicle for introducing new genes into yeast cells. In this case the HBsAg gene was the gene of interest. The resulting plasmid, in which the HBsAg is inserted, was designated pHBS56-GAP347/33 (Figure 6.4.). GAP indicates that the yeast glyceraldehyde 3-phosphate dehydrogenase (GAPDH promoter) gene is incorporated in the plasmid.

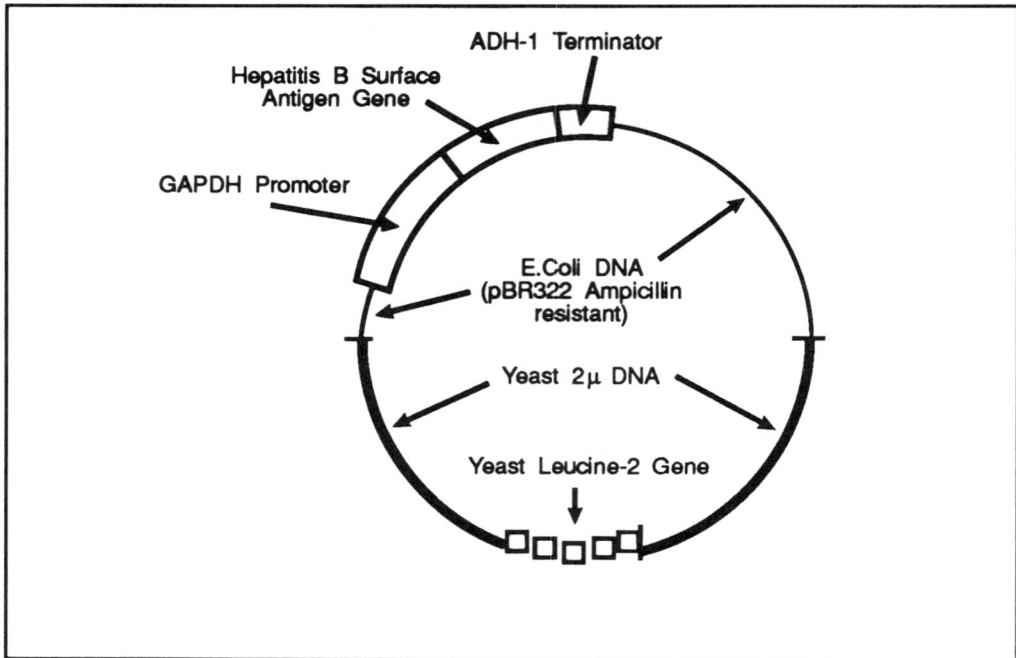

Figure 6.4. Construction of the plasmid pHBS56-GAP347/33

This plasmid contains:

- the complete yeast 2μ sequences which enables autonomous amplification to about 100 copies of a 2μ circular DNA molecule per cell;

- the Leucine-2 gene (LEU-2 gene), a selectable marker, which enables the host yeast strain to grow in the absence of leucine;

- the GAPDH promoter gene, which provides one of the strongest promoters for constitutive expression of HBsAg in yeast;

- the Hepatitis B surface antigen gene (S gene), which encodes for the 24 kilodalton (kd) polypeptide, the predominant protein in the 22 nm particles from human plasma;

- the final regulatory element, which is the transcriptional termination region from the ADH-1 (alcohol dehydrogenase).

SAQ 6.7.	Reproduce the cloning of the HBsAg coding sequence in the form of a flow diagram.

6.2.3. Outline of the method of manufacture

The main steps in the production of hepatitis vaccine (HB-Vax) are as follows:

1)   yeast fermentation

2)   culture harvest

3)   cell rupture and extraction

4)   absorption onto silica

5)   butyl agarose adsorption

6)   thiocyanate treatment

7)   sterile filtration

8)   formaldehyde treatment

9)   adsorption onto alum

10)   filling and packaging

Steps 4 through to 8 are unique for HB-Vax. Comparable steps are used in the production of other recombinant hepatitis B vaccines

6.2.4. Seed culture

The manufacture of biological products entails the use of a seed lot system. A recombinant clone was obtained through the isolation of a single colony of the transformed strain of *S. cerevisiae.*

∏   Why does one have to isolate one single clone?

The main reason is of course that by isolating a single clone, we will be working with a pure culture and can make certain guarantees about product quality and yield. How do we identify the colony we need?

master seed
vial

stock seed vial

The primary means of identifying the strain is its ability to grow on medium lacking leucine. In addition, the ability to produce HBsAg can be determined by radioimmunoassay or Western blot analysis (an electrophoretic method used to separate proteins in order to characterise them). The isolated clone can then be expanded to a large number of master seed vials, of which one or more are expanded to a large number of stock seed vials. Each vial of stock seed is the starting material for the large scale fermentation and manufacture of the vaccine. In this way, genetic identity of each fermentation is assured. The vials with the stock seed cultures are stored at -20°C or below.

6.2.5. Vaccine production

quality control
of cells

A stock seed vial derived from a single colony is grown in fermentation broth for a period of about two days. During this interval, the yeast continually manufactures and stores HBsAg. Because many cell divisions occur during the large-scale fermentation, several parameters are considered important in the evaluation and comparison of plasmids within the master seed vial with plasmids within cells at the end of fermentation.

These parameters are:

- overall structural stability;

- intactness of the S-gene sequence;

- mitotic stability;

- copy number.

structural stability

The overall structural stability is assessed by restriction endonuclease mapping. This assay will detect (more than 0.1 kilo base pairs) deletions or rearrangements. With a panel of restriction endonucleases, it is possible to confirm identity in the sizes of restriction fragments of purified plasmid DNA isolated from cells either in the master seed or at the end of full-scale fermentation.

intactness of the S gene in master seed

To confirm the intactness of the coding region for the S gene, DNA sequence analysis is performed both for the master seed and for cells from a final time sample of a full-scale fermentation.

mitotic stability

At mitosis, yeast plasmids can segregate asymmetrically to the daughter cells. This segregation process is influenced by gene products encoded by the 2μ plasmid. In the course of large-scale fermentation, some plasmid loss may occur. Because it is desirable that a large and consistent percentage of the cells should retain plasmids, the degree of loss is measured by replica plating assays. The fraction of cells which retain the plasmid can be determined by testing the cell growth on leucine-deficient medium; cells which have lost the expression plasmid are unable to grow on this medium.

copy number

A final parameter is the calculation of the number of expression plasmids in cells of the master seed compared with the number in cells after large-scale fermentation. After extraction of cells and restriction endonuclease digestion of DNA followed by agarose gel electrophoresis, the number of plasmids can be quantified relative to an appropriate internal control by densitometric scanning following staining or Southern blot hybridisation.

During the early stages of fermentation it is desirable to maintain a high plasmid number. This is achieved by using a defined selective medium, eg a leucine-deficient medium. During the later stages, in the larger fermentation vessels, it is desirable to use a more enriched medium which will favour the build-up of a high cell density and large quantities of HBsAg. The maintenance of plasmid numbers in the final fermentation as high as in the master seed is not critical, but it is important from the perspective of regulatory agencies that the copy number is consistent among all fermentations.

process control

In order to develop an efficient fermentation process, attention must be paid to:

- the method by which cells in the stock seed are propagated through initial stages of growth to the stage of growth in the large fermentation;

- the growth media, both selective and enriched;

- cell growth curves, for maintaining cells for significant periods of time in their log phase;

- defined volumes transferred during scale-up;

- physical parameters of fermentation, eg pH; dissolved oxygen; agitation rate.

All of the above must be controlled to ensure high cellular viability, good production of HBsAg, and reproducibility of the process.

SAQ 6.8.

1) In comparing the stock seed vial and the large scale fermentation used in the production of hepatitis B vaccine, select the most appropriate techniques from the list provided to achieve the following objectives:

   i) a comparison of the proportion of cells carrying the desired plasmids

   ii) a comparison of the number of plasmids per cell

   iii) a comparison of the coding region of the gene of the desired product

   iv) a comparison of the overall structure of the plasmids

   Which of the techniques described would be suitable to determine the quality of the protein product?

Techniques:

A - Nucleotide sequencing

B - Cultivation by replica plating on media with and without leucine

C - Circular dichroic spectroscopy

D - Electrophoretic patterns of restriction endonuclease fragments

E - Western blotting

F - Radioimmunoassay

G - Densitometric scanning of electrophoretically separated DNA fragments

6.2.6. Isolation of the expressed antigen

isolation of expressed antigen

In order to isolate the expressed antigen, the yeast is harvested by means of centrifugation, and disrupted under high pressure. The cell lysate is adsorbed onto fused silica gel and eluted with borate buffer. The eluate is then fractionated by hydrophobic interaction chromatography, which depends on selective retention of water immiscible molecules. While most proteins are hydrophillic, HBsAg is relatively hydrophobic and is easily separated. The antigen is treated with potassium thiocyanate, in order to convert the particles into a fully disulphide-linked form similar to that of the plasma-derived vaccine. The purified antigen is passed through a sterile filter and treated with formaldehyde, which inactivates infectious units of hepatitis B virus, preventing infectivity for man. Finally, the antigen is formulated by adsorption onto aluminium hydroxide, a commonly used adjuvant, and thiomersal is added as a preservative.

rejection criteria

The reproducibility of the purification scheme must be assured to satisfy the licensing requirements of the regulatory agencies. As part of this assurance, rejection criteria are set for individual stages during the process to stop the purification of poor or impure batches. Some of these rejection criteria are: a contaminating micro-organism in the fermenter; the total yield of HBsAg at a particular point in the process being less than a certain predetermined level; or a yeast impurity being present at too high a level in the final product.

6.2.7. *In vitro* analysis of antigen and stability tests of vaccine

Several assays are used to ensure a detailed physico-chemical analysis. These include measurements of the structure, purity and antigenicity of the HBsAg as well as the stability of the vaccine.

protein purity

For measurement of protein purity, assays are performed on samples during and at the end of the process. The samples are subjected to sodium dodecyl sulphate-polyacrylamide gel electrophoresis (SDS-PAGE), whereafter the gels are stained with either Coomassie Brilliant Blue or silver nitrate. By determining the limit of sensitivity of the assay for the contaminating protein with a series of control samples of known quantity, the purity of the in-process or final product can be determined by densitometric scanning of the stained gel.

Alternatively, the purity of a sample can be determined by high performance liquid chromatography (HPLC) and spectrophotometric scanning of individual fractions. By means of these assays, a final product purity of greater than 99% can be verified.

product identification

Tests to identify the product are performed to confirm that the 24 Kdalton species is the HBsAg polypeptide. Following SDS-PAGE the size of the monomeric HBsAG polypeptide can be verified in the stained gel. A sample of vaccine, resolved by SDS-PAGE, may also be subjected to immunoblot analysis with monoclonal antibodies to HBsAg. Other tests for identity are the demonstration of the *in vitro* reactivity of monoclonals with the vaccine and the amino-terminal sequence analysis of the 24 Kdalton polypeptide, which can help to rule out the presence of co-migrating contaminants.

impurities

Impurities in the final vaccine can be divided into extrinsic, which are introduced during purification, and intrinsic, which are derived from the yeast host cell.

The extrinsic impurities are minimised by an appropriate design of the purification process and monitored by sensitive chemical assays. Intrinsic impurities, principally residual yeast proteins, can be detected by immunoblot or HPLC techniques. Even though there is no known biological risk associated with low levels of residual yeast DNA, the purification process should also ensure the removal of as much DNA as possible. Guidelines for the preparation of recombinant-derived products suggests that a final dose of the product should contain no more than 10 picograms of residual host cellular DNA. As assayed by quantitative DNA hybridisation, this level is readily attainable in hepatitis B vaccines.

safety

Several tests which relate to the safety of the vaccine are carried out, namely:

- general safety test in guinea pigs and mice. These animals receive a large inoculum, 5 ml and 0.5 ml respectively, of the vaccine intraperitoneally. They are then observed for general health and continued weight gain relative to control animals;

- pyrogenicity test in rabbits. After inoculation with the vaccine any rise in temperature is followed in these animals. The product is considered non-pyrogenic if the temperatures rises less than 0.5°C over a four-hour period. (What other tests for pyrogens could be used? Look back at the insulin case study if you cannot remember);

- sterility test by standard microbiological assays.

The recombinant-derived HBsAg is formulated into a vaccine by adsorption to aluminium hydroxide. Completeness of this adsorption can be tested by assaying the supernatant of the alum-adsorbed vaccine for residual HBsAg. In addition, standard chemical tests are used for assaying the pH and osmolarity of the vaccine.

stability　　The assurance of stability is one of the most important determinations that needs to be made for the yeast-derived vaccine, as for any pharmaceutical product. Following the isolation and formulation process, vials are filled with doses of the vaccine. Each vial receives a label identifying the product and stating its shelf-life. Typically a period of two to three years at 4°C is assigned to many vaccines. A manufacturer must develop real-time stability data for the product in order to be able to assign it a realistic shelf-life. Typical assays which are performed to assure stability include SDS-PAGE analysis, *in vitro* reactivity with monoclonal antibodies, and mouse potency. (We will deal with this test later). For two year dating, it is expected that assays performed two years after filling the vials will give results statistically equivalent to those performed at the time of filling.

6.3. Pre-clinical studies

6.3.1. Tests in animals for potency of the vaccine

The recombinant yeast-derived vaccine has to undergo a wide range of tests to ensure its potency in eliciting biologically-active antibodies to HBV. The purified antigen, which is formulated into a vaccine by adsorption onto an aluminium hydroxide adjuvant to contain 10μg of antigen and 0.5 mg aluminium per 1 ml vaccine dose, is tested for anti-HBs production in mice, grivet monkeys and chimpanzees.

One of the key properties associated with the immunogenicity of a vaccine is the availability of epitopes (antigenic determinants) capable of eliciting the production of specific antibodies. In the recombinant hepatitis B vaccine the most important epitope is the 'a' epitope, which is present on all subtypes of HBV (adr, adw, ayr and ayw). The most abundant antibodies elicited by the HBsAg are against the 'a' epitope.

The mouse potency test

For this potency test, mice are given a single 1 ml injection intraperitoneally of serial four-fold dilutions of the vaccine. The mice are bled four weeks later and their sera individually tested for antibodies.

ED50　　The antigenic potency in mice is expressed as the 50% effective dose (ED50) and the geometric mean titre (GMT). The ED50 is the dose required to seroconvert (show
GMT　　antibodies against HBV) 50% of the mice. The GMT is the anti-logarithm of the arithmetic mean of the sum of the logarithms of the titres measured. That sounds complicated, so let us work through an example. If the titres were 100, 200, 250, 250, 100 IU/l we would have to take logarithms of each. These would, therefore, be:

2.0000; 2.3010; 2.3979; 2.3979; 2.000

Sum them = 11.0968

Therefore the mean = 11.0968 + 5 = 2.21936

Take the antilog = 165.7

This is the GMT = 165.7

The GMT and anti-HBs titres are expressed in International Units/litre (IU/l). The data have to be treated statistically. Consider the data in Table 6.1.

Vaccine source	Ag dose/injection (μg protein)	Anti-HBsAg response no.pos./total	GMT IU/l
Human plasma	10.0	9/10	563
Lot 799-2	2.5	10/10	2235
	0.625	4/9	32
	0.156	0/10	4
Yeast	40.0	10/10	5432
Lot 81-4	10.0	10/10	3400
	2.5	8/10	673
	0.625	8/10	967

Table 6.1. Antigenic potency in mice of HBsAg purified from yeast and from human plasma

If we begin with the data for Lot 799-2 produced from human plasma then when 10μg antigen dose was injected 9 out of 10 animals gave a positive response. The anti-HBsAg titres of these allowed for the determination of GMT for this sample (in this case 563 IU/l).

Notice that when the injection contained only 2.5μg antigen, all of the animals gave a positive response and that the GMT was higher.

The potency assay, as shown in Table 6.1. indicates that the yeast-derived vaccine to be at least as potent as the plasma-derived antigen based on comparison of the ED50 and the GMT. Although not shown here, the ED50 for the yeast-derived vaccine was less than 0.625μg compared to the ED50 for the plasma-derived vaccine (0.639μg).

The effective dose capable of seroconverting 50% of the mice is calculated for each batch of vaccine. The advantage of the one-dose mouse potency test over other animal assays for immunogenicity is the availability of mice of a single inbred strain, which assures statistical validity and reproducibility.

Protective efficacy study in grivet monkeys and in chimpanzees.

Seronegative grivet monkeys each received three 1 ml (10μg/ml) intramuscular doses of yeast vaccine four weeks apart. A single injection of the vaccine resulted in seroconversion with the highest titre always seen after the third injection.

Eight chimpanzees were selected for the efficacy study based on negative findings in tests for HBsAg, anti-HBs, anti-HBc, elevation of serum transaminases, liver histopathology and general health, i.e. the animals showed no evidence of previous infection by HBV. The animals were separated in two groups, four test animals and four controls. Each of the four test animals received three 10µg doses of yeast-derived HBsAg vaccine in 1 ml volume intramuscularly at four week intervals. All of the vaccinated animals showed strong antibody responses as compared to control animals (see Table 6.2.).

Chimpanzee	Ab-response to HBsAg	Challenge virus Subtype	HBsAG	Anti-HBcAg	Enzyme (ALT, AST) elevation	Liver path.
Yeast vaccine						
1	1830	adr	0	0	0	0
2	540	adr	0	0	0	0
3	18300	ayw	0	0	0	0
4	7200	ayw	0	0	0	0
Controls						
5	<8	adr	+	+	+	+
6	<8	adr	+	+	+	+
7	<8	ayw	+	+	+	+
8	<8	ayw	+	+	+	+

Table 6.2. Protective efficacy of purified yeast antigen vaccine (see text for detailed explanation). AST = aspartate aminotransferase, ALT = alanine aminotransferase

The above-mentioned chimpanzees were challenged intravenously, one month after the third dose was given, with heterologous adr or ayw subtype virus. All the vaccinated animals developed antibody following immunisation and all were solidly protected against the virus with all serological and histopathological markers remaining negative. The unvaccinated animals developed hepatitis B virus infection with positive antigenemia (presence of Ag in the blood), anti-HBcAg antibodies, elevation of the liver enzymes aspartate aminotransferase (AST) and alanine aminotransferase (ALT) and liverpathology (Table 6.2.).

The hepatitis B surface antigen vaccine contains the 'a' epitope common to all subtypes plus the d and w subtype determinants. Table 6.2. shows that protective immunity was induced in chimpanzees which were challenged with either the adr or the ayw HBV subtypes.

The preclinical immunological tests in chimpanzees were very important for they are one of the few species outside man which are known to be susceptible to HBV infection.

SAQ 6.9.

Label with an 'A' those of the tests listed below that are used to analyse the stability of hepatitis B vaccine.

Label with a 'B' those which can be used as *in vitro* tests of product stability.

Label with a 'C' those tests which can be regarded as tests specifically for safety.

Label with a 'D' those tests which can be used to measure the efficacy of a vaccine.

1) Pyrogenicity test

2) SDS-PAGE analysis

3) Mouse potency test

4) Microbiological sterility test

5) Titration with monoclonal antibodies

6) Protective value tests in monkeys

SAQ 6.10.

In a mouse potency test, the titres of antibodies against an antigen X after injection of the antigen into mice were as follows:

16; 300; 420; 100; 7; 90; 400; 10; 100; 100; 100 IU/litre

Calculate the GMT of the antigen sample used in this assay.

6.3.2. *In vitro* comparison of sera from plasma vaccinees versus sera from yeast vaccinees.

The two vaccines were compared by determining, in sera of vaccinated persons, the following items:

• the percentages of anti-HBs specific for a and d antigenic determinants by radioimmunoassay (The patterns of the a and d determinant-specific antibodies elicited in both vaccinee groups were indistinguishable);

- the total and relative proportion of IgM and IgG anti HBs (The anti-HBs antibodies were characterised as IgM or IgG by using ultracentrifugation sedimentation velocity analysis. The kinetics of development of IgM and IgG anti-HBs were virtually identical);

- the avidity constants of anti-HBs (Avidity is a term employed to express the interaction (binding) between an antiserum and a multivalent antigen molecule. The avidity constants of the anti-HBs antibodies produced to each vaccine were virtually identical - see Table 6.3.);

- cross-absorption of antibodies (Both vaccines were used to absorb antibody from the serum of persons vaccinated with either of the two vaccines. The cross-absorption patterns indicate that the spectra of anti-HBs antibodies elicited by the two products to be similar - see Table 6.3.).

	Avidity Constants			Cross-Absorption			
Vaccine	No.	Titre IU/l	Constant (10^{10})	No.	percentage absorption		
					ay plasma	ad plasma	ad yeast
Plasma derived							
	1	1350	4.0	11	86	100	99
	2	1350	8.0	12	97	99	95
	3	2080	4.0	13	94	100	97
	4	540	4.0	14	50	100	93
	5	1580	7.0	15	86	100	97
	6	920	8.0	16	87	100	97
Yeast derived							
	1	920	4.0	11	98	100	100
	2	1580	1.0	12	98	100	100
	3	1580	16.0	13	98	100	99
	4	720	5.0	14	94	100	99
	5	920	1.0	15	97	100	99

Table 6.3. Avidity constants of anti-HBs and cross-adsorption of antibodies (see text for details)

Note: The sera used in this study were obtained one to two months following the second vaccine dose.

6.4. Clinical studies

6.4.1. Design of clinical studies

When a drug or vaccine has successfully passed pre-clinical *in vitro* and animal studies it will be tested in clinical trials in humans. The number and design of the various trials undertaken with a new drug or vaccine depend largely on the type of agent under investigation. We remind you, every new drug passes through a clinical trial programme which may be divided into three main phases, each with its own objectives.

pharmaco-
kinetics

In phase I trials the emphasis lies on safety and clinical pharmacology, especially pharmacokinetics. However, pharmacokinetic studies are rarely, if ever, done with vaccines. The small number of subjects recruited for these studies are healthy volunteers. The volunteers receive a dose of a new drug which is only a small fraction (perhaps one-twentieth to one-tenth) of the predicted therapeutic dose. The latter is based on the data gained from the animal studies. When a drug is well tolerated at the initial dose, subsequent volunteers receive progressively higher single doses until the wanted pharmaceutical effect is reached or when adverse drug effects emerge.

optimisation

Phase II trials are dose-ranging studies in order to define the most appropriate clinical dose. The objective is to further optimise the use of a drug. This is done by gathering pharmacological, pharmacokinetic, metabolic and toxicological data in a small number of patients.

safety

Phase III trials are conducted in a large number of patients, when the overall safety of a drug is established as well as its therapeutic properties and its side-effects. In these trials the effectiveness of a drug is balanced against any harm it may cause. The phase III trials are mostly carried out in a double-blind mode, in which a drug is compared with the standard medication or with a placebo. Also the interaction with other drugs is investigated as well as the dosage range.

The above-mentioned trials are necessary for the registration of a drug by the authorities. The time occupied by these trials depends largely on the new drug and its indication, but can last as long as ten years. Of ten new drugs which enter phase 1 trials perhaps only one will achieve market authorisation. After registration a drug enters phase IV and V trials. These trials deal mainly with long-term safety, enlargement of the data, new applications and new groups of patients.

6.4.2. Clinical evaluation of a recombinant vaccine

Safety and tolerability of the vaccine

∏ The sera of recipients of the vaccine were tested for anti-yeast antibodies. Why is it also necessary to carry out these tests? (Think about the tests for *E.coli* polypeptides used in the insulin case study).

Samples from 133 of recipient vaccines were tested for anti-yeast IgG by radioimmunoassay. For this purpose, an extract of the parent strain, *S.cerevisiae*, lacking the HBsAg gene but containing plasmid was prepared by disrupting the cells. The homogenate was clarified by centrifugation and filtration and subsequently the extract was adsorbed onto polystyrene beads. Diluted sera were incubated with the coated beads, which were then washed and incubated with ^{125}I protein A. Protein A is a bacteria protein which binds very strongly with IgG. Thus Protein A will bind to any anti-yeast

IgG on the beads. After measuring the radioactivity on the beads, the titres of the anti-yeast antibodies in the sera were determined from a standard curve derived from dilutions of a guinea-pig serum with known amounts of anti-yeast antibodies.

In these studies pre-vaccination sera and post-vaccination sera samples were taken at 1,3,6 and 7-9 months and tested for anti-yeast IgG. All pre-vaccination sera had significant amounts of anti-yeast IgG, titres ranged from 12,000 to 104,111. This is because bakers' yeast is ingested as a component of bread, wine and beer. The anti-yeast IgG titre following vaccination fluctuated over time. Clinical reactions in subjects with increases in yeast antibody titres were no more frequent than in those with no increase titre.

In Table 6.4. we report the percentages of adults and children who had any complaint during the five-day period following each injection. A temperature of 38.3 °C was reported by 1% of the adults and in the children this percentage was 8% after the first vaccination and 3% after the second and third vaccination.

In clinical studies of more than 8000 people there have been no serious adverse experiences attributable to the vaccine.

Type of complaint	% of adults (10µg dose)			% of children (5µg dose)		
	1st inj (n=1144)	2nd inj (n=1056)	3rd inj (n=559)	1st inj (n=79)	2nd inj (n=75)	3rd inj (n=75)
Inj. site	18	13	20	3	3	1
Systemic	17	11	11	18	15	8

Table 6.4. Rates of complaint within 5 days of receiving an injection (inj = injection site)

Most frequent complaints include:

Injection site:

- adults: soreness, pain, tenderness;

- children: soreness.

Systemic:

- adults: headache, fatigue/weakness, nausea, malaise;

- children: fatigue/weakness, diarrhoea, irritability.

Immunogenicity.

Clinical trials to evaluate the immunogenicity and tolerability of yeast-derived hepatitis B vaccine in humans were initiated in 1983. Participants entering these studies had to be seronegative for HBsAg, anti-HBc and anti-HBs, had to have a normal level of serum alanine aminotransferase and not previously have received any hepatitis B vaccine (ie they showed no signs of previous HBV infection). The dose-ranging studies, in which four doses were tested, viz. 2.5, 5, 10 and 20µg for adults and 1.25, 2.5 and 5µg for children, were done early in the clinical studies programme. The vaccine was administered as a series of three injections. The first two injections were given a month apart followed by a third or booster injection at six months.

dose-ranging study

Blood samples were taken at 0,1,3,6 and 7-9 months following administration of the first dose of vaccine. Anti-HBs titres were determined by radioimmunoassay and expressed in International Units per litre (IU/l) or Milli-International Units per millilitre (MIU/ml). A responder was defined as an initially seronegative individual who developed, after vaccination, an anti-HBs titre of 10 IU/l or greater.

Figure 6.5. illustrates the degree of seroconversion and the height of antibody response for each dose given. At month three (two months after the second doses) between 57 and 76% of the vaccinated persons developed antibodies. Two months after the third vaccination, 89 to 97% of the vaccinated persons seroconverted. The geometric mean titre (GMT) of the responders varied substantially with dosage (Table 6.5.).

Adults		Children	
doses in µg	GMT in IU/l	doses in µg	GMT in IU/l
2.5	295	1.25	2059
5.0	349	2.5	6230
10.0	1286	5.0	15966
20.0	1022		

Table 6.5. The GMT of the responders among the adults and children

Both the 10 and 20µg doses of antigen gave similar serological responses, therefore most of the clinical trials in adults were done with the 10µg dose of vaccine.

Healthy children between one and twelve years of age were vaccinated with three different doses in order to establish the appropriate immunogenic dose. Following vaccination, children developed antibody more rapidly than adults even at a lower dose. After two injections with vaccine of all doses 82 to 100% of the children had an anti-HBs titre of 10 IU/l (Figure 6.5.). Two months after the third injection all the children responded with a titre ≥10 IU/l. As with adults, GMT following three injections of vaccine varied directly with the dose given (Table 6.5.). Although seroconversion rates were excellent with all doses, the highest antibody titres were obtained with the 5µg dose.

SAQ 6.11.

Select from the following list, the features which are emphasized in the safety assessment of a vaccine.

a) discomfort at the injection site;

b) the efficacy of the vaccine to protect the individual;

c) rise in body temperature;

d) whether or not IgG or IgM antibodies are produced;

e) the antibody titres achieved;

f) the general health of the individuals receiving the vaccine.

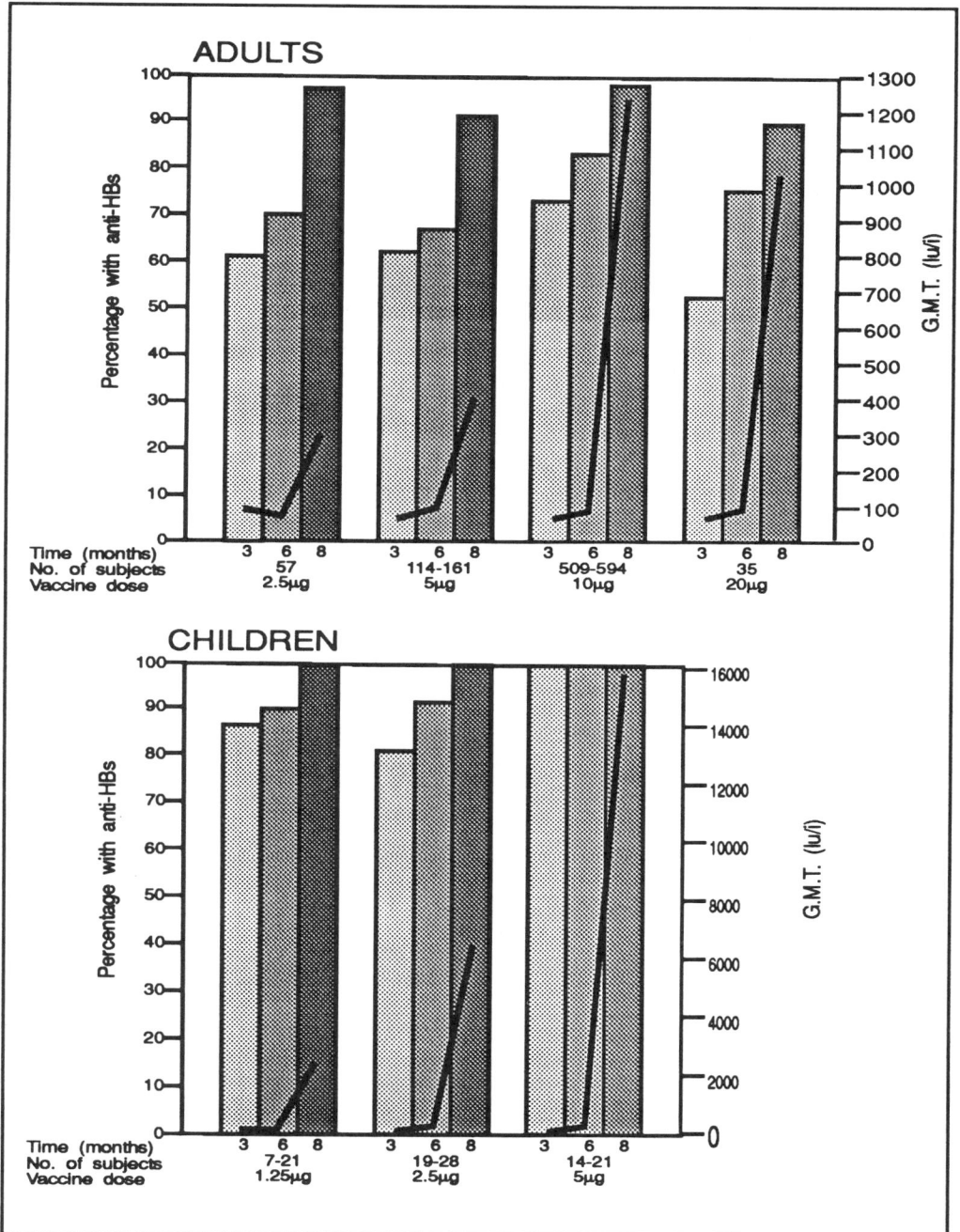

Figure 6.5. Seroconversion rates (bars) and GMT (lines) among humans vaccinated with varying doses of vaccine

| SAQ 6.12. | When participating in an immunogenicity study why do subjects have to be seronegative for HBV markers? |

| SAQ 6.13. | What is a seroconversion rate? |

6.4.3. Efficacy
Clinical trials with the effective dose of antigen

As mentioned above the effective dose (i.e. the dose which elicited the highest GMT) for adults is 10μg and for children 5μg. Healthy adults receiving 10μg of yeast-derived vaccine ranged in age from 20 to 70 years. The antibody response of adults by age is shown in Figure 6.6. The effect of age was apparent both in seroconversion rates and in the GMTs. The younger adults between 20 and 29 years of age showed a more vigorous antibody response to the vaccine than did the older (over 50 years of age) subjects.

At month eight, 97% of the adults had titres greater than 10 IU/l while 89% had titres greater than 100 IU/l. 58% had titres greater than 1000 IU/l (Table 6.6.).

Adults					Children			
Time after vaccination	titres in IU/l					titres in IU/l		
month	<10	10-99	100-999	1000	<10	10-99	100-999	1000
3	24%	45%	29%	2%	-	30%	60%	10%
6	14%	44%	39%	3%	-	32%	47%	21%
8	3%	8%	31%	58%	-	-	7%	93%

Table 6.6. Anti-HBs titres in IU/l among adults and children

Π What % were non- or hypo-responders (ie did not produce anti-HBs antibodies or produced anti-HBs titres of less than 10 IU/l)?

From the data presented in Table 6.6., did children respond more than adults to the vaccine in terms of producing anti-HBs antibodies?

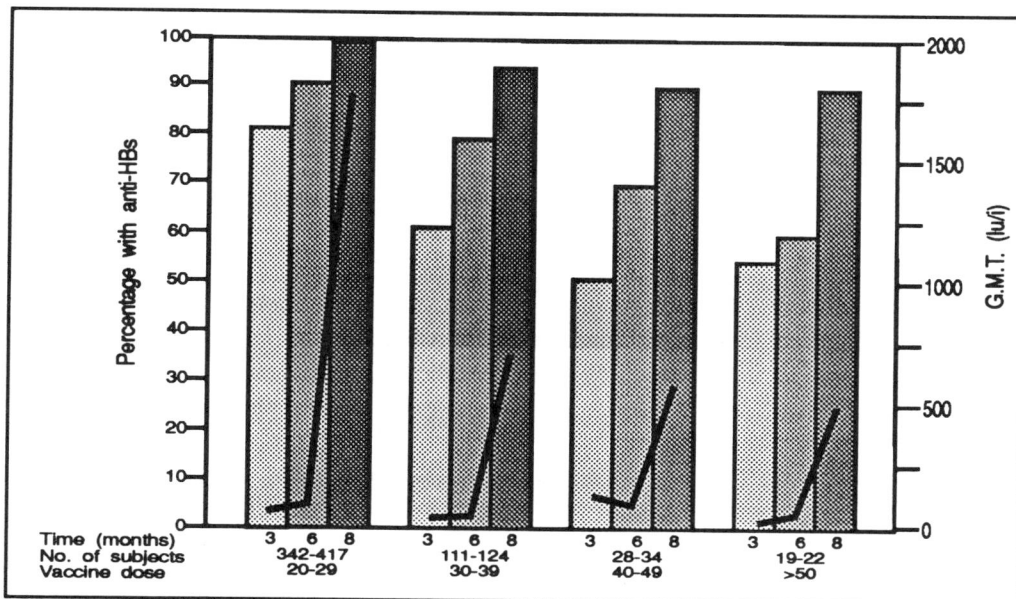

Figure 6.6. Seroconversion rates (bars) and GMT (lines) among healthy adults receiving 10µg doses of vaccine.

SAQ 6.14.

Vaccinees receiving hepatitis B vaccines may be described as responders, hyporesponders or non-responders. What is the difference between these three populations?

Protective efficacy

It is important to determine that the antibodies, induced by the vaccine, are protective against infection with the hepatitis B virus. This can be done in appropriately conducted clinical trials in high-risk people. The high-risk groups include, among others, intravenous drug abusers, dialysis patients, medical staff, dentists, homosexual men, and newborns whose mothers are carriers of the virus antigens. Of the latter group 86 to 96% become chronic HBsAg carriers when born to a HBeAg-positive mother and, of the newborns born to HBsAg-positive mothers, 40% become chronic HBsAg carriers.

To produce a statistical evaluation of the efficacy of the yeast-derived vaccine, children born to HBsAg-positive mothers were included in a protective efficacy study. Because the efficacy of the plasma vaccine had already been proven, this vaccine served as a control in the comparative study. Among infants in the study, who received hepatitis B immunoglobulin at birth and three 5µg doses of the yeast-recombinant hepatitis B vaccine, only 4.8% became chronic carriers. This is a better than 90% level of protection and a rate that is comparable with that seen with immune globulin at birth and three 10µg doses of the plasma-derived hepatitis B vaccine.

SAQ 6.15.

How has the protective efficacy of the hepatitis B vaccine been demonstrated in humans?

Summary and objectives

In this case study we have examined the biology and epidemiology of Hepatitis B virus and explained how the gene coding for a major viral antigen had been isolated, cloned and produced in yeast. We have also described how the antigen HBsAg has been evaluated as a vaccine in pre-clinical and clinical trials.

Now you have completed this chapter you should therefore be able to:

- explain why there is a need for a hepatitis B vaccine and why an alternative from the conventional plasma derived vaccine is desirable;

- explain why a particular hepatitis B antigen was targeted;

- explain the reasoning behind the selection of yeast cells as the host system of choice for the production of HBsAg;

- describe the stages required to clone the HBsAg sequences;

- describe the tests that enable comparison of the yeast carrying HBsAg gene sequences at the stock seed and large scale fermentation phases and to maintain the quality of the product;

- describe in outline, and distinguish between, the *in vitro* and *in vivo* tests conducted to determine product potency and stability and to determine the parameters which measure potency;

- describe the different emphasis placed on the pre-clinical studies of vaccines compared to that of therapeutic products such as insulin and erythropoietin;

- use terms commonly applied in immunology and vaccine evaluation.

7

Case Study: Orthoclone OKT3 - A therapeutic antibody produced by hybridoma technology

7.1. Introduction   166

7.2. Organ transplantation   166

7.3. Immunosuppressive agents   174

7.4. OKT3   179

7.5. Production process   181

7.6. Safeguards of the production process   188

7.7. Pharmacology   193

7.8. Clinical trials   194

Summary and objectives   198

Case Study: Orthoclone OKT3 - A therapeutic antibody produced by hybridoma technology

7.1. Introduction

hybridoma
technology

monoclonal
antibodies

In the three previous case studies, we have examined the production of materials using recombinant DNA technology. In this case study, we examine quite a different strategy, namely the application of hybridoma technology. By hybridoma technology we mean the fusion of two cell types to produce a new hybrid cell line carrying desired features of the two "parent" cells. This case study focuses on the production of monoclonal antibodies. Monoclonal antibodies are pure antibodies, produced by hybrid cells derived from the fusion of B-cells (antibody producing cells) and myeloma (cancerous B-cells) cells. The desired hybridomas produced by such fusions exhibit the rapid growth characteristics of the myeloma cells and the specific antibody producing characteristics of the B-cells.

The monoclonal antibody we have focused on is OKT3 which has application in organ transplantation. We begin this chapter by considering the general problem of organ rejection associated with transplantation and develop into a description of the technology associated with OKT3 production and evaluation.

7.2. Organ transplantation

multidisciplinary
aspects of
successful
transplantation

Transplantation is used to replace organ function or to overcome tissue defects. The expansion of this procedure over the past few years is based upon developments in five areas of knowledge:

• Immunology - the development of knowledge of the human leucocyte antigen system (HLA); further insights in rejection mechanisms;

• Pharmacology - the discovery and development of potent immunosuppressive agents (azathioprine, cyclosporin);

• Technology - the refinement of surgical techniques, especially in the microsurgical and vascular field. The development of organ harvesting and storage procedures, rinsing and storage solutions (eg Eurocollins and UW, solutions used for rinsing out the blood in the organs to be transplanted). They mainly contain physiological amounts of electrolytes and amino acids to prevent organ damage under avascular and hypothermic conditions. Note: UW = University of Wisconsin;

• Biotechnology - the development of monoclonal antibodies that interfere with the different steps of the rejection mechanism (eg anti-CD3 receptor, anti-IL2 receptor);

- Organisation - the foundation of international transplantation organisations which co-ordinate the exchange of available organs in a number of countries (eg Eurotransplant).

However, the underlying drive for the expansion of organ transplantation is the growing population of patients being kept alive despite organ failure. Medical technology has mainly been involved in artificially maintaining a large population of patients in organ failure. For example, where only a few decades ago all patients in renal failure died, they are kept alive today by means of the artificial kidney or other epuration techniques. Although these procedures supply a reasonably comfortable alternative for such patients, long-term survival with these devices often means a high morbidity rate. The most important complication, however, is the enormous psychological stress accompanied by the continuous requirement for four to five hours of intensive treatment two or three times a week.

renal transplant

In terms of numbers of patients in the Eurotransplant countries (The Netherlands, Germany, Belgium, Austria), there has been a steady increase in demand for renal transplantation. In 1979, 2500 patients were registered on this list rising to 9300 patients by 1989. Over the same period, the number of kidney transplants more than doubled from 1000 to 2500 in these countries. It is understandable that major efforts are being made regarding donor recruitment. Even governmental measures have been taken in Belgium to change the policy of organ donation from informed to uninformed consent. This means that every Belgian citizen is considered to be an organ donor unless he or she explicitly indicates the contrary in writing.

Let us consider the important advances made in immunology during the past two decades. These have given further insights into the immunological processes which develop when a normal human defence system is challenged by foreign antigens and will give us the background to this case study. The first defensive step the host's defence system will take is to remove the antigenic challenge. This results in rejection when it is directed towards a transplanted organ.

7.2.1. Rejection mechanisms

Before discussing the different mechanisms by which a transplanted organ is attacked in the human body, it will be useful to introduce basic terminology closely related to organ transplantation. In relation to the transplanted organ or tissue:

- allograft is an organ or tissue transplanted between individuals of the same species (e.g. human to human);

- homo- or auto-graft is an organ or tissue transplanted in the same individual (eg skin to cover large defects);

- xenograft is an organ or tissue transplanted between species (eg mouse-rabbit).

B-cell/T-cell mediated rejection

Allograft rejection is the most important problem almost every transplant patient has to deal with. Rejection can occur by two major mechanisms: humoral or antibody mediated (ie B-cell mediated) and T-cell mediated or cellular rejection.

7.2.2. Humoral or antibody mediated rejection

Humoral or antibody mediated rejection occurs when circulating antibodies directed against antigens present on the donor organ or tissue exist in the recipient.

Π   Are there any reasons why a person who has not received an earlier transplant
     can have antibodies directed against a donor organ? (Try to think of circumstances
     where an individual might receive human antigens other than by organ
     transplant).

There are several possible reasons:

*   at birth, many of us received minor blood transfusions with maternal blood during
    the disruption of the placenta. This minor antigen challenge is often enough to form
    antibodies against red blood cell (AB, rhesus) antigens.

*   many patients in renal failure are dependent on regular blood transfusions, resulting
    in the same problems. During these transfusions there might occur additional
    sensitisation against occasional leucocytes present in the transfused blood. These
    cells carry the human leucocyte (HLA) antigens.

human   Rejection can be directed against the A, B or rhesus antigens of the donor's red blood
leucocyte   cells or, more commonly, to the HLA antigens of the donor organ. The HLA (human
antigen   leucocyte antigen) system is a part of the major histocompatibility complex (MHC). Let
             us remind ourselves of some of the features of the MHC. The MHC is located on
major   chromosome 6 and consists of four major regions: HLA-A, HLA-B, HLA-C and HLA-D
histocompatibility all coding for the HLA antigens. Each person has a unique antigenic fingerprint of the
complex   MHC antigens. MHC antigens are divided into two major classes: Class I and Class II
             antigens. The occurrence of these different classes of MHC antigens on the surface of
             different cells is described in Figure 7.1.

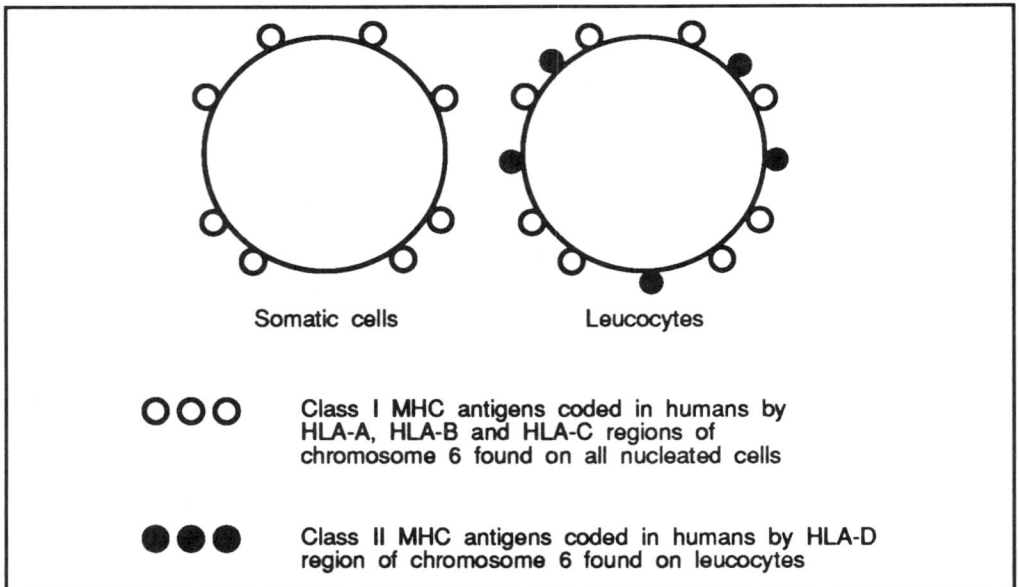

Figure 7.1. Occurrence of the major histocompatability antigens on cells

Class I antigens are the serologically defined antigens consisting of a 45,000 dalton
heavy chain located on the cell membrane in association with B_2-microglobulin. Class II

antigens consist of two glycoproteins, encoded by the HLA D-region, and are also located on the cell membrane (Figure 7.2.). Class II antigens are found on the cells involved in the immune response.

Π By examining Figure 7.2., work out which portions of the Class I and II antigens contain hydrophobic amino acids. The clue is to find which part of these molecules are associated with a lipid environment.

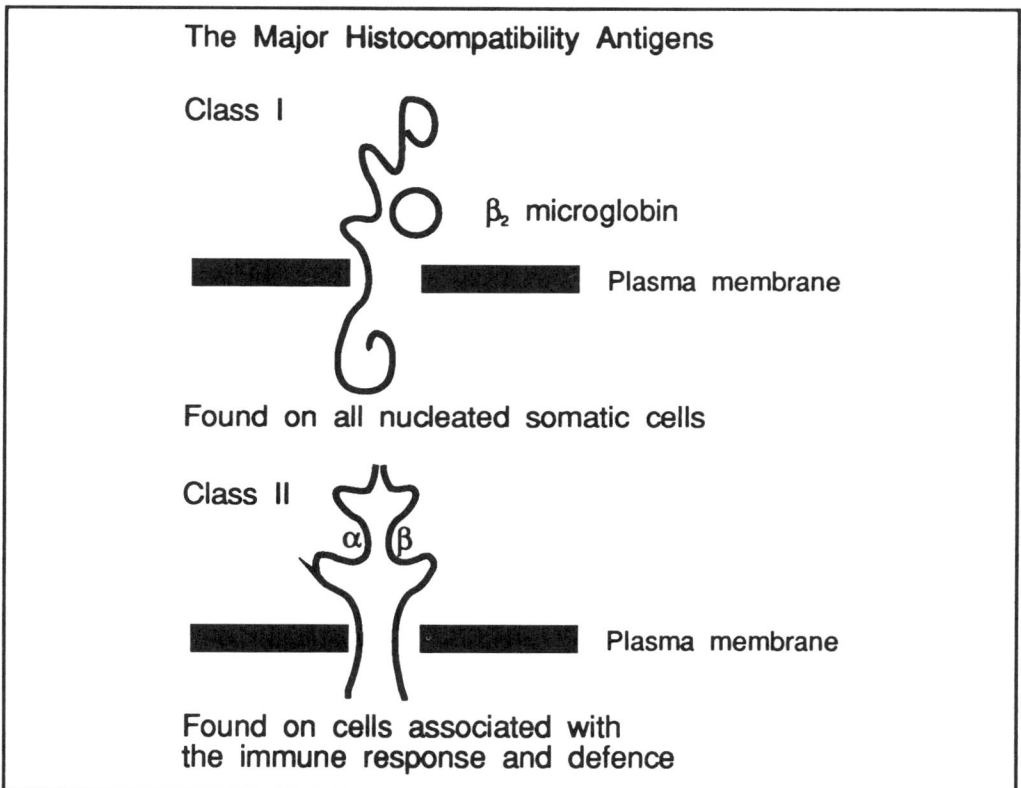

The Major Histocompatibility Antigens

Class I

β₂ microglobin

Plasma membrane

Found on all nucleated somatic cells

Class II

α β

Plasma membrane

Found on cells associated with
the immune response and defence

Figure 7.2. The major histocompatibility antigens

hyperacute
rejection

The antibodies against these antigens are produced by a B-cell mediated process, and can give rise to a hyperacute rejection with dramatic destruction of the transplanted organ within 24 hours after transplantation. Hyperacute rejection can be prevented by pretransplant A-B-O, Rhesus (Rh) and HLA matching of the donor and the recipient. A cytotoxic antibody matching test needs to be performed just prior to transplantation (cytotoxic antibodies are those which react with cells leading to cell death). This test is performed by incubating patient serum with lymphocytes of the donor as described in Figure 7.3.

plasma
exchange

If cytotoxic antibodies are present the lymphocytes will be lysed by adding complement to this test system. In such a case the organ is assigned to another compatible recipient. The only, but often temporary, solution to this kind of rejection is to remove all antibodies from the patient's serum by plasma-exchange. Plasma-exchange is a

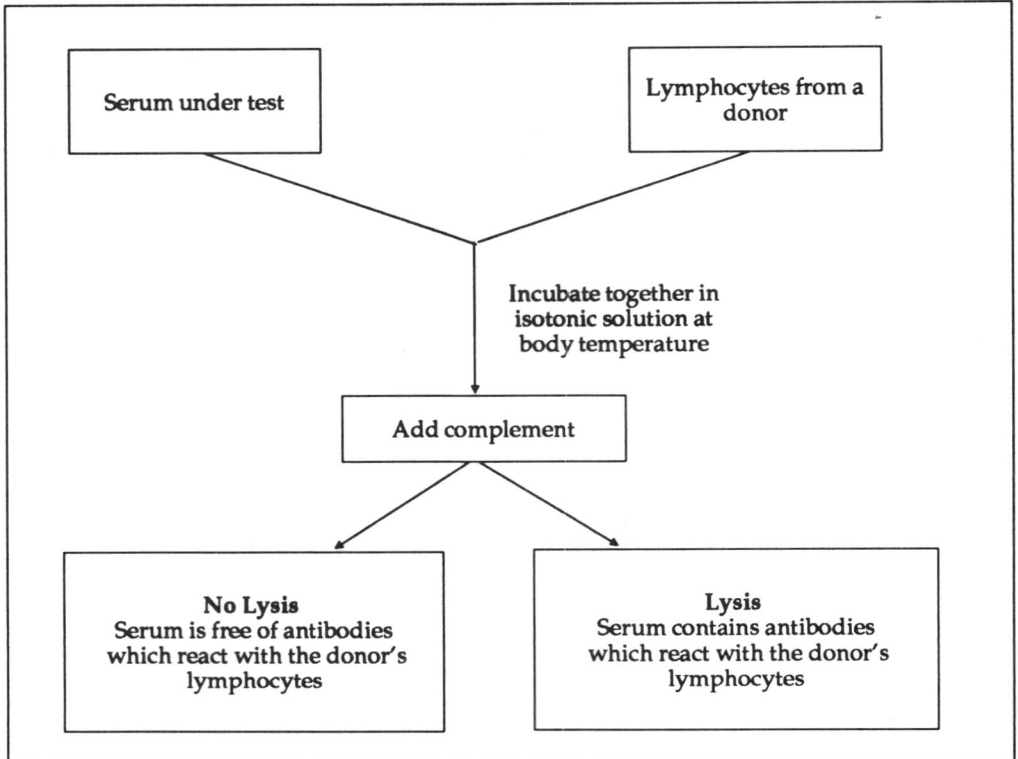

Figure 7.3. Testing for cytotoxic antibodies against the lymphocytes of a prospective donor

technique in which plasma is separated from the circulating blood cells by a cell separation procedure, after which the patient's plasma is replaced by fresh donor plasma.

Vigorous immunosuppression after plasma exchange diminishes B-cell proliferation from memory cells and subsequent antibody production.

∏ What could be a major complication after this technique? Does the body have any protection against infection?

Plasma-exchange means removing all the circulating immunoglobulins of the patient. If restoration of these immunoglobulins is reduced afterwards by immunosuppressive drugs, it will mean that the patient has lost not only antibodies against the graft, but also those against infective agents developed as a response to infections or after vaccination. The patient would, therefore, be susceptible to infection.

contact
triggering

mediator
release

The most common type of rejection occurring after transplantation is the T-cell mediated or cellular rejection mechanism. It is mainly in this field that most effort has been made to elucidate the different mechanisms of cell interactions. Since they are crucial to the understanding of the action of most immunosuppressive drugs, we need to look at these mechanisms in more detail.

7.2.3. Cellular rejection

The following summary of the mechanisms of cellular rejection represents current ideas. It is however, susceptible to change, because research in this field is very intense. The central role of these somewhat complicated immunological interactions is to protect the host from external aggression. Almost all antigens which the immune system of an individual does not recognise as its own may form a possible target. Under physiological conditions, it represents an effective way of dealing with bacterial, viral or fungal infections. Even ones own cells behaving abnormally (eg cancer cells) are continuously destroyed by this defence system. Rejection is, therefore, a normal reaction of a functioning immune system.

With this in mind, what will be the physician's major concern when he has a patient receiving immunosuppressive drugs to prevent or treat rejection? Preventing rejection or treating it means compromising the host with substantial hazards of infection or growth of neoplasms. This will remain true until a selective therapy is developed which can discriminate between immunological actions against the graft and the defence against infections or malignant degeneration.

summary of cell and antibody mediated immunity

Let us examine the sequence of events which leads to the rejection of an antigen. First let us deal with antibody mediated immunity. Use Figure 7.4. to follow the description given below. The first step is carried out by the antigen presenting cells (APC). These are cells of the monocyte macrophage lineage (ie macrophages). The antigen is recognised and processed by these cells, after which it is presented to the T-helper (T_H) lymphocyte. The APC recognises the appropriate T_H cell by sharing the same HLA-D or D-related antigens (Class II MHC antigen) on its cell surface. The recognition site thus consists of two major determinants: the antigen and the proper Class II antigen determinant. This means that an APC will not work with T_H cells from individuals with different Class II antigens. The contact itself and the secretion of interleukin-1 (IL-1) by the APC induces proliferation of lymphocyte subsets. In this step, B-cells are stimulated and differentiated to produce antibodies by secretion of lymphokines by the T_H cell. The lymphokines include B-cell growth factor (BCGF) and T-cell-replacing factor (TCRF). The B cells are stimulated to grow and produce antibodies.

Now let us turn our attention to cell mediated immunity. Again refer to Figure 7.4. We begin with the activated T_H cells. Another important lymphokine, interleukin-2 (IL-2) is also secreted by the T_H cell. IL-2 induces a cascade of proliferation and differentiation steps. First, it stimulates the formation and avidity of IL-2 receptors on the T_H cell itself, which in turn stimulates the proliferation of the appropriate T_H cells. Furthermore, it stimulates the proliferation and maturation of effector cells responsible for the final destruction of the foreign antigen-bearing cell. These cells are called cytotoxic T-cells or T_C cells. These T_C cells can be antigen specific or non-antigen specific. Indeed IL-2 promotes proliferation of all T-cells, including the non-specific T-lymphocytes, and the suppressor T-cell subset (T_S). The T_S subset represents an autoregulatory mechanism suppressing further expansion of this destruction process.

The efferent limb of rejection is based on antigen recognition by sensitised T_C cells or by specific B-cell derived antibodies. We will focus on rejection by T_C cells. The recognition of the antigen-bearing cell by the T_C cell is based upon the recognition of, first, an HLA antigen (Class I) and, second, the foreign antigen, to which it was sensitised. This contact results in the activation of several enzyme-dependent cascades (complement, coagulation, renin) which ultimately kill the target cell.

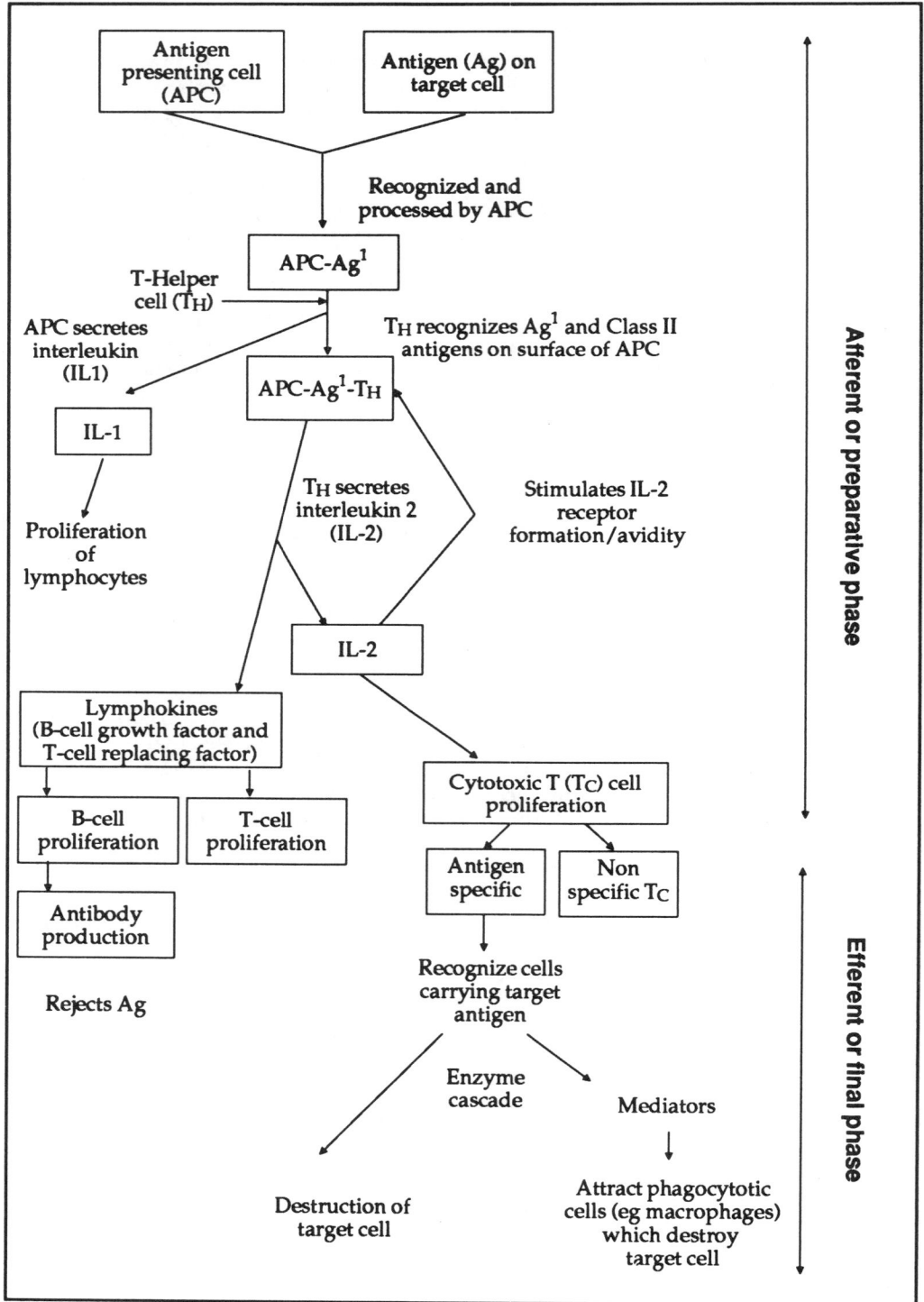

Figure 7.4. An outline of cell and antibody mediated immunity

In addition, other mediators are secreted resulting in attraction of other sensitised or unsensitised T-cells, macrophages, platelets and polymorphonuclear leucocytes. These direct and indirect mechanisms result in destruction of the allogenic target cell.

T-cell

sub-populations

These mechanisms have been discovered largely through the development of monoclonal antibodies recognising specific T-lymphocyte subsets, thus specifying their role and action. Indeed most T-cells are characterised and classified by their cell surface antigen determinants. Some of these are shared by different subsets, some of them are specific for just one subset. Table 7.1. illustrates the characterisation of the T_H and T_S cell subsets by their membrane markers. Most of these surface antigens were first recognised by monoclonal antibodies developed by the Ortho Pharmaceutical Corporation and were named the OKT series. Until recently, the surface antigens were called after the monoclonal by which they were recognised (e.g. OKT4, OKT8). These surface antigens, however, are not a single molecule but are each a cluster of molecules with distinct proteins of varying molecular weight. Therefore, a more appropriate terminology has been chosen to designate these clusters, namely cell differentiation cluster or CDs (e.g. CD1, CD3).

Sub-population	Surface Antigens
T-Helper cells	CD4, CD3, CD2, CD5
T-Suppressor	CD8, CD3, CD2

Table 7.1. Antigenic clusters on the surface of sub-populations of T-cells

SAQ 7.1.

Answer true or false to the following statements

1) All nucleated cells of the body carry Class II MHC antigens on their surfaces.

2) The presence of antibodies against Class II MHC antigens in a serum may cause the lysis of lymphocytes from a donor.

3) Interleukin 2 (IL-2) is produced by activated T-Helper cells.

4) Interleukin 2 causes T-Helper cells to produce receptors for Interleukin 2 with high avidity for this ligand.

5) Inhibition of Interleukin 2 production will greatly inhibit antibody production.

6) Antigen presenting cells (APCs) from one individual will interact with T-Helper cells from another.

7) T-Helper cells can be distinguished from T-Suppressor cells by the presence of the surface antigen CD3.

7.3. Immunosuppressive agents

The ideal immunosuppressive agent would interfere with the recognition of antigen(s) of the donor without compromising the patient's defence system. This idea has not yet been achieved but in theory a specific way to deal with or prevent rejection would be to remove all the antigens which could trigger an immune response. However, this "antigen stripping" of the donor organ still remains a future prospect. The main restriction to this approach is the limited access to donor antigens. Only those that can be reached by the circulatory system can be dealt with. After transplantation, the turn-over of cells will again expose new antigen. Moreover, cellular replication cannot be inhibited and all surface antigens will eventually be re-expressed on cell membranes. Until now most efforts have been concentrated on interfering with the host's immune system. To this end the pharmaceutical industry has developed several drugs designed to suppress the immune system. We have listed these under the headings of antimetabolites, alkylating agents, corticosteroids, cyclosporin and antibodies.

antigen stripping

7.3.1. Antimetabolites

interference with metabolic activity

These drugs are often used in cancer treatment. They interfere with cell division making any dividing cells susceptible to its action. Of course the faster the growth of a population of cells, the greater the impact these drugs have on the cells. Since rejection implies a rapid multiplication of cells, it is understandable that it is also vulnerable to these agents. Antimetabolites closely mimic the structures of normal metabolites. By this structural analogy, they inhibit enzymic pathways or are built into proteins making them useless for normal metabolic activities.

These drugs usually have a close resemblance to purine, pyrimidine or folic acid. The structures of antimetabolites commonly used in transplantation are depicted in Figure 7.5.

Π   Antimetabolites interfere with rapidly dividing cells. Considering this mode of action, write down what you think might be their major side effects (Do this before reading on).

Due to their mode of action the most important side effect of the antimetabolites is on bone marrow where they induce leucopenia (deficiency in white blood cells). Their hepatotoxic (toxicity to the liver) effect is still controversial. Their side effects on bone marrow can easily be explained by the fact that this is one of the most productive site in the human body for cellular replication. Other high-turn-over areas include the gut, the liver and the gonadal tissue (testis and ovaria) and these too suffer damage during the treatment with such antimetabolites.

leucopenia

hepatotoxicity

7.3.2. Alkylating agents

blocking of DNA replication

Alkylating agents are also borrowed from cancer therapy. Their action mainly consists of combining with DNA and blocking replication in the S-phase (DNA synthesis) of the cell cycle. Although these drugs are used in bone marrow transplantation, their therapeutic application in solid organ transplantation is limited due to their severe pancytopenic effect (ie wide ranging cell deficiency leading to leucopenia, anaemia). The most commonly used drug in this category is cyclophosphamide.

pancytopenic effect

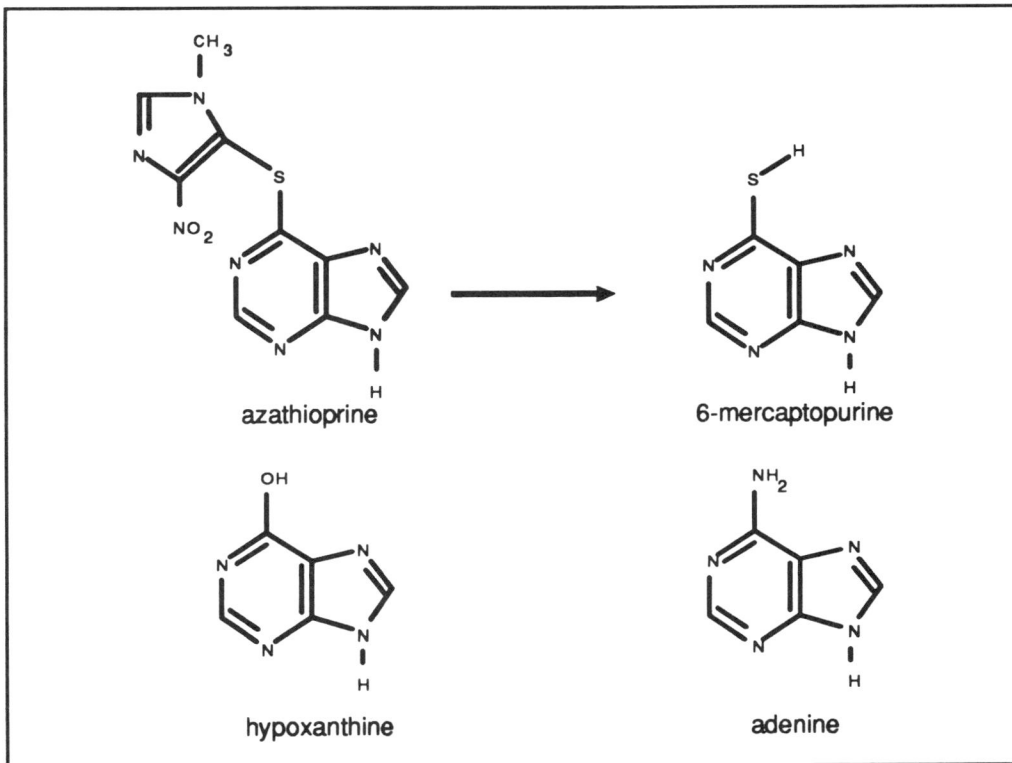

Figure 7.5. Azathioprine and 6-mercaptopurine, purine analogues used as immunosuppressor agents

7.3.3. Corticosteroids

corticosteroids
and
inflammation

These drugs form the basis of most immunosuppressive protocols in organ transplantation. Corticosteroids inhibit all mechanisms of acute or chronic inflammation, a process similar to rejection. Since the majority of human cells have steroid receptors in their cytoplasm, the mode of action of corticosteroids is a result of a complex interference of intra- and extra-cellular pathways. In summary, they inhibit release of cell mediators such as lymphokines, chemotactants, bioactive peptides and vasoactive substances. Chemotactants are released at sites of inflammation and attract different cell types such as macrophages. An example of such chemotactants is the Macrophage Activating Factor (MAF). In fact an inflammatory reaction is a process in which the body tries to eliminate substances (eg bacteria in infection) which it considers as "foreign" and consequently harmful. Inflammation is often associated with infection but can also be linked with auto-immune disease or other pathological conditions. It is characterised by a classic series of symptoms such as rubor (redness), calor (elevated temperature), dolor (pain) and tumour (swelling).

redistribution of
circulating
leucocytes

In addition to inhibiting the inflammatory response, corticosteroids also cause a redistribution of circulating leucocytes with a depletion of circulating lymphocytes, monocytes, eosinophils and basophils and a peak increase of neutrophils shortly after administration.

The preparations most commonly used are hydrocortisone, prednisolone and methylprednisolone. Although they are relatively inexpensive and easy to produce, many efforts have been made to reduce their use to the absolute minimum, mainly because of severe and multiple side effects. Corticosteroids are normally produced by the cortex of the adrenal gland. Their natural function involves many metabolic pathways including water regulation, regulation of glycaemia, cerebral function etc. When excessive production of steroids occurs, as is the case in endocrine tumours of the adrenal gland, Cushing Syndrome develops. By external administration of corticosteroids an iatrogenic Cushing Syndrome can occur. (Iatrogenic means induced in a patient by a doctor's action or treatment). Patients with Cushing Syndrome develop hypertension, obesity, fluid and salt retention, osteoporosis, infections and, if there is impaired glucose tolerance, it may produce severe diabetes.

7.3.4. Cyclosporin

Since its discovery in the mid-1970s this drug has had a major impact on organ transplantation. One year graft survival in renal transplantation increased by about 20% and it roughly doubled the survival of liver allografts. Due to this medication, heart and lung transplantation became feasible. Its mode of action is mainly by inhibiting IL-2 secretion of the T_H-cell.

inhibition of IL-2 secretion

∏ Look back to Figure 7.4. and see if you can explain how cyclosporin acts as an immunosuppressor.

Since cyclosporin inhibits IL-2 secretion, adminstration of this drug results in an inhibition of the proliferation of T-cells to an antigenic trigger, thus preventing the major pathway of cellular rejection. It is obvious, however, that B-cell proliferation and antibody synthesis will mostly escape from these actions since they do not involve IL-2. Also the direct cytotoxicity of the existing T_C-cells will not be inhibited.

Therefore, cyclosporin in combination with corticosteroids is considered to be the best maintenance therapy. However, once the cascade of rejection has started, additional cyclosporin administration will not be helpful to treatment because the rejection effectors (T_C cells, Ab) will already exist. We can perhaps represent the effect of cyclosporin as described in Figure 7.6. Although the discovery and development of cyclosporin can be considered as a milestone in the history of organ transplantation, its application is still limited by its nephrotoxicity. This toxicity can be acute as well as chronic. It may lead, and it often does, to a situation in which it is difficult to decide whether a renal graft is not functioning due to rejection or to cyclosporin nephrotoxicity. Despite this major drawback, cyclosporin, together with corticosteroids, forms the basis of most current immunosuppressive protocols.

7.3.5. Polyclonal antibodies, antisera and monoclonal antibodies

In principle, if we could destroy/remove the white blood cells of the recipient, then the graft could not be rejected. Several devices have been used to try to achieve this. Antisera directed against lymphocytes (ALS) have been used. A purified preparation of the active globulin fraction of this ALS, the anti-lymphocyte globulins (ALG), and the active globulin fraction of an antiserum raised against T-cells (ATG) have also been employed (ALGs remove lymphocytes, ATGs destroy T-cells). ATG especially is a potent immunosuppressive agent. Administration results in a fast and almost complete depletion of all circulating lymphocytes and those located in the lymphoid organs. Despite their potency, severe side effects such as serum sickness, severe thrombocytopenia (low platelet count) and leucopenia (low white blood cell count) often occur with these preparations, resulting sometimes in fatal infections.

ALS

ALG

ATG

∏ What do you think are the reasons for these adverse effects?

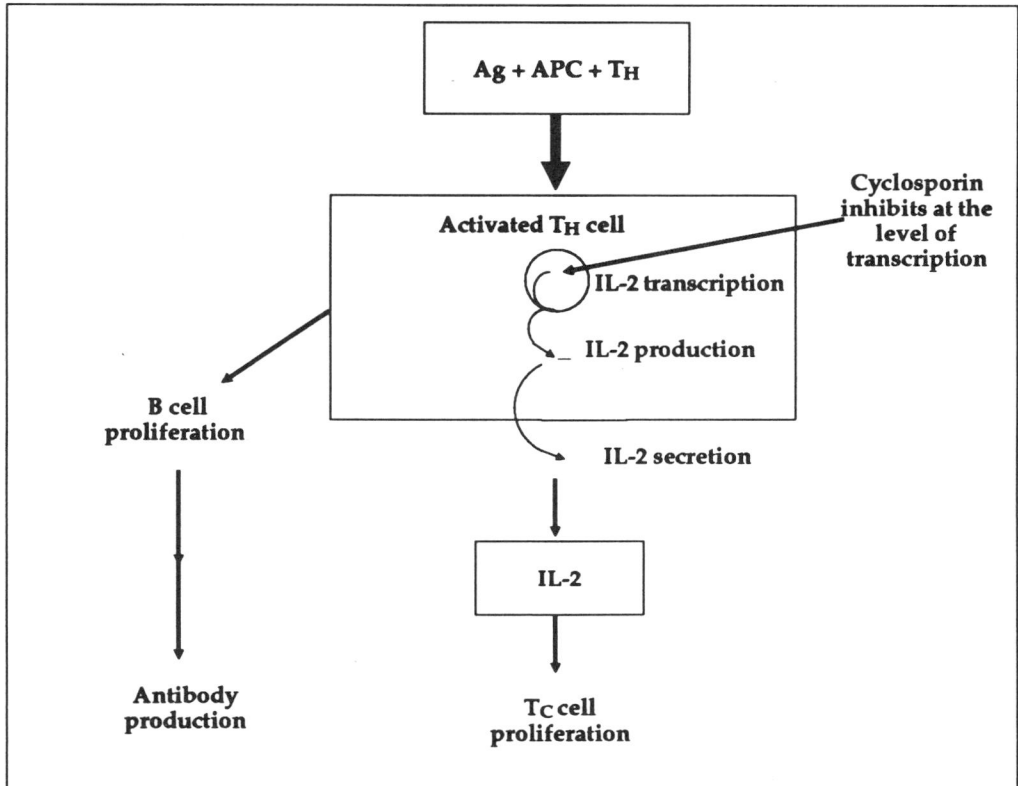

Figure 7.6. The site of action of cyclosporin

ATG antisera are raised by immunising animals with human lymphocyte preparations. Since these often contain other cells such as platelets or granulocytes it is obvious that the procedure will also result in antibodies against other components of the blood. It is the effect of these antibodies on other host cells which can give rise to many side effects (eg loss of blood platelets or thrombocytopenia). Moreover, horse (the usual animal used to raise such sera) serum contains potent immunising proteins itself. Using horse serum therefore induces anti-horse antibody production often making the individual sensitive (allergic) to further horse serum addition (a condition called anaphylaxis). These allergic reactions can be severe and can prove fatal.

However, until recently ALS, ALG and ATG were often the only alternative in severe rejection which did not respond to high-dose steroid treatment. Renal patients may return to dialysis and wait for retransplantation but in heart or liver transplant cases no alternatives were available. This duality of major, sometime life threatening side effects, or returning to dialysis for renal transplant patients, is often a major problem.

In an attempt to change this situation, a more selective therapy has been offered by the introduction of monoclonal antibodies. These are designed to selectively deplete the transplant recipient of a certain T-cell subset, without interfering with other cells. Thus some major disadvantages of polyclonal antibodies, such as severe thrombocytopenia, can be avoided. The first monoclonal antibody to be used for this purpose was OKT3, directed against all mature T-lymphocytes carrying the membrane antigen CD3.

As a summary, Figure 7.7. depicts the modes of action of the immunosuppressive agents discussed above.

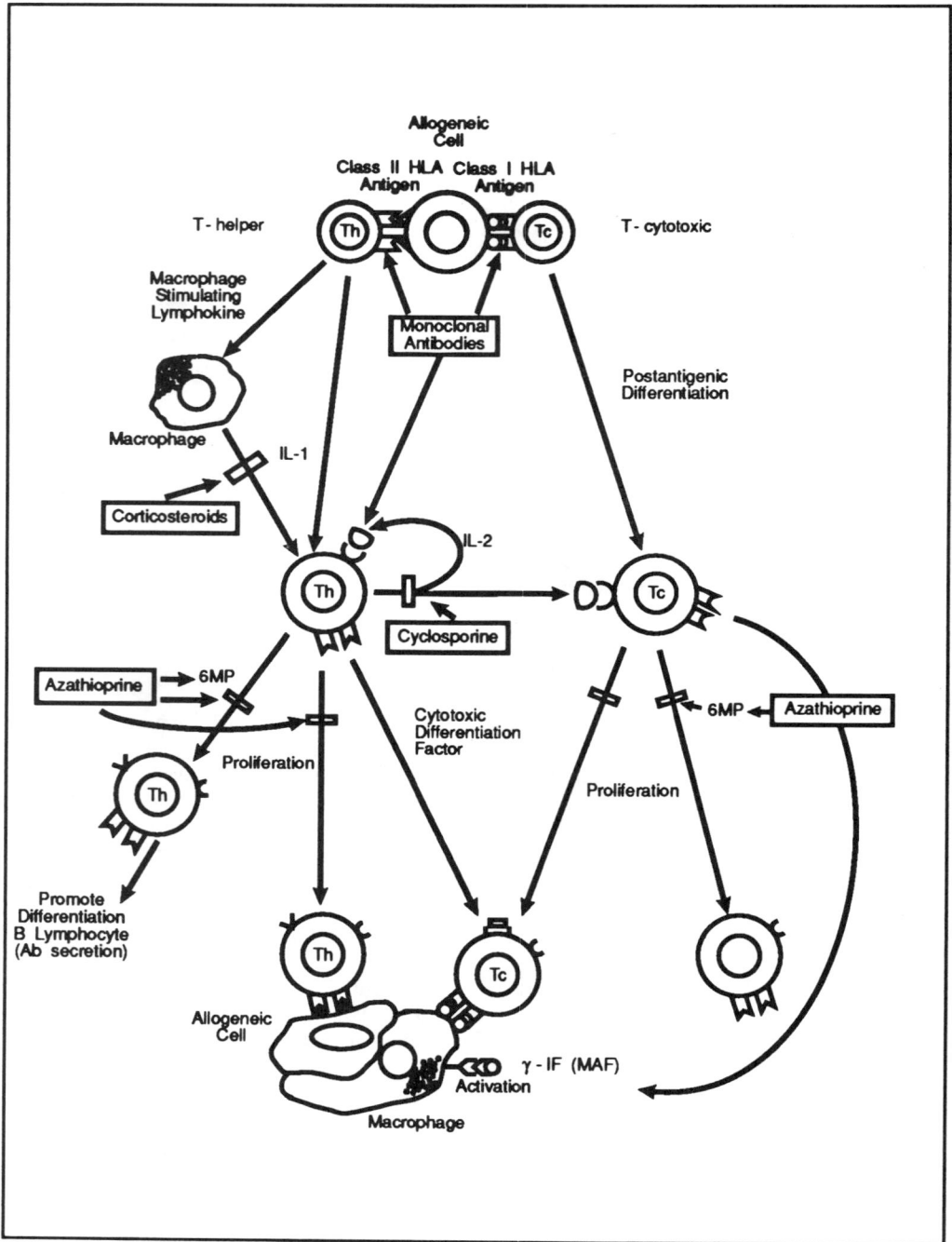

Figure 7.7. Site of action of most currently used immunosuppressive agents in the rejection process

SAQ 7.2.

From the list below select the reagent(s) that is best suited to:-

1) stop interleukin-1 action

2) stop the proliferation of T-cells, but not B-cells

3) stop the proliferation of B- and T-cells

4) the removal of T-cells from the blood stream

Monoclonal antibodies against CD3; azathioprine, cyclophosphamide, corticosteroids, cyclosporin, polyclonal antibodies against lymphocytes (ALS), polyclonal antibodies against T-cells (ATG).

SAQ 7.3.

Choose one of the options below which you think might be a more selective alternative to OKT3 (monoclonal antibody against CD3) as an immunosuppressive agent.

a) Monoclonal antibody (MoAb) against all thymocytes (T-cells)

b) MoAb against all TH-cells (CD4+)

c) MoAb against all TS-cells (CD8+)

d) MoAb against all TC-cells

7.4. OKT3

7.4.1. The concept

Monoclonal antibodies are an attractive way of selectively blocking certain biological pathways. The general idea is to develop an antibody directed against an antigen, which may be exposed on the cell membrane or freely circulating in body fluids. Antibody-antigen coupling subsequently inhibits normal cell function or neutralises circulating enzymes, messengers or medication. However, this "magic bullet" concept has some major restrictions.

As in many new developments, monoclonal antibodies are being extensively investigated in cancer diagnosis and treatment. It is in this field that many, sometimes frustrating difficulties have been encountered. The next section will deal mainly with experiences in cancer diagnosis and therapy. The reason for this is that many of the findings are applicable in the field of anti-rejection immunotherapy.

specific
surface
markers

First of all, if specific tumour cell lysis has to be induced it is of special importance that the target to which the monoclonal antibody is directed, is uniquely expressed on the tumour cell membrane. Until now, no absolutely specific tumour marker has been found. All markers such as CEA, CA 19.9, CA 125, HMFG, HPLAP etc are also expressed in normal, non-cancerous tissues. This means that these normal tissues will also be affected eventually by cell lysis due to binding of monoclonal antibodies against such markers.

barriers

A second problem which has arisen is the fact that some cells cannot be reached by the antibody. There are physiological barriers such as the blood-brain barrier and tumour specific barriers and tumours often develop without vascularisation, rendering intravenously administered antibodies ineffective.

non-specific uptake

Another problem is that many monoclonal antibodies are of murine (mouse) origin. These "foreign proteins" are often taken up non-specifically by the reticuloendothelial system (RES) if administered to other animals where they are broken down and thus inactivated. The reticuloendothelial system (RES) or mononuclear phagocyte system is a host defence system consisting of cells able to assimilate and digest (phagocytose) foreign substances. The cells are localised in blood (monocytes) or in organs or tissues such as the lymph nodes (macrophages), the lung, the spleen, the liver (Kupffer cells), kidney, pancreas, brain and connective tissue (histiocytes). The breakdown of foreign antibodies is often seen in imaging experiments where a radioactively labelled monoclonal antibody is injected intravenously. There is an almost immediate uptake in liver, spleen, lymph nodes and lungs which are the sites that have an extensive RES lining.

complement

opsonisation

internalisation

Suppose a monoclonal has reached its target. At this point some other problems arise. First of all the affinity of the antibody for the antigen must be high enough to prevent early dissociation. What will happen after this coupling? The antigen-antibody complex may induce the complement cascade after which the tumour cell is lysed. A second mechanism by which the cell may be removed is through opsonisation.

This means that cells covered with antibody can be phagocytosed by the cells of the RES system such as macrophages or Kupffer cells in the liver. A third mechanism can be that the Ag-Ab complex is internalised into the cell after which the cell can remain viable but without its membrane marker. If, in this process, some major recognition sites of the cell closely associated to the antibody recognition site are also internalised, the cell will also lose its normal function, although this may not always be the case. The membrane fragment with the antigen-antibody may also be shed from the cell surface; this often happens with tumour cells.

The concept for the production of a monoclonal antibody to act as an immunosuppressor is that by producing an antibody against a specific cell surface antigen of T-cells, administration of the antibody should lead to the inactivation or restriction of T-cells by one of the mechanisms indicated above for the inactivation of tumour cells. An antibody directed against CD3 surface antigen and called OKT3 has been produced. Before we examine how this was achieved we will examine the action of OKT3.

7.4.2. Specific action of OKT3

∏ What would you expect to happen if a monoclonal antibody against CD3 (ie OKT3) is introduced into a body? (Think about the consequences of antibodies binding to tumour cells and write down two mechanisms by which these cells may be inactivated).

OKT3 mechanisms of action

OKT3 is administered by intravenous injection. The gross clinical effect is that all CD3 antigen-bearing cells are removed from circulation and from the graft. There is strong evidence that there are two mechanisms by which these cells are removed. First by opsonisation and phagocytosis by the RES, secondly by internalisation of the complex disabling the cell for normal functions.

There is no complement activation. The specificity of the antibody is absolute because the CD3 antigen is only exposed on mature lymphocytes.

It is understandable that by depleting the patient of all mature T-cells, cellular rejection of the transplanted organ will promptly be abolished. Counting of CD3 positive (+) T-lymphocytes in the blood forms a good parameter by which the effectiveness of OKT3 therapy can be monitored. Some problems, however, do occur.

side effects   Since OKT3 interferes with the patient's immune system, serious infections can occur during therapy. Another complication is the production of antibodies to the murine IgG. These anti OKT3 antibodies are often anti-idiotypic (ie they bind with the antigen recognition site of OKT3). If these antibodies occur in low titre, they can be overcome by higher doses of OKT3 but high titres render it completely inactive. Upon injection of OKT3, most patients show side effects such as fever, nausea, vomiting and diarrhoea. It is believed, but not proven, that these side effects are related to the release of lymphokines.

∏   What does the release of lymphokines after OKT3 administration suggest? (Remember that lymphokines are released after interaction of T_H cells with antigen and antigen presenting cells - refer to Figure 7.4. Remember also that these lymphokines stimulate cell proliferation).

OKT3 is mitogenic in mixed lymphocyte cultures. This means it is able to activate T-cells for IL-2 secretion and proliferation, mimicking the response of T-cells to antigen presentation. This feature suggests that the CD3-antigen complex is located near the antigen recognition site of the T-cell. The binding of OKT3 by T-cells, at least in part, mimics the binding of antigen.

Addition of IL-1 (secreted by the antigen presenting macrophage) starts the cascade of IL-2 release and T_H-cell proliferation in which a number of lymphokines such as IL-2 and γ-INF (gamma interferon), are released. However, due to removal of all CD3+ cells this activation will not be effective for the specific allogenic cells. In summary, the rejection process is induced but the efferent loop is inactive. The activation of the efferent part of the rejection process (see Figure 7.4.) is still a major concern since a smaller number of side effects might result if it could be avoided.

7.5. Production process

7.5.1. Introduction

Orthoclone OKT3 was the first murine monoclonal antibody to be licensed for therapeutic use. The license was approved by the FDA, USA on June 19, 1986. The antibody is a highly effective therapeutic agent which reverses acute renal allograft rejection. The OKT3 monoclonal antibody is produced by a hybridoma cell created by the techniques of Köhler and Milstein. The antibody is a murine type IgG_{2a}.

7.5.2. Overview of production of Orthoclone OKT3

We remind you of the basic strategy used to produce monoclonal antibodies. First suitable animals (usually mice) are immunised with the antigen of interest. This causes the mice to produce an increased population of B-cells producing antibodies which react with the antigen. The spleens, which are rich in B-cells, are collected and after disruption, the B-cells are isolated and fused with myeloma cells.

The desired fused cells (called hybridomas) are those which produce the required antibody and have the growth characteristics of the myeloma cells - that is they grow rapidly. After isolation of the hybridoma cells, they can be cultivated either *in vitro* or as ascites producing tumours in the peritoneal cavities of mice.

The application of this process to the production of a monoclonal antibody (OKT3) which reacts with CD3 is described in Figures 7.8. and 7.9.

Note from Figure 7.9. that the OKT3 production process has five production phases and three quality assurance testing stages.

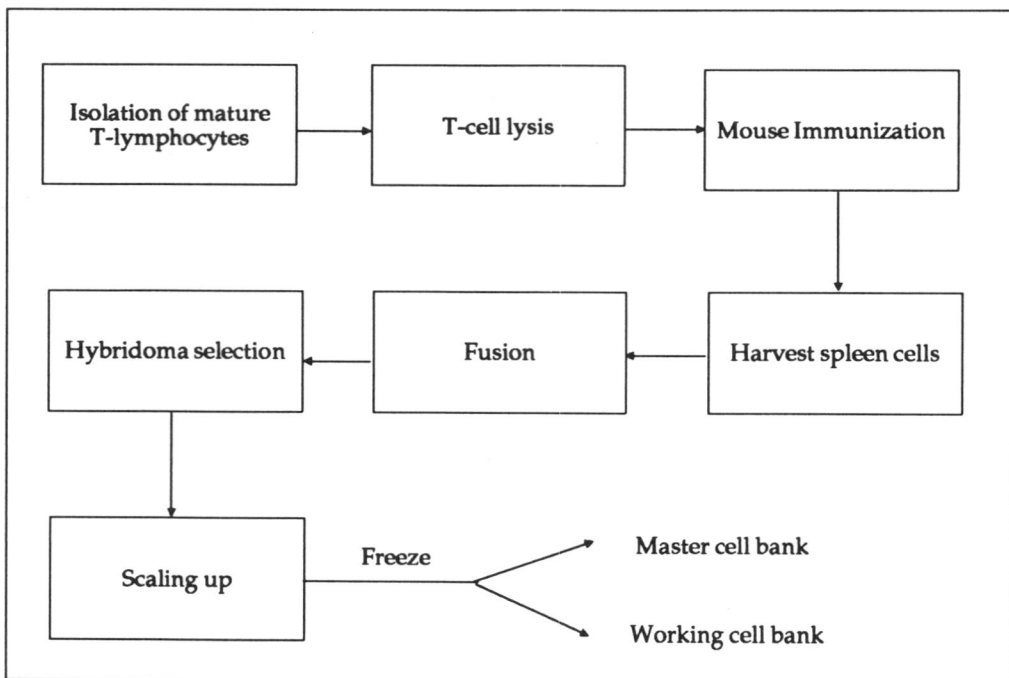

Figure 7.8.   Generation of hybridoma cells producing anti-T-cell antibodies

Generation of hybridoma cells

Hybridoma technology has become a common procedure for the production of specific antibodies. To raise a specific antibody to the desired antigen, it is important that the immunising agent is enriched with the antigen, in this case CD3.

∏   From what you have already learnt, what do you think might be the best immunising agent in this case?

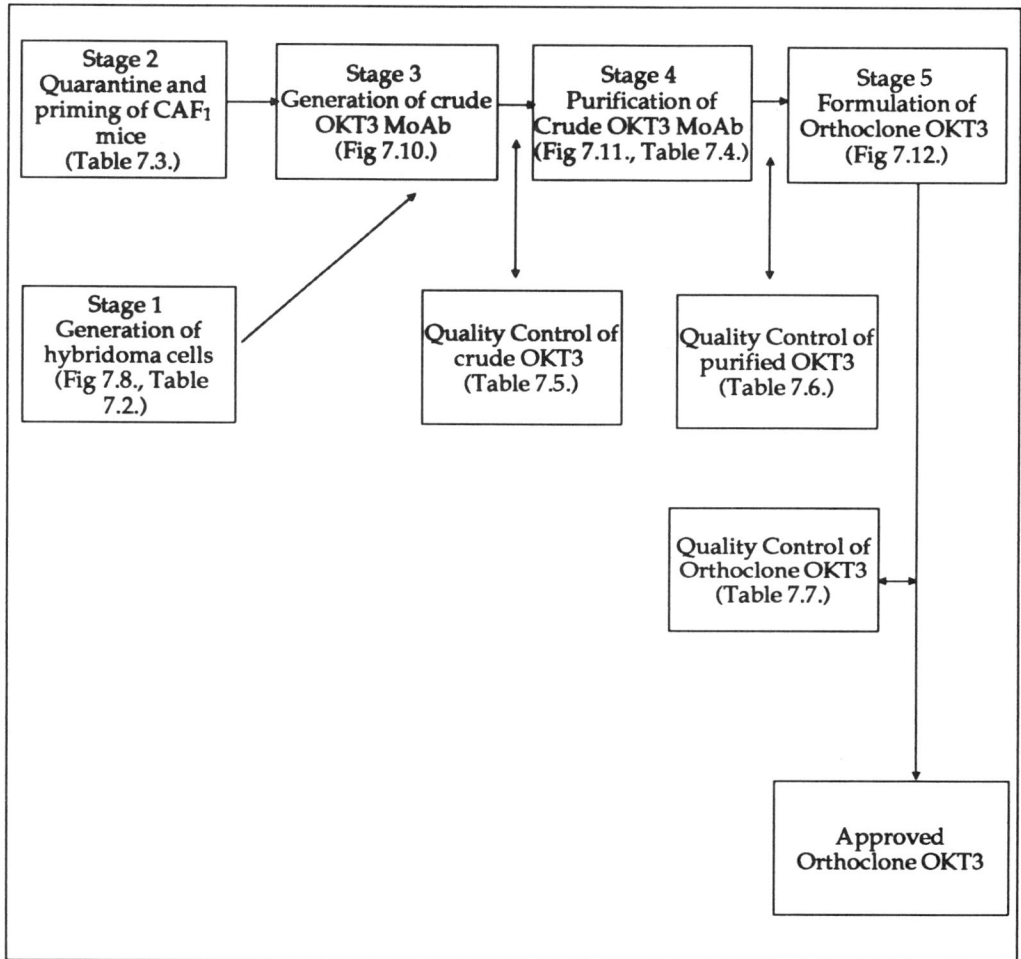

Figure 7.9. OKT3 production process (Data relating to each of these stages are given in the figures and tables referred to in the boxes) MoAb = Monoclonal Antibody. CAF₁ mice are a special strain of mice certified free of certain viruses (see text).

producing the
antigen

The answer is mature lymphocytes, which all contain the CD3 antigen on their surface. Since CD3 is localised on and not in the cell, cell lysis is necessary to increase its concentration by harvesting cell membranes from the lysate.

The antigen is injected into an animal to stimulate the production of the appropriate B-cells. The spleen from such animals is collected and the lymphocytes (B-cells) isolated by density gradient centrifugation.

Π   After hybridisation how should the right hybridoma clone be selected? (Think
     about the specificity of the desired antibody).

anti-CD3
producers

affinity
measurements

The "right clone" is selected on its ability to bind selectively only mature, T-cells. Testing should include challenging with other non-CD3 positive cells such as granulocytes and blood platelets. After this procedure there will be several cell lines as candidates. To make a selection between these, competition assays are used to determine which cell line is producing antibodies which have the greatest affinity for the antigen. This can be performed by first preparing antibodies from the cell lines and then by labelling one antibody with a fluorescent or radioactive label and allowing it to bind to mature T-cells. By adding a non-labelled antibody in the same amount, a new equilibrium will be set up. If the non-labelled antibody, in equimolar concentrations, is able to reduce the binding of the labelled antibody by more than 50%, it means that the former has a higher affinity for the antigen.

in vitro and in vivo antibody production

The next stage of the process is to generate large numbers of the appropriately selected hybridoma cells. There are basically two options here. We can either grow the hybridoma cells in the peritoneal cavities of mice where they produce ascites producing tumours or we can grow them *in vitro* in rich culture media. The choice is not clear cut. If the ascites producing tumour route is taken the yields of antibody are quite high (typically 12-15mg/ml). Using the *in vitro* cell cultivation technique yields are much lower (in tens of μg/ml). But the product from the ascites producing tumour route has to be extensively purified since the product is recovered in peritoneal fluid recovered from mice. In contrast, antibodies produced by *in vitro* cell cultivation have to be recovered from the culture medium.

∏ Why is it more difficult to purify the antibody from peritoneal fluid than from culture medium?

The main reason is that the peritoneal (ascites) fluid contains many mouse-derived proteins which need to be separated from the desired product, since these will be immunogenic when introduced into the patient. Essentially the problem is to separate one set of proteins (antibodies) from a wide variety of others (mouse proteins). The composition of culture medium is much simpler and well defined compared with mouse peritoneal fluid and, therefore, poses few problems.

Despite these difficulties, however, the ascites producing tumour route was chosen since the much greater yields of this route more than compensated for the greater efforts needed to achieve the desired degree of purity.

Once a desired hybridoma cell line has been identified, a culture is divided into ampoules and frozen in liquid nitrogen. Some of these are stored securely and are referred to as the master cell bank, whilst the remainder make up the working cell bank. When a vial of the OKT3 working cell bank is thawed, the recovered hybridoma cells are expanded in tissue culture by standardised methods. Cells are cultured to generate sufficient inoculum for injecting 400 mice.

CAF₁ mice

The second stage of the process is to prepare specific pathogen free CAF₁ male mice, which are used to generate the crude monoclonal antibody, in the production process. The CAF₁ mice are received and quarantined for one week. They are then primed with 0.5 cm³ of pristane (2, 6, 10, 14, - tetramethylpentadecane) by intraperitoneal injection 7-28 days prior to injection of the hybridoma cells. Pristane irritates the peritoneal cavity which encourages the multiplication of the hybridoma cells and the production of peritoneal fluid or ascites.

Generation of crude monoclonal antibody

The third stage of the production process is the generation of the crude monoclonal antibody (ascitic fluid) which is shown diagrammatically in Figure 7.10.

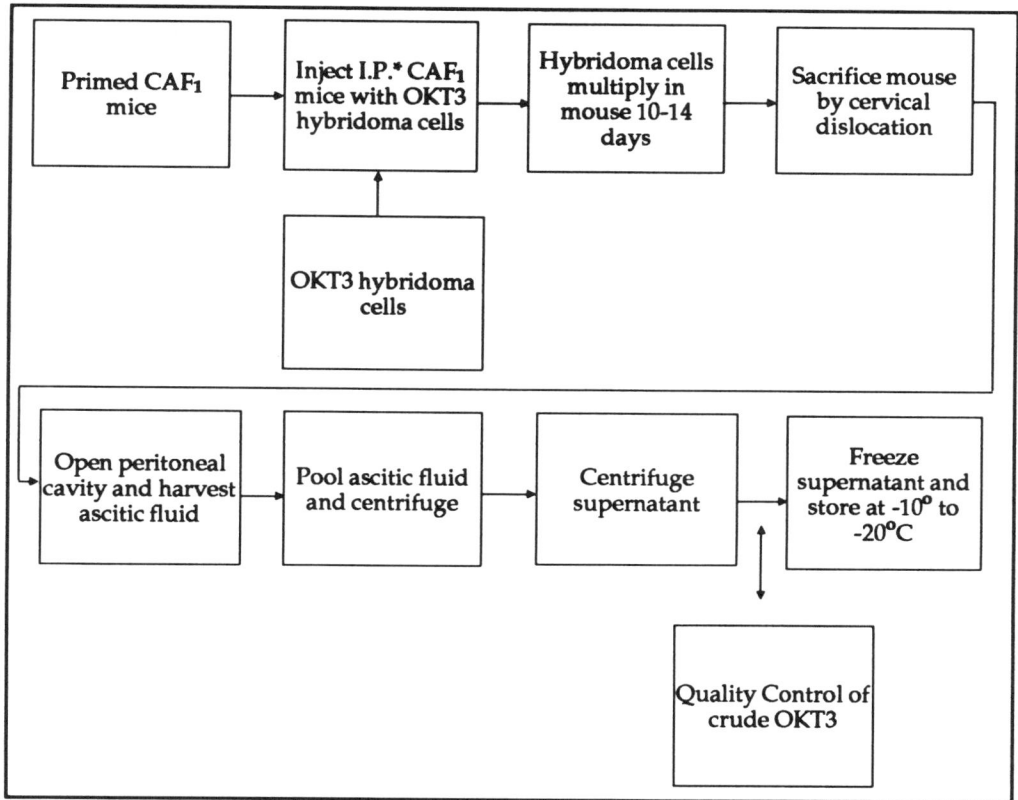

Figure 7.10. Generation of crude monoclonal antibody (I.P. = Intraperitoneal)

harvesting
ascites fluid

The CAF_1 mice are injected intraperitoneally with approximately 10^7 hybridoma cells, which multiply and secrete the OKT3 monoclonal antibody into the peritoneal cavity.

They cause the permeability of the peritoneum to increase, allowing serum proteins (albumin, transferrin) to enter the cavity. Ten to fourteen days after the injection of the hybridoma cells, the mice, which have swollen abdomens due to ascitic fluid, are sacrificed by cervical dislocation and the peritoneal cavity is opened for the removal of the ascitic fluid. The fluid is pooled and centrifuged for 10 minutes at $2°$ to $8°$C to remove red cells and hybridoma cells. The supernatant is recentrifuged at low speed for 30 minutes at $2°$ to $8°$C to remove any residual cells and cellular debris. High speeds could cause disruption of cell membrane, contaminating the supernatant with intracellular material. The crude monoclonal antibody (CMA) is sampled for quality assurance testing, frozen and stored at $-10°$ to $-20°$C. Upon completion of the quality assurance testing, the CMA is released for further processing.

An alternative method for production of the crude monoclonal antibody is to passage (inject) 0.2 cm^3 of the unprocessed ascitic fluid, which contains the hybridoma cells, into each CAF_1 mouse. Two *in vivo* passages are allowed.

The crude monoclonal antibody has an OKT3 concentration of about 12-15 mg/ml.

Purification of crude monoclonal antibody

The fourth stage of production is the purification of the crude monoclonal antibody, illustrated in Figure 7.11.

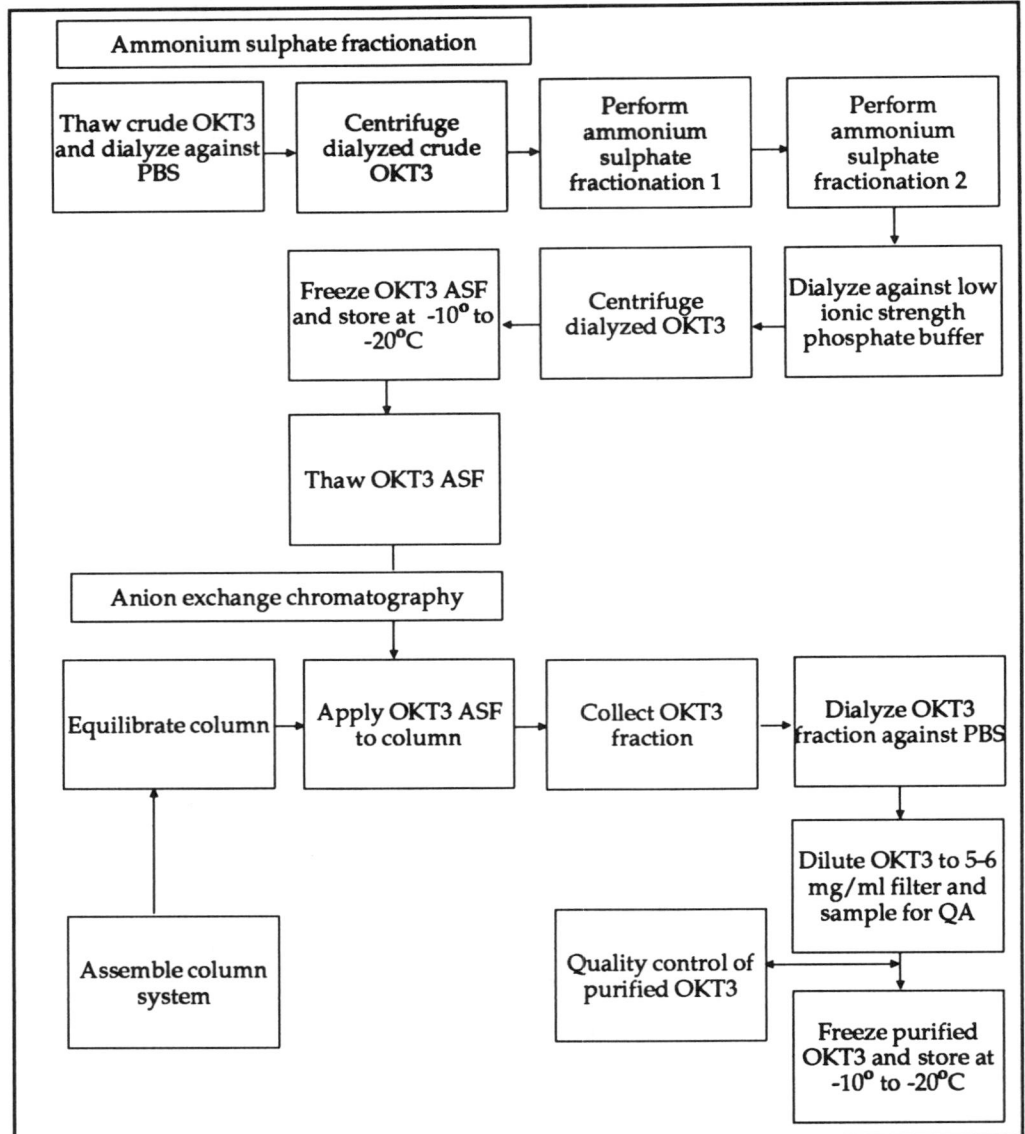

Figure 7.11. Purification of crude OKT3 monoclonal antibodymonoclonal antibody (PBS = Phosphate Buffered Saline, ASF = Ammonium Sulphate Fraction, QA = Quality Assurance)

salt
fractionation
chromatography

Use the flow diagram given in Figure 7.11. to follow this description. The purification is a combination of classical salt fractionation and anion exchange chromatography. Approved batches of OKT3 are thawed and dialysed at 2° to 8° C against phosphate buffered saline for 18-24 hours. The dialysed solution of antibody is then centrifuged at 2° to 8° C for 30 minutes. The supernatants are pooled and two sequential ammonium sulphate fractionations are performed at 0° C. The IgG precipitated, at 50% saturation, is centrifuged for 30 minutes at 2° to 8° C and resuspended in saline. After the completion of the second fractionation, the OKT3 is dialysed against a low ionic strength phosphate buffer in preparation for anion exchange chromatography.

∏ Why is there an ammonium sulphate precipitation step in this process? Are all compounds present in the mixture precipitated by ammonium sulphate?

After the completion of the dialysis step the OKT3 is centrifuged to remove any insoluble immunoglobulin. It is then frozen and stored at -10° to -20° C. The second phase of the purification process is anion exchange chromatography, which is carried out at 2° to 8° C. The OKT3 fraction is dialysed against phosphate buffered saline at 2° to 8° C for 16-24 hours. The dialysis step is repeated against two additional batches of buffer. At the end of the process, the OKT3 is diluted to approximately 5-6 mg/ml with phosphate buffered saline, filtered to minimise microbiological contamination, sampled for quality assurance testing, frozen, and stored at -10° to -20° C. Upon completion of the quality assurance testing, the purified OKT3 is released for formulation.

In summary, the ascitic fluid undergoes the following steps: centrifugation, dialysis, precipitation, chromatography, dialysis and filtration. The crude monoclonal antibody is obtained at a concentration of 12-15 mg/cm^3 which means it does not have to be made more concentrated.

Formulation and filling of Orthoclone OKT3

The fifth stage of the production process is the formulation and filling of Orthoclone OKT3 into five ml ampoules.

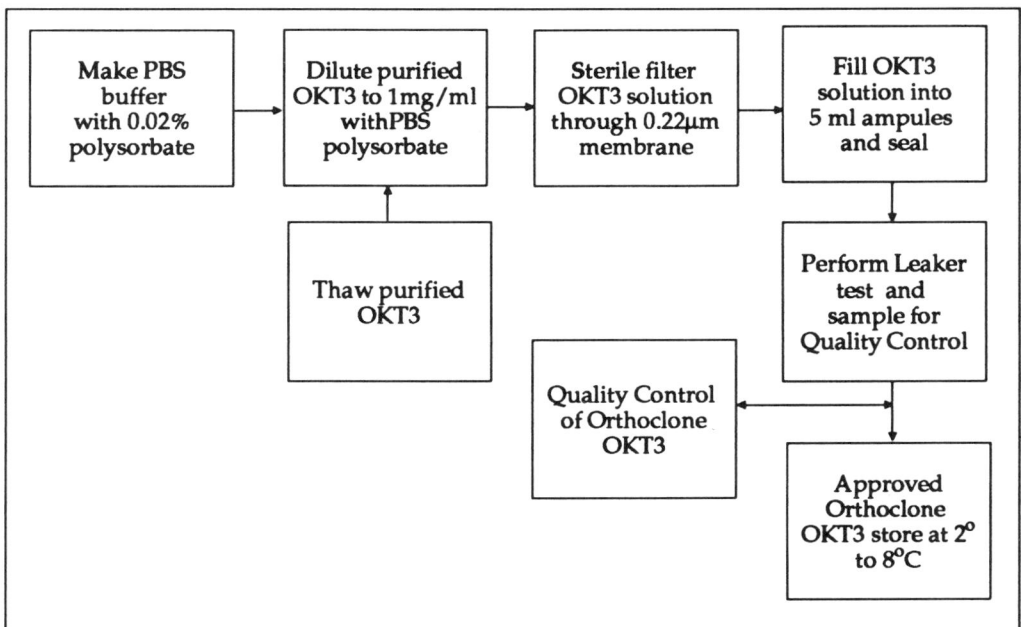

Figure 7.12. Formulation and filling of Orthoclone OKT3 (PBS = Phosphate buffered saline)

sterilisation

asceptic filling

A batch of purified OKT3 is thawed at 2° to 8° C and diluted to 1mg/ml in a phosphate buffer saline (0.15 M NaCl/0.01 M phosphate, pH 7.0) containing 0.02% polysorbate. The solution is gently but thoroughly mixed and then sterilised by filtration through a 0.22μm membrane filter. The filtered solution is then aseptically filled into Type I pyrex glass 5 ml ampoules and sealed. The ampoules are inspected for leaks by a dye test and stored at 2° to 8° C. Samples are taken for quality assurance testing. Upon successful completion of the quality assurance checks, the ampoules are labelled and returned to cold storage.

SAQ 7.4.

1) Which of the following would be a good source of the antigen CD3?

a) *Escherichia coli* b) liver cells c) lymph cells d) B-cells

2) Which of the following techniques would enable selection of hybridoma cells producing antibodies against CD3?

a) Use of radioactively labelled CD3 to label the cells

b) The addition of lymphocytes to the hybridoma cells to see if they are bound

c) The addition of blood platelets to the hybridoma cells to see if they are bound

d) The addition of a selection of viruses to the hybridoma to see if they are inactivated

3) Which of the following reasons can be applied to the use of 0.22μm membrane filters used in the final stages of the manufacture of OKT3?

a) to remove bacteria and other microbes from the product

b) to prevent microbiological spoilage of the product

c) to prevent the chances of transmitting viruses through the use of the product

d) to aid dispersion of the product

7.6. Safeguards of the production process

Good Manufacturing Practices

The production of a therapeutic murine monoclonal antibody for human use has some similarities and many differences to normal pharmaceutical production methods. As with standard pharmaceutical products, the production process is performed under Good Manufacturing Practices (GMPs) using quality control approved reagents and chemicals. The process is performed by highly trained personnel and is controlled by strict adherence to batch records and standard operating procedures. All equipment in contact with the product is sterile or is pyrogen-free. Buffers for injection are tested by the Limulus Amoebocyte Lysate (LAL) assay for endotoxins. The product also undergoes quality control testing at various stages in the production process. All of these measures are to ensure the safety, purity, potency, efficacy and stability of the product, and all are redone before the final release of the product. Purity of the solution is assayed by electrophoretic methods. The potency is tested in a competition assay similar to the one described in the selection method for the hybridoma. The stability of the product is determined by testing its potency after it has been stored under different conditions of temperature, exposure to UV light, etc.

risk of viral
infection

The production of a therapeutic monoclonal antibody has some major differences from a standard pharmaceutical because it has a biological origin. Specific difficulties relate to the starting materials for the process, the hybridoma cells, the CAF_1 mice, and various animal serum additives for tissue culture. The use of these materials in the process presents the risk of the final product being contaminated with micro-organisms and viruses. The possibility of **viral contamination** is a particular concern with immunosuppressed transplant patients. Several safeguards have been built into the production process to reduce potential risks to negligible levels. These safeguards include the following: the use of a dedicated facility, characterisation of cell banks, CAF_1 mouse serology profiles and virus validation studies (we will meet similar issues in the Myoscint case study in Chapter 9).

Dedicated facility

separation of
areas

The OKT3 facility is dedicated to the production of Orthoclone OKT3. The facility has limited access and has three separate isolation zones each requiring special precautions (e.g. face masks, gloves, washing procedures, jump suits etc) corresponding to the three production areas:

- tissue culture and animal production rooms;

- purification;

- formulation and sterile filling.

All critical steps of the process occur in laminar flow hoods or in closed systems. Personnel are dressed in disposable jump suits, face masks, head covering, shoe coverings and sterile gloves. Personnel are assigned to only one area.

Characterisation of cell banks

master and
working cell
banks

The hybridoma cell banks provide the critical starting material for the OKT3 process. There are two categories of cell banks: the master cell bank and the working cell bank. The master cell bank is used to generate working cell banks, which in turn are used to generate the monoclonal antibody. Working cell banks are typically composed of approximately 100-200 vials. The use of this master cell bank/working cell bank concept assures a consistent, high quality supply of the OKT3 monoclonal antibody well into the twenty-first century.

confirmation of
safety and
identity

The master cell bank is derived from the OKT3 hybridoma initially generated in 1978 by the technique of Kohler and Milstein. Both master and working cell banks have undergone extensive characterisation to ensure their safety and consistency. Each is tested for absence of murine viruses, bacteria, mycoplasma and fungi. They are also examined for karyotype and isoenzyme profile to confirm their identity. The cell banks have been examined, by transmission electron microscopy, for the presence of virus-like particles. It is worth noting that all hybridomas, which have been examined by electron microscopy and reported in the literature, have had virus-like particles present. (Those used for OKT3 production are virus-free). The testing for OKT3 cell banks is summarised in Table 7.2.

Bacteria and Fungi		
Mycoplasma		
Murine Viruses		
Intracerebral Inoculation for:	Lymphocytic Choriomeningitis Virus (LCM)	
Isolation tests for:	Mouse Salivary Gland Virus	
	Mouse Thymic Virus	
	Epizootic Diarrhoea Virus of Infant Mice	
	Lactic Dehydrogenase Elevating Virus	
Mouse Antibody Production tests for:	K Virus	Reovirus 3
	LCM Virus	Pneumonia Virus of Mouse
	Sendai Virus	Mouse Adeno Virus
	Minute Virus of Mouse	Mouse Hepatitis Virus
	Ectromelia virus	Polyoma Virus
	Mouse Encephalomyelitis Virus	Hantaan Virus
Assay for Murine Leukaemia Viruses	S+L-	Xenotropic Murine Leukaemia Virus
	XC Plaque	Ecotropic Murine Leukaemia Virus

Table 7.2. Infective agents tested for in master and working cell banks

CAF$_1$ mice

quality assurance of mice

Mice used for the production process always present the potential for contaminating the crude monoclonal antibody with murine viruses. To assure freedom from contamination of Orthoclone OKT3, the CAF$_1$ mice used are specific pathogen free mice. The mice are raised in a virus-free and specific-pathogen-free environment. The quality of the mice is assured by the producer by frequent quality control testing for specific pathogens and antibodies to murine viruses. The absence of antibodies against murine viruses assures that mice have not been exposed to these murine viruses. The absence of pathogens and viruses thus provide a high degree of assurance that mice are virus-free and specific-pathogen-free.

Mice are transported to the Ortho production in gauze filtered shipping cases. On arrival the cases are wiped down with a viricidal agent on the loading dock. The cases are then transported into a quarantine animal production room. Mice are held in quarantine for one week. A sample of ten mice from the incoming shipment is sent to a testing laboratory and evaluated for absence of antibodies to murine viruses to confirm the clean health status of the mice.

∏ The animal production rooms also house sentinel mice. Why is this?

proof of virus
free status

These mice are periodically tested for antibodies to murine viruses to provide assurance that the animal production rooms remain virus free throughout the entire production process for crude monoclonal antibody. The testing of the incoming mice and the sentinel mice is summarised in Table 7.3. Antibodies against the viruses listed can be detected by ELISA or haemagglutination assays. The CAF_1 mice used for OKT3 production produce no antibodies against these viruses.

Testing by both the vendor and Ortho ensures that mice are free of pathogenic murine viruses. However, the mice and the hybridoma cells from the cell banks have the genome of xenotropic and ecotropic murine leukaemia viruses incorporated into the mouse genome. The virus may be expressed at low levels at any time. This potential problem has been addressed by virus removal validation studies.

Antibodies to the following viruses are detectable by ELISA and haemagglutination assays	
K Virus	Reovirus 3
LCM Virus	Pneumonia Virus of Mouse
Sendai Virus	Mouse Adeno Virus
Minute Virus of Mouse	Mouse Hepatitis Virus
Ectromelia Virus	Polyoma Virus

Table 7.3. Murine serology profile for CAF_1 mice

Virus removal validation studies

spiking studies

Virus removal validation studies have been performed to assess the ability of the purification process to remove viruses of selected representative groups. The successive stages of the purification process, the ammonium sulphate fractionation and the anion exchange chromatography, were evaluated separately for their ability to remove viruses.

Studies were performed by scaling the ammonium sulphate fractionation process to one tenth scale and the anion exchange process to one twentieth scale. In the ammonium sulphate fractionation, the crude monoclonal antibody was spiked with high titres of viruses. Samples were also taken at various stages of the process for virus testing. Based on the virus levels detected, clearance factors for the individual and overall process were calculated. Selected viruses and clearance factors are summarised in Table 7.4. The overall purification process reduced the high viral titres to non-detectable levels. These studies demonstrate that there is a large margin of safety in the process.

Virus	Reduction in Virus
Murine Xenotropic Leukaemia Virus - retrovirus	10^6
Rubella	10^6
Herpes Simplex	10^6
Simian Virus 40 (SV 40)	10^6
Minute Virus of Mouse (MVM)	10^6

Table 7.4. Virus spiking validation studies (Reduction in viral numbers after ammonia sulphate precipitation and anion exchange chromatography).

7.6.1. Quality control testing

Quality control   Quality control testing is performed at three stages of the production process: crude monoclonal antibody, purified monoclonal antibody and the final dosage form.

The testing on the crude monoclonal antibody ensures that the parameters - appearance, identity and concentration of the immunoglobulin - meet specifications and that the antibody is suitable for further processing. The testing is summarised in Table 7.5.

Appearance:	Pale yellow to pale red solution
Isoelectric focusing:	Sample conforms to standard
OKT3 concentration:	$3 \geq 4mg/ml$

Table 7.5. Quality control testing for crude OKT3 monoclonal antibody

The purified monoclonal antibody is tested for appearance, pH, molecular integrity, potency and purity to make sure that it is suitable for further processing. The testing is summarised in Table 7.6.

Physical Parameters:	
Appearance:	Clear, colourless solution
pH:	7.0 ± 0.5
Protein Concentration:	5.0 - 6.0 mg/ml
Identity:	
Isoelectric focusing	Sample conforms to standard
Immunodiffusion	Sample must have characteristics of IgG2a
Molecular Integrity:	
SDS/PAGE (reduced)	Sample conforms to standard
SDS/PAGE (unreduced)	Sample conforms to standard
HPLC (GPC) (gel permeation)	Monomer greater than 95%
HPLC (ion exchange)	$OKT3 \geq 93\%$
Potency:	
Competition Assay	Potency $\geq$ 79% standard
Dilution Curve Assay	Similar to standard
Micro-organisms:	
Sterility	Negative
Mycoplasma	Negative
Murine Viruses	Negative
Purity:	
Murine DNA	≤ 10 pg of DNA/dose

Table 7.6. Quality control testing for purified OKT3 monoclonal antibody

Similar parameters are checked on the final dosage form, Orthoclone OKT3 (Table 7.7.). Quality assurance testing on more than a hundred lots of Orthoclone OKT3 has demonstrated that the product is a reproducible, highly purified monoclonal antibody with consistent potency and stability.

You will have noted that in the biotechnological production using *in vivo* systems that particular attention is paid to ensure that the host system does not harbour infective agents.

7.7. Pharmacology

In vitro studies have shown that an OKT3 concentration of 1 µg/ml is required to block all T-cell function. By giving 5mg daily, a steady state concentration of about 1µg/ml is reached after 3 days.

The first injection depletes the blood of all mature T-cells within 1 hour. At this point, the OKT3 concentration is about 1 µg/ml, after which it falls to about 0.1 µg/ml 24 hours after the first administration.

Physical Parameters:	
Appearance:	Clear, colourless solution
pH:	7.0 0.5
Protein concentration:	0.99 - 1.25 mg/ml
Identity:	
Isoelectric focusing	Sample conforms to standard
Molecular Integrity:	
SDS/PAGE (reduced)	Sample conforms to standard
SDS/PAGE (unreduced)	Sample conforms to standard
HPLC (GPC) (gel permeation chromatography)	Monomer greater than 95%
HPLC (ion exchange)	OKT3 93%
Potency:	
Competition assay	Potency 79% standard
Dilution curve assay	Dilution same as standard
Micro-organisms:	
Sterility	Negative
Safety:	
General safety test	Passed
Pyrogen test (Rabbit)	Passed

Table 7.7. Quality control testing for Orthoclone OKT3 monoclonal antibody

After cessation of therapy OKT3 becomes undetectable within 72 hours. Meanwhile CD3+ T-cells reappear in circulation. This can be considered as a major advantage, since severe infection, necessitating withdrawal of OKT3 therapy, is more easily combatted by a rapid restoration of the patient's immune responsiveness, which is often not the case with polyclonal antibody preparations.

Π   The plasma half-life of OKT3 is estimated at about 20 hours. But on first injecting OKT3 it only takes 24 hours to fall from 1 µg/ml to 0.1 µg/ml. How can this difference be explained?

After the first injection, most of the administered OKT3 binds CD3 receptors on the circulating T-lymphocytes, which are removed from the circulation (ie OKT3 is quickly removed from the plasma). The next injections create a steady state condition in which OKT3 is only removed by metabolism and by binding to weakly CD3 positive T-cells newly released by cell division and maturation.

7.8. Clinical trials

To prove the safety and effectiveness of a new medication, clinical trials have to be organised. On the basis of these results, it is decided in which situation(s) a new drug may be used and what the restrictions are. This section describes the trials pointing

---

SAQ 7.5.

Which of the tests listed below should be used for:

1) testing the quality of the purified antibody preparation?

2) confirming the safety of the master and working cell banks?

3) testing the status of the mice used to produce the ascites fluid?

Tests

a) test for the presence of mycoplasma

b) test for antibodies against Sendai virus

c) test for murine DNA

d) test for pyrogens

e) SDS PAGE

f) Isoelectric focusing

g) test for Mouse Adeno viruses

towards the major indications for OKT3 therapy. This does not yet mean that OKT3 is approved for all these indications by official registration agencies such as the FDA.

The first indication for which OKT3 therapy is approved in most cases is the treatment of renal allograft rejections. In this indication the monoclonal is injected at a daily dose of 5 mg for 10 days. However, a distinction has to be made for this broad indication. At first, OKT3 was used for "rescue" therapy, that is after steroid and polyclonal treatment had failed. OKT3 as a last resort therapy, was quite successful, since it was able to save about 70 to 90% of the grafts that otherwise would have been lost. These impressive results lead to a so-called compassionate approval of OKT3 in the treatment of severe cardiac or liver allograft rejection. The same results were obtained in these indications, making OKT3 at this stage a "life saving" drug.

rescue therapy

early treatment

This rescue treatment is, however, often accompanied by severe infections. This is understandable since the patient at this stage is usually receiving maximal immunosuppression with steroids, cyclosporin, azathioprine, ATG or ALG and OKT3. It is for this reason that investigators became aware of the fact that earlier treatment could avoid these situations. Since then OKT3 has been used in large multicentre trials for the early treatment of kidney rejection, the so-called first line treatment. This means that the monoclonal is administered at the first sign of allograft rejection.

The largest of these studies was performed by the Ortho Multicentre Transplant Study Group (1985). It showed that OKT3 is more powerful in treating rejections than steroids and that the 1-year kidney survival significantly improved after OKT3 treatment. The results are summarised in Figure 7.13.

prophylactic treatment

Efforts are concentrated on the use of OKT3 in preventing rejection (prophylactic treatment). In this indication OKT3 is used for 10 to 14 days at a daily dose of 5mg.

It is indeed an interesting idea to start treatment with OKT3 at the moment a patient is challenged by foreign antigens. The final goal is to prevent the onset of rejection, thus

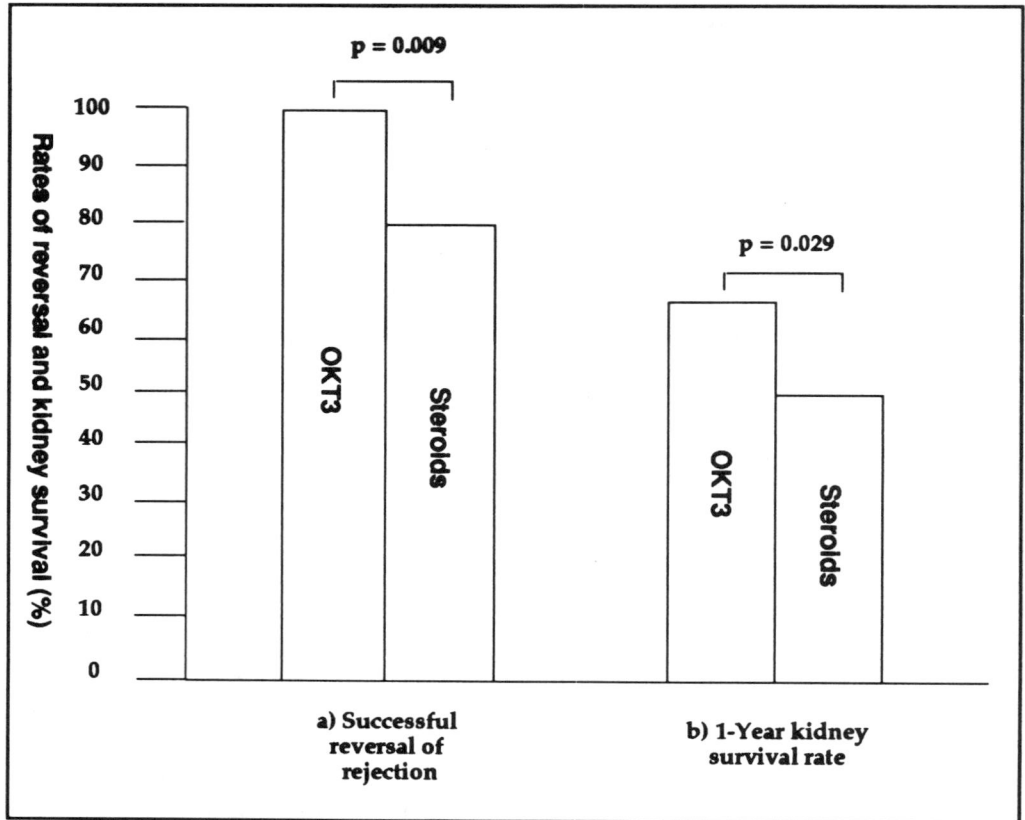

Figure 7.13. Rejection reversal rate and 1-year kidney survival rate in patients receiving OKT3 (n = 63) or steroids (n = 60) for rejection treatment. (Ortho Multicentre 1985)

avoiding irreversible early graft damage by which later renal function can be compromised. A further argument to back-up the early use of OKT3 is that cyclosporin is especially toxic to a kidney recovering from the treatment it receives being transported and handled between donor and recipient. By starting with a non-nephrotoxic, potent immunosuppressive agent and obviating the use of cyclosporin during the first two weeks, the renal function might benefit at long term, since the kidney will get the appropriate time to recover from "storage injury" without having to defend itself against rejection or nephrotoxicity of cyclosporin.

A major objection to this strategy is the development of anti-OKT3 antibodies, rendering it useless for further administration as a rescue treatment. However, concomitant immunosuppression with azathioprine and steroids has reduced this problem substantially.

A prophylactic study is currently running in the USA. Many reports have been published showing significantly less rejection episodes during the first six months in renal, cardiac and hepatic allograft patients. However, graft survival after 18 months in the longest study tends to be comparable to standard protocols (without OKT3 prophylactic treatment). A major advantage could be superior renal function at long-term follow-up, for the reasons discussed above, although this has still to be proved. Similar studies have been conducted in other organ transplantation programmes, with comparable results.

In conclusion, OKT3 treatment is a logical step towards future therapies. Its introduction to the medical world has been a major development in medicine, not only for its effectiveness in the field of transplantation, but also as a radically new way of treating disease conditions using monoclonal antibodies.

Summary and objectives

In this case study we have examined the human immune system and how it responds to transplanted organs and tissues to induce rejection. We have explored a variety of strategies that have been used to suppress the immune system and prevent rejection. In particular we have focused on the use of the monoclonal antibody, OKT3 directed towards a T-cell antigenic determinant (CD3) as a means of disabling the process of rejection. We have discussed the process of manufacture of OKT3 and the validation of its quality and therapeutic activity.

Now that you have completed this chapter you should be able to:-

- describe the distribution of Class I and Class II MHC antigens and the role of Class II antigens in the interaction of antigen presenting cells (APCs) and T-helper (T$_H$) cells;

- describe the roles of interleukins 1 and 2 in the stimulation of the immune response;

- distinguish between sub-populations of T-cells in terms of their surface antigens;

- explain why antimetabolites, alkylating agents, corticosteroids, cyclosporin and certain types of antibodies can act as immunosuppressive agents including their modes of action and the likely consequences of their use;

- identify sources of the antigen CD3 used to stimulate the production of B-cells producing antibodies against CD3;

- explain how hybridomas producing CD3-antibodies may be identified;

- describe the reasoning behind the tests used to check the quality of the hybridoma cells used to produce OKT3, the mice used to cultivate these hybridoma cells and the quality of the purified antibody;

- explain why OKT3 is produced *in vivo* rather than in *in vitro*.

8

Case study: A vaccine for Aujeszky's disease

8.1. Introduction   200

8.2. History of Aujeszky's disease   200

8.3. The pathology of Aujeszky's disease   200

8.4. Control of Aujeszky's disease by vaccination   201

8.5. The molecular characterisation of PRV   202

8.6. Vaccines: The choice of vaccination strategy   206

8.7. Development of a deletion mutant of PRV   208

8.8. Biological characterisation of the deletion mutants   214

8.9. Construction of an additional deletion in the TK gene   216

8.10. Biological characterisation of the 2.4 N3A TK⁻ 783 mutant   217

8.11. The safety of genetically engineered vaccines   220

8.12. Development of the gI test kit: towards eradication of Aujeszky's
      disease   222

Summary and objectives   224

Case study: A vaccine for Aujeszky's disease

8.1. Introduction

gene
inactivation

In the case study on hepatitis B, the use of recombinant DNA to produce a vaccine was examined. In this case study, we again focus on the production of a vaccine. The objectives of each of these are essentially similar in so far as each is intended to produce protection against infection. There are however some quite important differences. In the case of hepatitis B, recombinant DNA technology was used to produce a specific antigen. In this case, recombinant DNA technology was used to deliberately inactivate certain genes. This is not however the only difference between the two. The vaccine against Aujeszky's disease described here, not only provides protection against the disease but also provides a strategy for the complete eradication of the disease. The final important different between hepatitis B and Aujeszky's disease is that the latter is a disease of domestic animals rather than humans.

Thus in this case study you will learn how the deliberate inactivation of genes can find practical application in vaccine production, as well as broaden your experience of vaccine development into the area of veterinary as well as human medicine.

8.2. History of Aujeszky's disease

pseudorabies
virus

Aujeszky described the disease, now named after him, in 1902. It is caused by pseudorabies virus and is particularly serious in pigs, although the virus can be transmitted to other animals (dog, cat, rabbit, guinea pig, ox). The serological characterisation of pseudorabies virus by Shope in 1931 was soon followed by its cultivation in tissue culture, laying the foundation for future research into Aujeszky's disease (AD).

In recent years, there has been an increase in the number and severity of outbreaks of AD in pigs, particularly in Europe and the U.S.A. Much effort has therefore been expended in developing an effective vaccine against pseudorabies virus (PRV), and success has been achieved by combining detailed knowledge of the virus with biotechnology to produce adequate quantities of a safe vaccine.

8.3. The pathology of Aujeszky's disease

8.3.1. Aetiology

The causative agent of AD is Herpesvirus suis or pseudorabies virus (PRV).

8.3.2. Epidemiology

PRV is found throughout the world. It has been an important cause of disease in pigs for many decades in Eastern Europe. In Western Europe and the U.S.A. the disease took a mild form until the 1960s when more virulent strains of PRV began to emerge. Worldwide, the number and severity of outbreaks of AD have shown a steady increase since 1980.

The transmission of viruses between pigs occurs mainly through nasal secretions, which contain high titres of virus in animals 2-4 weeks after the primary infection. Foetuses can be infected *in utero* and transmission to new-born piglets may occur through milk. Spreading from farm to farm is likely to involve aerosols.

reservoirs of
infection

Other sensitive species may be infected via the same route or, in the case of farm animals, by being fed contaminated offal. Pigs are not always killed by PRV and may become latently infected, providing a reservoir of the virus. In contrast, infection of other species is usually fatal within a short time.

8.3.3. Pathogenesis

respiratory
tract
central nervous
system

Infection via the respiratory tract is the most common route, and the upper respiratory tract is the primary site of viral replication. Viruses then pass along nerve fibres and the central nervous system is invaded. Here, further replication of virus takes place in the medulla and pons.

8.3.4. Clinical signs and pathology

The severity of symptoms depends on the age of the animal but excessive salivation, fever, depression and convulsions are characteristic signs of PRV infection. Death is most common in young animals.

Macroscopic lesions in the nasal mucosa, larynx and trachea may be found in the late stages of a terminal infection. Microscopic lesions occur in the central nervous system, particularly in the cerebrum. Lesions with necrotic spots may also be detected in the respiratory tract.

8.4. Control of Aujeszky's disease by vaccination

harmonisation
of policy

In Europe at least three different policies are followed with respect to the use of live or inactivated vaccines against AD The intention is that they will need to be harmonised by 1992 (the Single Community Act) as all vaccination strategies are supposed to be the same in all participating countries. Eradication of the disease seems to be the favoured option; vaccination alone only serves to prevent major outbreaks.

In what ways do the policies concerning vaccination against Aujeszky's disease differ in different EC states?

In Great Britain (except Northern Ireland), Sweden, Norway and Finland, vaccination against AD is forbidden by law, whilst Germany and Italy forbid immunisation with live vaccines. In contrast, in the Netherlands, Belgium, France, Spain, East and South European countries there are no restrictions on the use of either live or inactivated vaccines.

All conventional live vaccines possess some degree of virulence; a live vaccine without some virulence would not be effective, because it would not be immunogenic. The disadvantage of the remaining virulence is that a vaccine may not be suitable for all the species for which it may be required. Some commercially-available live vaccines can be dangerous for young pigs, dogs, sheep and cats.

The specific benefits and problems of live versus inactivated vaccines will be discussed in more detail later.

8.5. The molecular characterisation of PRV

8.5.1. The virus

PRV characteristics

Pseudorabies virus belongs to the Herpesviridae, a family which is widely disseminated in higher animals. Herpesviridae are generally defined by:

- the presence of double-stranded, linear DNA in the core of the virion;

- an icosahedral capsid containing 162 capsomers, assembled in the nucleus of the host cell;

- an envelope derived from the nuclear membrane of the host cell in which the virus was produced.

PRV is a member of a sub-family (Herpes viruses) characterised by:

- variable host range;

- efficient destruction of infected cells;

- rapid spread in cultured cells;

- the possibility of establishing latent infections, in most cases, in ganglia (nerve cells). Human Herpes simplex (1 and 2), Varicella zoster and Equine herpes viruses belong to the same sub-family.

8.5.2. Morphology

The core of PRV, which contains the DNA, has a toroidal structure and is surrounded by 162 capsomers to make up the capsid with a diameter of 110 nm. The tegument, the structure between the capsid and envelope, is variable in size. The envelope has a three-layered appearance in the electron microscope and spikes protrude from it. Virions vary in size from 120 to 300 nm (Figure 8.1.).

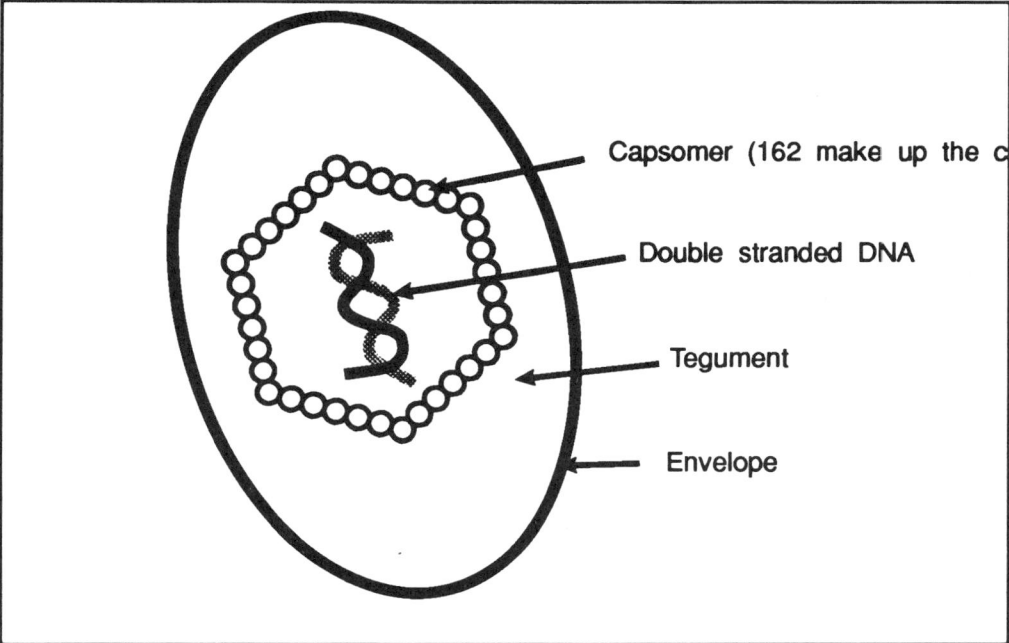

Figure 8.1. Stylised drawing of pseudorabies virus (PRV)

8.5.3. The viral genome and the infectious cycle

The basic organisation of the PRV genome is shown in Figure 8.2.

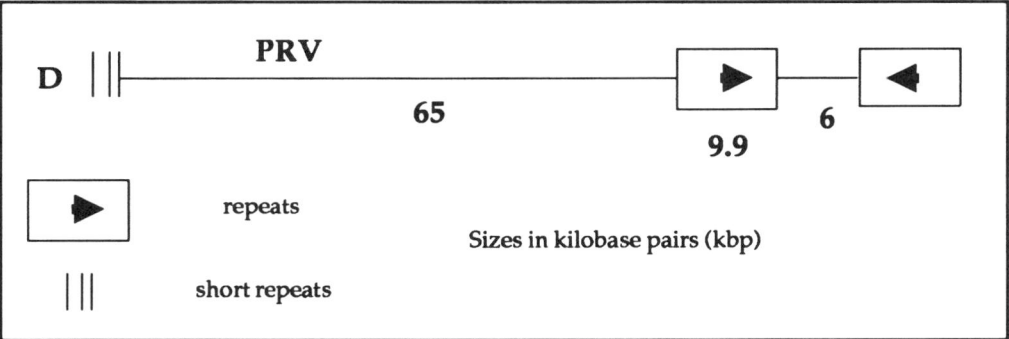

Figure 8.2. Genome of PRV

repeats and
inverted
repeats

The virus contains short segments of nucleotides at one end (D) of the genome which are repeated several times. Then there is a long segment (65 kilobase pairs (kbp) which carries structural genes including the gene coding for thymidine kinase (TK). Then there is quite a long segment of nucleotides (9.9 kbp), which is repeated at the further end of the molecule, but in the opposite direction (ie inverted repeats). To explain exactly what is meant by inverted repeats, consider the following nucleotide sequence:

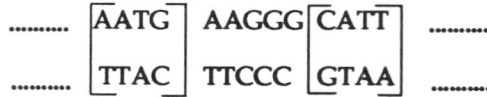

```
..........  AATG   AAGGG  CATT   ..........
..........  TTAC   TTCCC  GTAA   ..........
```

Note that these bracket sequences are repeats of each other, but are read in the opposite direction to each other (ie they are inverted repeats). The PRV genome contains such inverted repeats but they are much longer than those just described above.

Between the inverted repeat sequences, there is a 6 kbp segment described as the Us (Unique short) region. We will be examining this area of the genome of PRV a little later because this region carries some important genes (especially gI) concerned with the virulence of the virus.

Figure 8.3. describes the phases in the development of the virus.

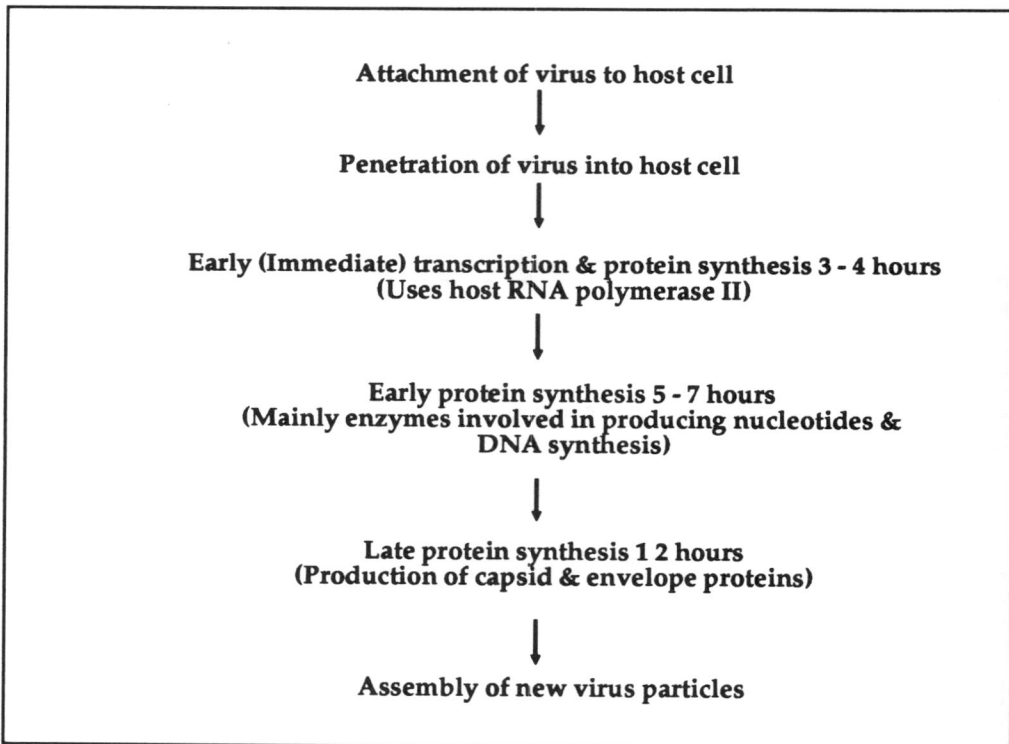

Attachment of virus to host cell

↓

Penetration of virus into host cell

↓

**Early (Immediate) transcription & protein synthesis 3 - 4 hours
(Uses host RNA polymerase II)**

↓

**Early protein synthesis 5 - 7 hours
(Mainly enzymes involved in producing nucleotides &
DNA synthesis)**

↓

**Late protein synthesis 1 2 hours
(Production of capsid & envelope proteins)**

↓

Assembly of new virus particles

Figure 8.3. Phases of development of PRV

PRV infectious cycle

The infectious cycle begins with the attachment of the virus to the cell surface, whereupon the viral envelope fuses with the cellular membrane and the capsid is released. The capsid interacts with the nucleus where the viral DNA is transcribed.

Transcription and accompanying protein synthesis can be divided into three phases as shown in Figure 8.3.

The phases of transcription and translation can be described as:

- immediate early (between 3-4 hours) after infection using the host RNA polymerase II. The mRNAs produced specify regulatory proteins involved in the transcription of the next phase

- early (between 5-7 hours). Proteins produced in this phase include thymidine kinase and DNA polymerase (ie concerned with nucleic acid synthesis)

- late (from 12 hours onwards). The structural proteins are synthesised from this group of mRNAs (ie concerned with final viral construction). In the mature virion, over forty distinct proteins have been found, many having undergone post-translational processing such as phosphorylation and glycosylation.

8.5.4. The glycoproteins of PRV

The genome of PRV codes for seven glycoproteins, six of which are found in the virus particle. The glycoprotein genes are usually labelled with a prefix g. Thus gI, gp50, gp63 and gX are coded for in the Us region of the genome; gII and gH are coded for within the 65 kbp region. Glycoprotein gI proved to be important in the design of a new, live vaccine. It is involved in the release of virus from certain cell types, but it is not absolutley essential for viral replication. The deletion of the gene for gI does not block the growth of virus *in vitro*. Now attempt SAQ 8.1.

SAQ 8.1.

Using the labels listed below, complete the labelling of the diagram of the genome of PRV. (You will need to use Figure 8.2. and the text to help you answer this).

Labels

Us region	Thymidine kinase gene
Short repeats	65 kbp
Long repeat	6 kbp
Inverted repeat	9.9 kbp

The genome of PRV:

```
 ||          gII     gH              gX gp50 gp63 gI
 ||  _____
```

8.6. Vaccines: The choice of vaccination strategy

Vaccination is designed to elicit a successful immune response in an animal in the virtual absence of clinical signs of disease. Subsequently, infection should not lead to perceptible illness. There are many options in the design of a vaccine. We have already met one in the case study about hepatitis B vaccine, namely the production of a purified antigen.

In this case study we will focus a little more on the choice of options by exploring the advantages and disadvantages of different vaccines.

conventional
vaccines
Conventional vaccines are based on the causative organism in either an attenuated or inactivated form.

∏   How do inactivated and attenuated viruses differ from each other? (Think especially in terms of their interaction with the immune system).

We can cite the following differences:

• a live vaccine can be administered in a very low dose, because viruses or bacteria will replicate before being inactivated by the immune system. With inactivated viral vaccines, substantially larger doses are required;

• a live vaccine challenges the immune system in the same way as a wild-type infection. This is rarely the case with inactivated vaccines and explains why attenuated, live vaccines are often more efficacious than their inactivated counterparts;

• attenuated vaccines have usually resulted from selection of slow-growing mutants, the nature of the mutation being unknown. The risk here is that the effect of a point mutation could be altered or even reversed by a further point mutation.

8.6.1. Live, attenuated vaccines

deletion
mutants
site directed
mutations
The argument for the deliberate construction of deletion mutants, the strategy adopted for PRV, is that it should give the advantages of live vaccines without the uncertainty of an unknown mutation arising. Virulence is in general determined by several genes. Alterations at different sites may influence the overall virulence. The techniques of recombinant DNA technology allow site-directed mutations/deletions to be made in one or more virulence-determining genes. The possibility of deleting specific information from a genome is an important development in the design of attenuated vaccines that are both safe and efficacious.

8.6.2. Inactivated vaccines

adjuvants
A drawback of inactivated vaccines is that only macrophages are capable of processing proteins from inactivated organisms since inactive viruses will not invade cells and activate T-cells. Furthermore, the stimulus of the immune system is less efficient than is the case with a live vaccine. Adjuvants (materials that enhance the immune response) have to be added to inactivated vaccines. The latter has to be used in sufficient quantity to obtain a good immune response.

Inactivated vaccines do have some advantages: they are generally more stable and heat resistant than their live counterparts and they do not induce any pathological effects. This last property is useful in immune-compromised patients.

8.6.3. Subunit vaccines

subunit vaccines

An immunological response is always directed towards several antigenic determinants displayed by an invading organism. In principle, it is possible to use just one of the antigenic determinants (protein or carbohydrate) as a vaccine. So-called subunit vaccines are often safer than inactivated vaccines, because inactivation can be incomplete, and a subunit vaccine contains only the antigen without contaminating material. The case study on hepatitis B discusses an example of a subunit vaccine, produced in this instance by genetic engineering.

8.6.4. Peptide vaccines

epitopes

A pure peptide, produced by chemical synthesis, should make a safe antigen. However, many antigenic determinants (epitopes) involve different regions of a protein brought together in the folding that creates the tertiary structure. Therefore a peptide vaccine, to be useful, must have the right configuration. In addition, it must be coupled to a suitable carrier protein to become immunogenic.

8.6.5. Live recombinant carriers

A live recombinant carrier is a combination of a live, attenuated virus and a subunit vaccine. It is created by adding the required genetic information to the genome of a vaccinating organism. In principle, the additional genes should be expressed, with the vaccinating organism acting as a carrier. Possible carrier viruses include:

- vaccinia and other pox viruses;

- adenoviruses;

- retroviruses;

- adeno-associated viruses;

- herpes viruses.

Π Can you think of ways in which such a live recombinant carrier vaccine could be produced? (Think especially about which viral genes you might replace with new genetic material).

replication defective

Two possible approaches for the construction of a virus-based live recombinant carrier are:

- to exchange a non-essential viral gene for a foreign gene;

- to exchange an essential viral gene for a foreign gene, in which case progeny virus can only be produced in cells supplied with the missing viral gene. Such a virus would be infectious but defective for replication, that is the virus in this system can be regarded as a 'semi-live' recombinant carrier. This is an advantage because it means that a semi-live recombinant virus is confined to the individual animal into which it is injected as a vaccine.

Attenuated forms of PRV could be used as a carrier for genes from other viruses and microorganisms pathogenic for pigs. It must be realised that such a 'live' recombinant virus which has not had an essential gene exchanged, could spread from animal to animal. The use of such a virus as a vaccine would therefore represent the environmental release of a genetically-modified organism. The release of such modified strains is subject to a considerable debate. Currently there are strict regulatory controls governing the release of such strains.

SAQ 8.2.

Let us assume that virus X causes a major disease and that you are seeking to produce a vaccine for the disease. You are attempting to decide whether or not to produce a vaccine composed of attenuated viruses, or one composed of inactivated viruses or to produce a single antigen as a vaccine.

Below is a series of factors which may contribute to your decision. Label those which favour the choice of an attenuated virus with an A, those which favour an inactivated virus with a B and those which favour using a purified antigen with a C.

1) Special requirement for an adjuvant

2) A risk of 'reversion' of the virus into a virulent form

3) A greater immunological response

4) Lack of knowledge of the specific antigens produced by the virus

5) Greater vaccine stability

6) Greater ability to characterise the vaccine

8.7. Development of a deletion mutant of PRV

From the preceding section, it should be clear that a good vaccine would be a deletion mutant of PRV, able to function as both:

* a safe PRV vaccine

* a carrier for antigen-determinants of other pig pathogens

The key to producing a safe vaccine is to ensure that there is little chance of the deletion mutant being replaced and thereby restoring virulence.

8.7.1. Construction of a gI deletion mutant

The virulent NIA3 strain of PRV (Northern Ireland Aujeszky 3) was chosen to begin the selection. The rationale for this choice was that a virus that is already attenuated may become too avirulent with the introduction of further mutations and the advantages of using an attenuated strain would be lost.

unique short
region (Us)

On the basis of earlier research, the unique short (Us) region in between two inverted repeat sequences was chosen as the starting point for creating deletion mutants that might affect the virulence of PRV without reducing viability. Figure 8.4. shows the location of the Us region and its relationship to genes specifying viral glycoproteins.

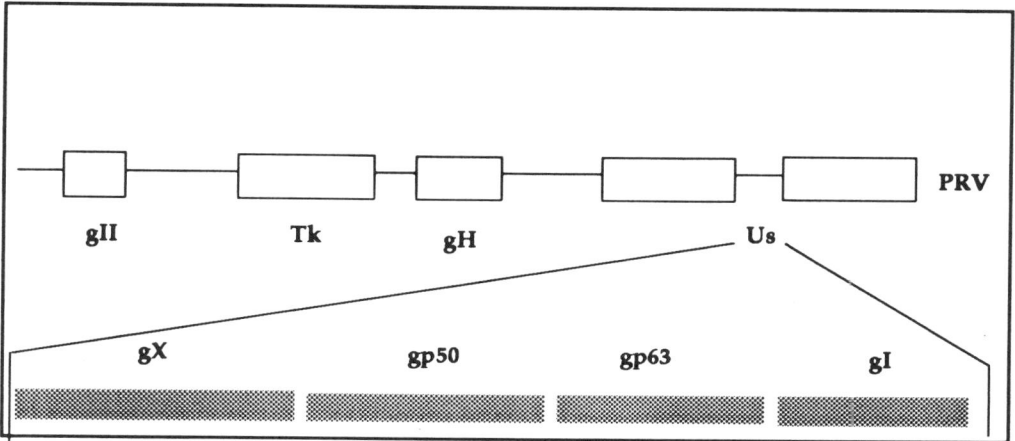

Figure 8.4. The location of several important (glycoproteins) on the PRV genome

The production of the deletion mutants was quite a complex procedure.

Let us first consider the restriction enzyme sites in the Us region and in the adjacent inverted repeat regions. We especially draw your attention to the *Hind* III site (see Figure 8.5.).

treatment with
restriction
enzymes

You will notice that there are *Hind* III sites in the inverted repeat sections. Treatment of PRV DNA with *Hind* III allowed for the isolation of a fragment carrying the whole of the Us region with a small part of the inverted repeats at each end. This is shown in Figure 8.6.

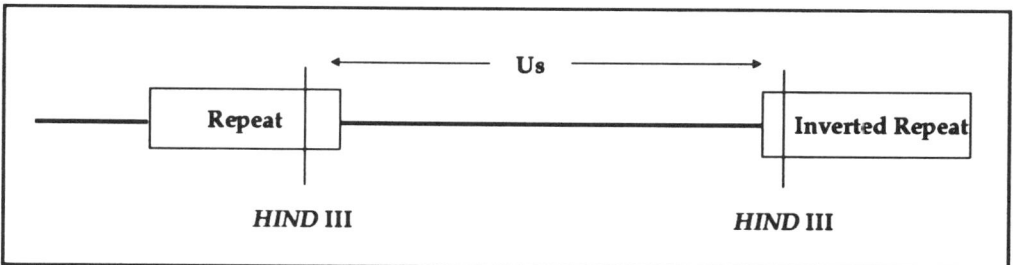

Figure 8.5. The *Hind* III sites adjacent to the Us region of PRV

The *Hind* III fragments were cloned in pBr322 to produce a new plasmid pN3HB. The fragments produced after *Hind* III treatment carried sites for the restriction enzymes *Mlu*1 (Ml), *Bg*III (Bg), *Bal*1 (Bl) and *Bam*HI (B) (Figure 8.6.).

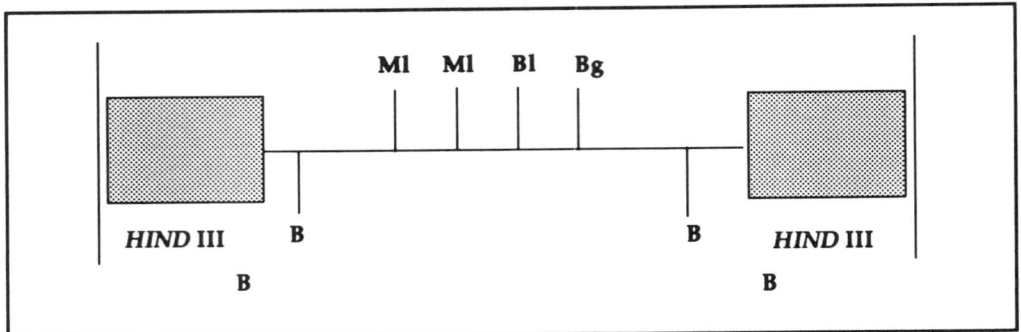

Figure 8.6. Restriction sites in the Us fragment isolated after *Hind* III treatment

The treatment of the *Hind* III fragments with Bg and Ml means that quite a large section of Us region could be deleted as shown in Figure 8.7.

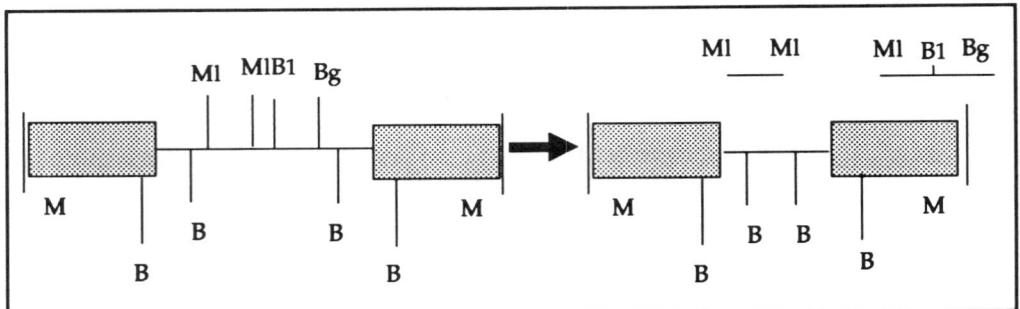

Figure 8.7. The *Hind* III fragment of PRV after treatment with Bg and Ml.

In principle this fragment contained only a short part of the Us region (ie much of the glycoprotein coding region had been removed). Such a fragment, however, is not ideal for re-incorporating into the virus because too much of the viral genome has been removed.

What was really required was for there to be only a limited section removed since it was not the intention to produce a totally inactivated virus. Part of the removed section of the Us region was replaced in the following way.

Treatment of viral DNA with *Bam*HI allowed isolation of the central portion of the Us region (Figure 8.8.).

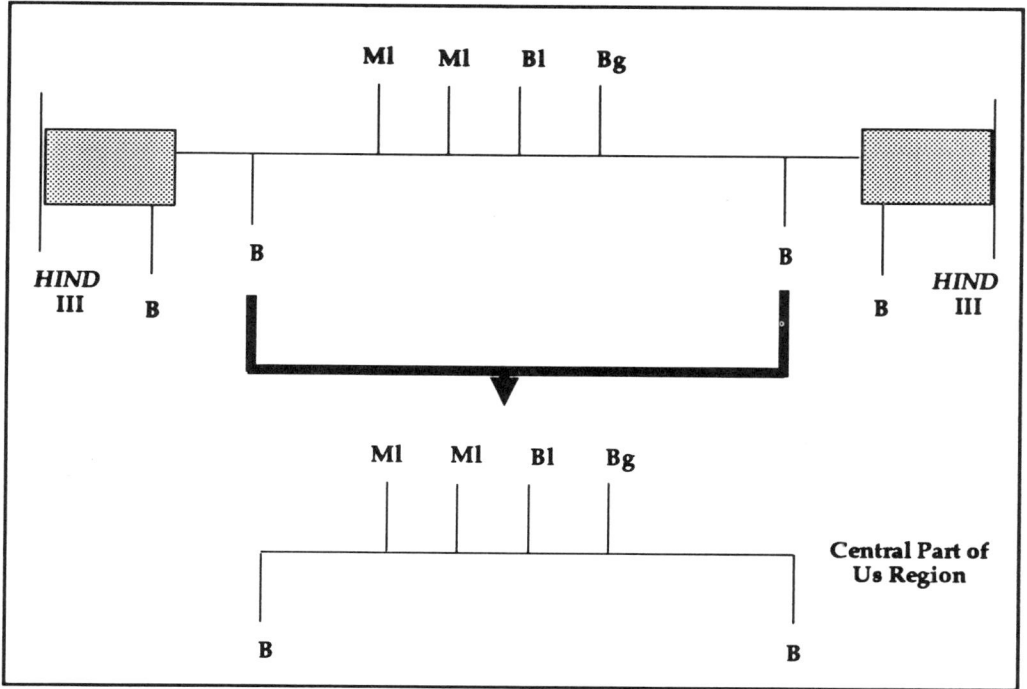

Figure 8.8. Fragmentation of PRV DNA with *Bam*HI

The central part of the Us region was cloned in pBR322 to produce a new plasmid pN387. This plasmid was then opened with the restriction enzyme *Bal*1 and partially hydrolysed with exonuclease *Bal* 31. This sequence is represented in Figure 8.9.

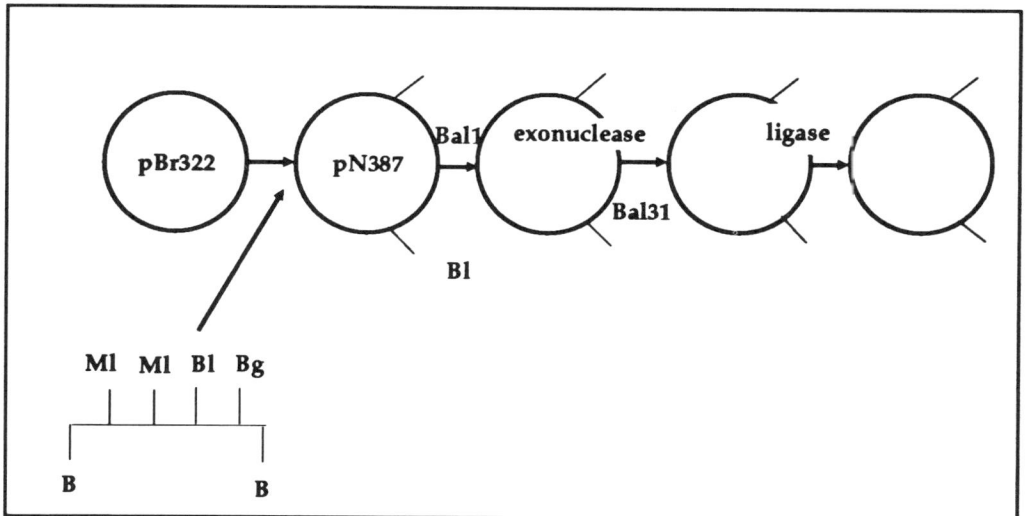

Figure 8.9. The removal of nucleotides adjacent to *Bal* 1 sites using *Bal* 1 and exonuclease *Bal* 31

After this treatment, three clones were isolated one which had 2.1 kbp removed, one 2.4 kbp and one with 2.8 kbp (kbp - kilobase pairs).

Thus we can write this overall sequence as:-

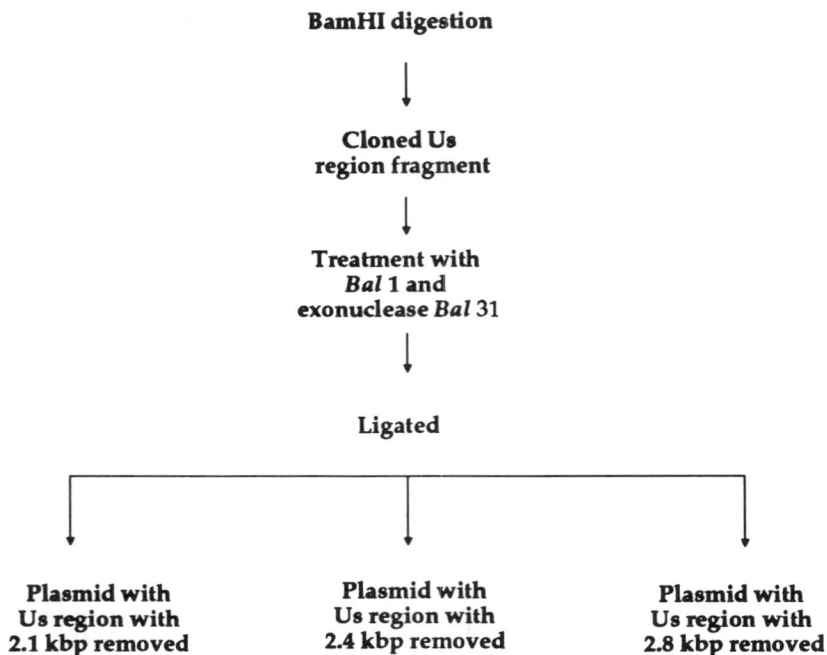

BamHI digestion

↓

**Cloned Us
region fragment**

↓

**Treatment with
Bal 1 and
exonuclease *Bal* 31**

↓

Ligated

| **Plasmid with
Us region with
2.1 kbp removed** | **Plasmid with
Us region with
2.4 kbp removed** | **Plasmid with
Us region with
2.8 kbp removed** |

These plasmids were cloned into the *Hind III* fragments described earlier to produce three types of plasmids. These are pHB2.1 which contains a Us region with a 2.1 kbp deletion, pHB2.4 containing a Us region with a 2.4 kbp deletion and pHB2.8 containing a Us region with a 2.8 kbp deletion.

This whole sequence is summarised in Figure 8.10.

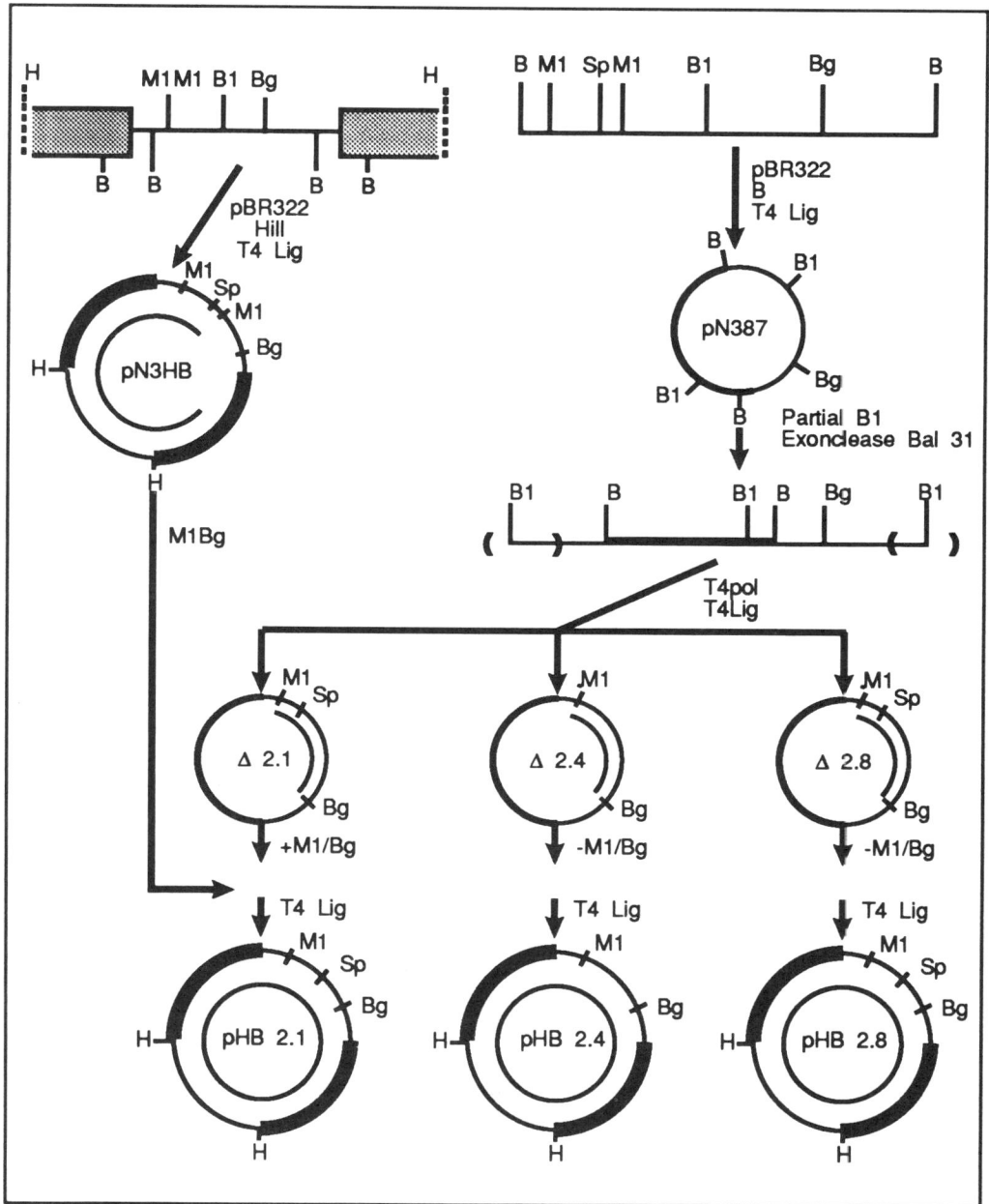

Figure 8.10 The cloning of deletion mutants in the Us region. Restriction enzymes used: B = *BamH* I, BI = *Bal* I, Bg = *Bgl* II, MI = *Mlu* I, Sp = *Sph* I, H = *Hind* III

The key now was to put these Us regions carrying the deletions back into PRV. Think back to the viral genome. We remind you that if this is treated with *Hind* III it is fragmented in the following way (Figure 8.11.).

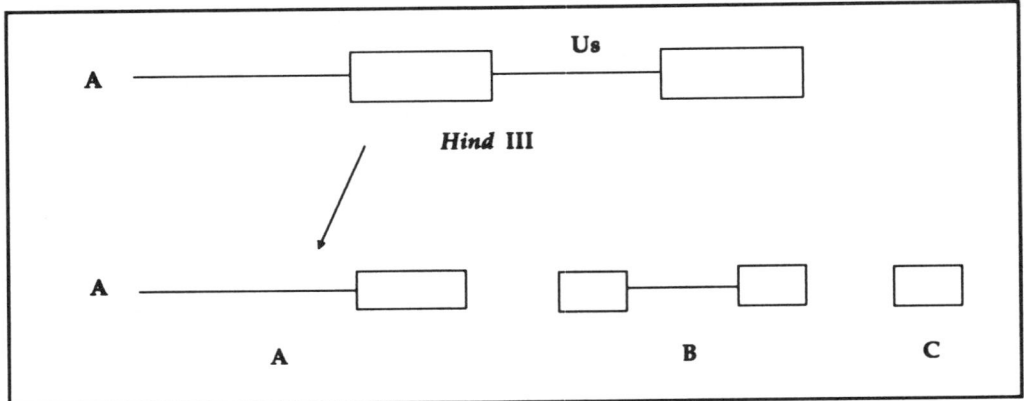

Figure 8.11. Fragmentation of PRV DNA with *Hind* III

∏ In Figure 8.11. we have labelled the fragments as A, B, C. Use this figure to answer the following questions:

a)   If porcine kidney cells are infected with fragment A will the virus develop?

b)   If porcine kidney cells are infected with fragment B will the virus develop?

c)   Would virus develop if porcine kidney cells are transfected with both A and B?

The answers to a) and b) are 'no' because in a) essential genes in the Us region would be absent, in b) essential genes in the A fragment would be missing. In c) we can basically infect cells with all of the essential genes by using an appropriate selection of fragments.

virus
re-construction

The technique that was used to reconstitute viruses carrying deletion mutants from the various restriction fragments was as follows. Fragment A (Figure 8.11.) was mixed with an excess of each of the plasmid constructs (pHB2.1, pHB2.4, pHB2.8) and the mixtures used to transfect porcine kidney cells. The two *Hind* III fragments must reconstitute before yielding infectious viral particles, which means that any virus obtained must be reconstituted PRV containing one of the deletions.

The reconstructed viruses containing the 2.4 kbp deletion has been labelled 2.4 N3A while that carrying the 2.8 kbp deletion has been called 2.8 N3A.

8.8. Biological characterisation of the deletion mutants

The parental virus, (NIA3), an avirulent strain, (NIA4) and the deletion mutants, (2.4 N3A and 2.8 N3A) were examined for their biological properties in mice and pigs. Note that we have not met strain NIA4 before. It is a natural occuring strain of PRV which shows less virulence than strain NIA3.

mean-time-
to-death

In general, PRV is lethal to mice and the mean-time-to-death (MTD) is a good indicator of any change in virulence in modified strains of PRV. Table 8.1. shows results obtained

SAQ 8.3.

In the construction of PRV with deletions in the Us region several enzymes were used and several plasmids constructed. Answer the following questions concerning these enzymes and plasmids.

1) Where does the restriction enzyme *Hind* III cut the genome of PRV?

2) Using the drawing of plasmid pN3HB below, which restriction enzymes should be used to remove the major part of the Us region of PRV?

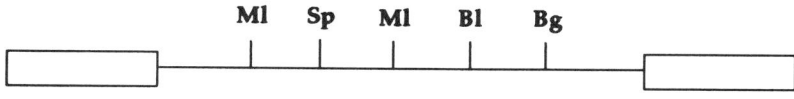

Ml   Sp   Ml   Bl   Bg

3) Using the map described in 2) which restriction enzyme(s) could be used to cut the sequence in just one place?

4) Which of the following enzymes could be used to remove nucleotides from the end of a linear molecule of DNA: *Bal*31, *Bam*HI, T4 polymerase, T4 ligase, *Hind* III?

5) Which of the following plasmids is the largest: pHB2.1; pHB2.4 or pHB2.8.?

with 6-7 week-old BALB/C mice (8 mice per group) inoculated subcutaneously with 10^6 plaque-forming units of PRV.

Virus	Mouse MTD (hours)
NIA3	50±4
NIA4	170±29
2.4 N3A	73±7
2.8 N3A	60±8.5

Table 8.1. Virulence tests of PRV mutants in mice (see text for details).

The results show that both 2.4 N3A and 2.8 N3A give a significantly longer MTD than wild-type PRV, indicating that both deletions influence virulence. This justified a more complicated and expensive experiment with the natural host, the pig.

Groups of five ten-week-old pigs were inoculated intranasally with 10^5 plaque-forming units of the different derivatives of PRV and Table 8.2. summarises the results obtained.

Virus	Deaths (out of 5)	Growth Arrest	Fever (days)	Virus POS	Excretion Mean Titre	Antibody Titre
NIA3	2	9	4	90	3.64	ND
NIA4	0	0	0	ND	ND	1.13
2.4 N3A	0	0	0	80	2.65	1.65
2.8 N3A	0	0	0	50	2.20	1.62

Table 8.2. Effects of inoculating pigs with various strains of PRV (see text for details of column headings)

growth arrest

Growth arrest is the number of days needed to regain the groups mean weight of post-inoculation day 3 (PID). Virus was assayed daily in oropharyngeal fluid (OPF) and POS indicates the percentage of positive samples. The mean titre is expressed as Log_{10} pfu/ml. The neutralising antibody titre on PID 21 (post inoculation day 21) is given as Log_{10} of the reciprocal of the final serum dilution that inhibited cytopathic effects in 50% of the cell cultures. Again, it is clear that the two deletion mutants of NIA3 exhibit reduced virulence.

The key question that follows such a study, is whether or not exposure to the attenuated mutant strain provides immunity against wild-type PRV. Do the antibody titres obtained provide adequate protection? The answer to that question was yes for pigs vaccinated with 2.4 N3A or 2.8 N3A.

8.9. Construction of an additional deletion in the TK gene

∏ Why is a second deletion required?

deletion of
thymidine
kinase gene

Despite the success of 2.4 N3A and 2.8 N3A in vaccinating pigs, it became clear that young piglets (less than one week old) experience too many adverse symptoms for either strain to be acceptable. It was therefore decided to construct an additional deletion in the thymidine kinase (TK) gene of 2.4 N3A. Thymidine kinase is not absolutely essential but virus lacking the enzyme is less virulent *in vivo*. In such cases the virus is dependent on the host cells' thymidine kinase.

latent state in
nerve cells

A second reason for deleting the TK gene is that it should render the virus much less likely to become reactivated from the latent state in nerve cells. A general problem in eradicating Herpes viruses is their ability to become reactivated after remaining latent in nerve tissue, sometimes for many years. Future reactivation of the virus is always a possibility, as with Herpes simplex 1, the causative agent of cold sores in man.

∏ Why should deletion of the TK gene render the virus less likely to become reactivated from the latent state?

Nerve cells do not replicate and so the demand for the substrates for DNA synthesis is extremely low (ie they have low levels of enzymes involved in DNA synthesis). In consequence, PRV lacking thymidine kinase should not be able to replicate in nerve tissue, greatly diminishing the possibility of reactivation of latent infection with its unpredictable consequences for the future.

8.9.1. Construction of the TK mutant

A restriction enzyme fragment containing almost the whole of the TK gene was isolated from wild type (NIA3) virus. An *Eco*RI site was introduced into this fragment. This fragment was purified and transfected with 2.4 N3A DNA into porcine kidney cells.

∏ How were recombinant viruses carrying both deletions in the TK gene and in the Us region isolated? The key is to think of the activity of thymidine kinase and to make use of this.

bromodeoxy-
uridine
selection Thymidine kinase will also phosphorylate the nucleoside base bromodeoxyuridine which is then incorporated into DNA during DNA synthesis. The incorporation of bromodeoxyuridine into DNA leads to incorporation of wrong bases during second strand replication of DNA. This ultimately results in the random inactivation of genes. Bromodeoxyuridine can therefore be used to select against systems with active thymidine kinase. The selection procedure is illustrated in Figure 8.12.

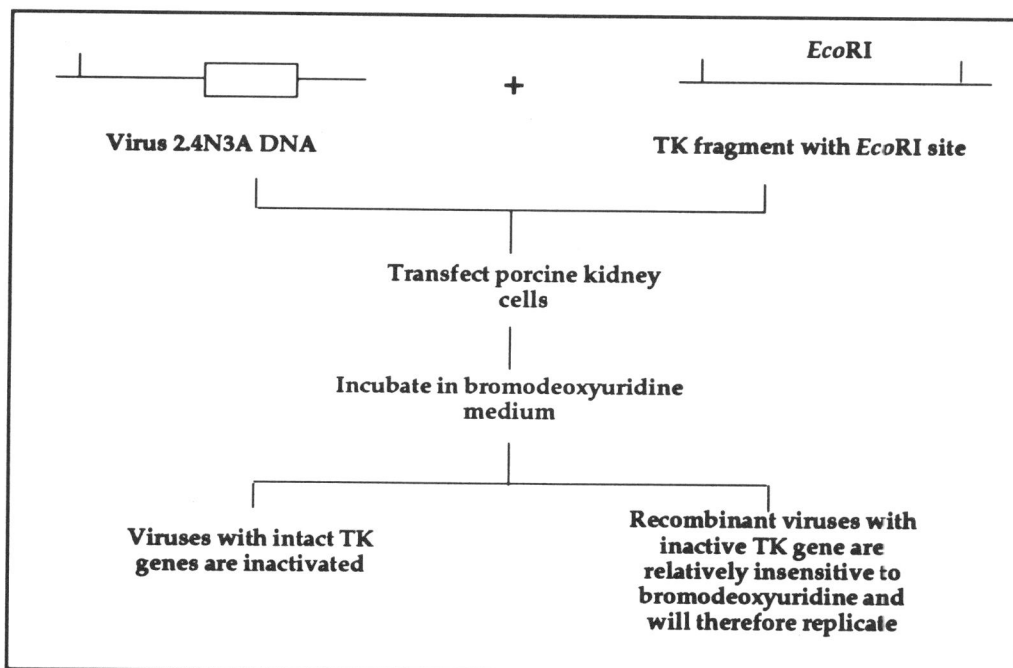

Figure 8.12. Selection of PVR carrying a double deletion

One TK mutant was further manipulated by cutting the DNA with *Eco*RI and transfecting cells with both fragments. In some cases fusion of the two fragments after co-transfection creates short deletions around the fusion site. This has to be tested by sequencing the DNA around the fusion site. A mutant (TK-783) with a deletion of 19 bp in the TK gene was chosen for use as a possible vaccine

8.10. Biological characterisation of the 2.4 N3A TK⁻ 783 mutant

This section illustrates two important things about the 'design' of a virus to be used as a vaccine: the amount of work that has to be done after the construction of a recombinant virus and the extent of the difference there can be between a wild-type virus and a recombinant.

8.10.1. Absence of the gI gene

no gI antibodies Restriction analysis of 2.4 N3A had suggested that the deletion had removed the gI gene. This was confirmed by showing that pigs vaccinated with 783 (ie 2.4 N3A TK⁻ 783) developed a good immune response to the virus but without any antibodies to gI.

8.10.2. Absence of thymidine kinase

∏ Can you suggest two ways of demonstrating the absence of an active thymidine kinase gene in 783? (Think about the way in which viruses with inactive TK genes can be isolated. You could also consider the metabolic activity of thymidine kinase).

BdUr The researchers checked for the absence of thymidine kinase (TK) activity in this strain by titration of the virus in the presence of bromodeoxyuridine (BdUr) in TK-negative mouse fibroblasts. The thymidine analogue is incorporated into the DNA of TK-positive viruses and reduces the titre because of its mutagenic effect. Table 8.3. shows that the titres of 783 is unaffected by BdUr whereas a TK-positive control strain in significantly reduced.

Virus	Virus Titre ($\log_{10}$)	
	Without BdUr	**With BdUr**
TK⁺ control	4.20	3.00
783	5.60	5.65

Table 8.3. Effects on virus titres of incubating TK⁺ and TK⁻ viruses with and without Bromodeoxyuridine

[¹⁴C] thymidine In the second method the virus was grown in the presence of [¹⁴C] thymidine. Incorporation of ¹⁴C into the DNA of replicating virus requires TK activity and this is assessed by the autoradiography of cells infected by PRV. The result for 783, by comparison with a TK⁺ strain, was negative.

8.10.3. Virulence, latency and spreading

∏ Using the three properties of virulence, latency and spreading, write down what you think would be desirable features of a virus to be used as a vaccine. Now compare your answer with the following:

virulence The virulence of a potent vaccine strain is an important consideration in determining how useful it could be. If it induces clear pathological effects, it would be unacceptable; if too avirulent it would not challenge the immune system sufficiently to confer protection against infection by wild-type virus. A 'good' vaccine strain must achieve a balance between the two extremes.

reactivation Furthermore, a strain suitable for vaccination should have lost the ability to become reactivated from the latent state.

Finally, a vaccine strain should only spread from animal to animal to a very limited extent, if at all.

Virulence, latency and spreading of 783 were tested together using young piglets from a pathogen-free herd. Results were as follows:

- virulence: no serious clinical symptoms were observed;

- latency: this was tested for by using prednisolone to suppress the immune system, thereby creating the appropriate environment for reactivation of any latent virus to occur. 783 could not be reactivated by this treatment, suggesting it is incapable of being reactivated from the latent state in young pigs;

- spreading: infection of control animals which had been allowed to come into close contact with vaccinated animals, occurred only to a limited extent, and in some experiments not at all.

8.10.4. Reversion to virulence

As will be made clear later, it is virtually impossible for a double deletion mutant like 783 to become as virulent as the wild type virus or even a single mutant virus. Nevertheless, a test for reversion to virulence must be carried out. After serial passage of 783 through four animals, there was no evidence for any increase in virulence.

8.10.5. Safety of 783 in various conditions

The safety of strain 783 was checked in pigs of different ages and in other farm animals. The results obtained may be summarised as follows:

- 783 did not cause any adverse effects in 11 out of 12 sows vaccinated during pregnancy and there was no evidence for transmission of the virus across the placenta

- yorkshire pigs at an age of 4 weeks were vaccinated with a 32-fold overdose of PRV 783 (note: the European Pharmacopoeia only requires a 10-fold overdose) and their progress compared to twelve control animals. Four of the vaccinated pigs showed very minor, transient swelling at the site of injection but otherwise the overdose had no obvious effects on the health and growth of the animals;

- the response of cattle to the vaccine strain 783 was tested in calves. It was not infectious when administered intranasally. Animals injected intramuscularly did develop antibodies;

- analogous results to those described for cattle were obtained in experiments with sheep and the same conclusion drawn (i.e. the virus is safe).

SAQ 8.4.

1) Which of the following viruses will be sensitive to bromodeoxyuridine (BdUr): Wild type PRV (strain NIA3), strain 2.4 N3A, strain 2.8 N3A strain, 2.4 N3A TK-783?

2) Which of the strains indicated in 1) is likely to lead to the least incorporation of ^{14}C-thymidine into DNA?

3) If a single member of a herd of pigs is inoculated with NIA3, what would be the likely result? What would be the result if a single member of a similar herd was inoculated with 2.4 N3A TK 783?

8.10.6. Efficacy of the 783 strain

vaccine testing
efficacy = low
doses
safety = high
doses

A 'safe' virus would be of little use as a vaccine if it did not trigger an immune response sufficient to protect the host against infection with the wild type strain. It is important to note that the efficacy of a vaccine is generally tested with low doses of virus, in contrast to evaluation of safety which often involves high doses.

The efficacy of 783 was tested in 10-week old pigs; three weeks after vaccination these animals and ten others which had not been vaccinated were challenged with 10^5 pfu of the virulent PRV strain NIA3. The results are summarised in Table 8.4. and they clearly indicate the value of 783 as a vaccine even at the low dose used (10^5 TCID 50 per animal).

Note that only slight symptoms (vomiting/diarrhoea) were observed over a limited period with vaccinated pigs compared with the control group. Respiratory and nervous symptoms were virtually absent from the vaccinated group.

∏ Did vaccinated pigs show any depression of growth or any elevation of temperature? How did this compare with the control group?

8.11. The safety of genetically engineered vaccines

The selection of viruses with reduced virulence (attenuated strains) has been the traditional route for producing vaccines. The process of attenuation depends on mutations in genes which are not essential for replication of the virus.

Until recently, it was not possible to control the production of potentially useful mutants but the techniques of molecular genetics now make this possible, as we have seen in the construction of PRV 783. Since this is a new development we should explore the likelihood of reversion to virulent forms. The likelihood of reversion to wild-type virulence must always be evaluated whatever the origin of the mutation. The basic question relates to the nature of the mutation responsible for attenuation. Was it a point mutation or a deletion?

∏ Are vaccines based on point mutations safe? (Answer yes or no and think up a reason before reading on).

reversion   A point mutation is the replacement of one base by another. Some do not result in any phenotypic changes, although many do. There is always a (low) probability of direct reversion to wild-type, that is a second mutation restoring the original base (reversion). If a gene is a thousand nucleotides long, the probability of a mutation occurring at a defined (ie the mutant) site must be 10^3 times lower than the probability of the original point mutation. Nevertheless with the very large number of viral particles being produced such an event may occur.

However such a 'reverted' virus would be only one amongst many avirulent viruses. Before such a virulent revertant could take over, the avirulent viruses would outgrow the revertant and stimulate an immune response. If reversion to virulence occurs during the process of growing the virus *in vitro*, the revertant may out-grow the vaccine strain.

					Number of days scored				
Group	Pig No	SN titre	Temp 40°C	SE*	RS*	NS*	VE*	GD*	Death at Day
Vaccinated Pigs									
	942	11	1	1	0	0	4	0	
	946	16	2	2	0	0	nd	0	
	947	11	3	1	0	0	nd	1	
	948	22	4	1	0	0	5	0	
	952	45	1	1	0	0	3	0	
	955	3	5	1	1	0	nd	0	
	956	3	6	2	0	0	5	0	
	957	11	4	1	0	0	nd	0	
	960	11	4	1	1	0	4	0	
	Mean		3.3	1.2	0.2	0	4.2	0.1	
Control pigs									
	943		8	11	2	2	8	>14	
	944		5	12	0	0	nd	0	
	945		4	4	3	1	5	na	7
	949		5	5	1	8	nd	>14	
	950		8	7	3	4	8	na	12
	953		5	4	1	3	nd	na	9
	954		4	12	3	3	nd	11	
	958		9	11	3	3	9	>14	
	961		5	4	2	2	nd	na	7
	962		5	4	3	2	6	na	6
	Mean		5.8	7.4	2.1	2.8	7.2	>10.6	

Table 8.4. Results of challenge-infection experiments

* SE = Side effects including inappetance, lassitude, vomiting, diarrhoea

RS = Respiratory symptoms

NS = Nervous symptoms

VE = Virus excretion

GD = Growth depression

This would, of course, be a very dangerous situation since the vaccine prepared from this would consist of virulent viruses!!!

When a virus is selected for mutations in two genes, the chance of reversion to virulence is equal to the chance of reversion of one gene multiplied by the chance of reversion of the other, that is practically zero. Thus we conculde that vaccines based on a single point mutation carry some risk.

Π Are vaccines based on deletion mutants safer? (Again answer yes or no before moving on).

Genetic material, even one nucleotide, once deleted cannot be regained; hence the attraction of using deletion mutations in the design of vaccine strains. The introduction of a second deletion (as in 783) provides an even greater degree of safety.

Π Can viruses with deletions recombine with wild viruses and thereby re-establish their virulence? If so, does this present a health risk? (Again try to reach your own conclusion before reading on).

coinfection   There is one mechanism that could allow the deleted region of a gene to be restored and that is homologous recombination between a vaccine virus and the wild-type. For this to happen, both viruses must infect the same cell at the same time. The chance of this happening is clearly very small. A gap of just a few hours between infection with the two viruses would be sufficient to block any interaction because of the alteration in the metabolism of a cell after the first infection. Even if homologous recombination does occur, it cannot generate a virus more virulent than the wild-type that is required to be present in the first place!

The overall conclusion is that double deletion mutants like PRV 783 are safe to use as vaccines.

8.12. Development of the gI test kit: towards eradication of Aujeszky's disease

8.12.1. Why a gI test kit was developed

We have seen that an attenuated virus can be used as a vaccine, stimulating the production of antibodies against antigenic determinants of the virus. Some of these neutralising epitopes can protect against subsequent infection by the wild-type virus.

It is currently forbidden by law to export pigs with antibodies to PRV, because pigs may carry wild-type virus for many years in a latent form. If reactivated, the animal concerned becomes a focus for infection and reintroduction of the disease. However, vaccination with 783 also creates a seropositive response.

Π Is it possible to distinguish animals that have survived wild-type PRV infection from those vaccinated with 783? (Think about the antigenic differences between wild type PRV and the vaccine strain 783).

wild-type   A way has been found to do this using the effect of the deletion mutation introduced
specific antigen   into the gI gene. The gI protein is a surface protein and is an antigen of wild-type virus. This is not the case for the deletion mutant and so pigs vaccinated with the deletion

mutant, PRV 783 can be distinguished from animals that have been infected with wild-type PRV by the fact that they do not possess antibodies against the gI protein. By testing serum samples for antibodies against PRV in general and for antibodies specifically against gI, it is possible to tell if an animal has been vaccinated or has suffered a field infection.

8.12.2. An eradication programme based on the gI test

As it is possible to discriminate between vaccinated and field infected pigs, it is possible to design a programme for eradicating AD This would involve large-scale vaccination with PRV 783; after a few years only pigs infected with a field strain of the virus will be gI positive and these animals could be eliminated. A fail-safe test for gI antibodies is an absolute pre-requisite for such an eradication to be successful and considerable effort has been put in to making sure a reliable test is available. The combination of a gI-negative mutant and a diagnostic gI test kit provides the potential for eradicating Aujeszky's disease.

SAQ 8.5.

A disease is known to be caused by viruses which can exist in a variety of slightly different forms. All are enveloped viruses and their capsids consists of approximately 160 capsomers. They are also known to be double stranded DNA viruses. Genetic and biochemical analysis of these viruses have revealed the following:

a) They all produce an identical glycoprotein (Gly 1) which appears to be involved in the maturation of the viruses within the host cells. Gly 1 is not a component of the envelope or the capsid. Gly 1 does not appear to be released by the host cell and it does not appear to be essential;

b) The capsomer proteins from different viruses show many similarities but antibodies raised against some of these proteins show little affinity for the proteins derived from the capsids of others;

c) All of the envelopes of these viruses contain a mixture of glycoproteins. SDS-PAGE analysis of these glycoprotein mixtures indicates many similarities between the envelopes of the various viruses. These glycoproteins have not however been isolated and characterized;

d) The gene for Gly 1 has been mapped onto a short (3 kbp) fragment of DNA produced by treating viral DNA with the restriction enzyme EcoRI. Incubation of viral DNA with this enzyme only produces 3 fragments;

Using this information, choose from the list below, the strategy most likely to lead to the production of a vaccine for the disease.

Strategy I - Clone the EcoRI fragment carrying the Gly 1 gene into an expression vector and use this to produce a lot of the glycoprotein which can be used as a vaccine.

Strategy II - Continue the genetic mapping to trace the capsomer protein genes. Then isolate these and clone them in an expression vector in order to produce capsomer proteins to use as vaccines.

Strategy III - use EcoRI to specifically delete Gly 1 genes. Use a re-constructed virus using the remaining two EcoRI fragments as the vaccine.

Summary and objectives

This case study illustrates the close relationship there must be between different biological disciplines for progress to be made to controlling a virus like the causative agent of Aujeszky's disease. The case study has illustrated the need for a detailed knowledge both of the natural history of the virus and its molecular structure. It also described how deletion mutants may be produced by recombinant DNA technology and how the vaccine composed of viruses carrying deletion mutants was tested.

Now that you have completed this chapter you should be able to:-

* describe in general terms the structure of PRV and the arrangement of its genome especially in relation to inverted repeat sequences, Us region and the genes which code for glycoproteins;

* explain the advantages of using attenuated viruses rather than inactivated viruses as vaccines;

* describe the sequence of steps which led to the production of derivatives of PRV which carry deletion mutants;

* explain why the deletion of thymidine kinase genes from PRV resulted in greater attenuation of the virus and decreased the chance of the vaccine carrying this deletion taking up or being activated from a latent state;

* explain how antibodies against gI provide a mechanism for identifying animals carrying wild type (field) PRV.

9

Case Study: Myoscint - A monoclonal antibody preparation used for cardiac imaging

9.1. Introduction   226

9.2. Development of Myoscint   228

9.3. Manufacture of Myoscint   231

9.4. Quality control   236

9.5. Toxicological and pharmacological tests   240

9.6. The clinical trials with Myoscint   243

Summary and objectives   247

Case Study: Myoscint - A monoclonal antibody preparation used for cardiac imaging

9.1. Introduction

In this case study, we explore the development and evaluation of another system based on the production of a monoclonal antibody. In this case, however the antibody is used as an *in vivo* diagnostic tool. We first explore the reasons why such a system is needed before examining the development and evaluation of the product.

9.1.1. Why is a monoclonal antibody needed for *in vivo* diagnosis?

scintigraphic
image

target specific

In nuclear medicine, a trace amount of a gamma-ray-emitting radioactive substance is administered to the patient and allowed to distribute within the body. The location of the distributed radioactive substance is then determined by obtaining what is called a scintigraphic image, by means of a gamma camera (similar to an X-ray camera). The radioactive substance is called a radiopharmaceutical. It is usually administered intravenously, and it is chosen for its particular characteristic of localising at a site of disease or physiological abnormality. In this way the image of the distribution of the radiopharmaceutical in the body can give a physician useful diagnostic information regarding the presence, location or extent of the pathology of interest. In order to provide diagnostically useful images, the radiopharmaceutical must concentrate to a greater extent in the target tissue than in surrounding or background tissues. In other words, it must show specificity for the target.

Antibodies demonstrate exquisite specificity for the antigen against which they are raised. Polyclonal antibodies are the immunoglobulin fraction of serum from an animal immunised with a particular antigen. Because they comprise multiple different antibodies they may show some cross-reactivity with non-target tissues, ie non-specificity. In comparison, a monoclonal antibody is the only immunoglobulin synthesised and secreted by a particular hybridoma clone (cell-line) and therefore retains maximum specificity for the target antigen.

The antigen specificity of radiolabelled polyclonal antibodies was early exploited in the *in vitro* diagnostic procedure called radioimmunoassay (RIA). Radiolabelled polyclonal antibodies were subsequently examined as radiopharmaceuticals in anticipation of their specificity for their antigen *in vivo*. However with the advent of hybridoma technology it was quickly seen that radiolabelled monoclonal antibodies offered superior prospects as diagnostic imaging agents. Current hybridoma technology allows the relatively routine generation of hybridoma cell-lines which has allowed the preparation of antibodies with specificities for a wide range of antigens.

Monoclonal antibodies, radiolabelled with gamma-emitting isotopes such as iodine-131, indium-111 or technetium-99m have been investigated in the detection of various pathologies such as cancer and venous thrombi. Studies, both in animals and in man, have shown that only a small portion of the injected dose actually localises and binds to the target. The majority of the injected dose is distributed throughout the body through non-specific uptake (not necessarily non-specific binding) of the protein in extra-vascular spaces and excretory organs. However, in most cases the target-uptake has been sufficient to provide target-to-background contrast and subsequent scintigraphic images.

Π See if you can list three or four requirements of a radiopharmaceutical based on a monoclonal antibody? (Do this before reading on. Then check your list with that we have produced).

The main requirements include:

• the selection of an antibody;

• choice of antibody type;

• choosing whether to use the whole immunoglobulin or a fragment;

• selecting a suitable radionuclide and a method for its attachment to protein.

IgG
F(ab)₂
Fab

The selection of a suitable antibody against a particular target involves assessment of affinity for the desired target and screening for cross-reactivity for non-target tissue. Gammaglobulins (IgGs) are usually selected for radiopharmaceutical development; IgMs are too large and other types are associated with undesirable physiological reactions. The selection of whole immunoglobulin or fragment depends on whether bivalency (with respect to antigen-binding) is required, in which case whole IgGs or F(ab')₂ fragments are selected, or whether molecular size is a limitation (50 000 dalton Fab fragments diffuse faster than larger fragments), or whether blood clearance needs to be fast (to lower blood background activity) or slow (to allow more of the circulating antibody to accumulate at the target). The relationship between whole IgGs and F(ab')₂ and Fab fragments is shown in Figure 9.1. In selecting a suitable radionuclide, the gamma emission(s), should be in an energy range that maximises gamma camera imaging efficiency and the half-life should be matched to the anticipated time required for target-to-non-target contrast to be adequate for imaging. We will return to this aspect in later sections.

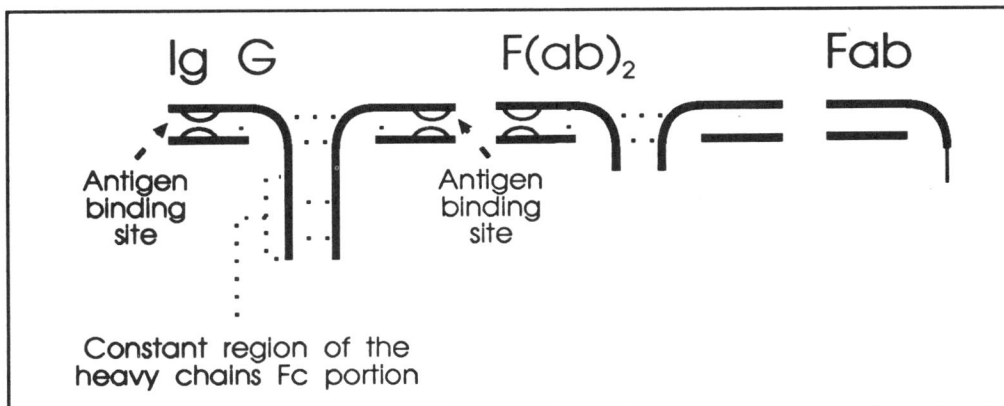

Figure 9.1. Stylized IgG, F(ab')₂ and Fab fragments

These are the major factors which were considered in the development of an antibody that would localise and provide an image of myocardial infarction.

9.2. Development of Myoscint

traditional
diagnosis

Diagnosis of myocardial infarction is commonly based on initial clinical presentation, characteristic chest pain, electrocardiographic changes and elevations in serum creatine kinase. However each of these methods has its limitations with respect to defining accurately the location and extent of the infarct. Non-invasive nuclear medicine procedures have been developed to identify regions of compromised myocardium based on the localisation and external scintigraphic imaging of radiotracers. For example, (Tc-99m) technetium-stannous pyrophosphate has been used to image myocardial infarcts and (Tl-201) thallous chloride has been used to define regions of deficient myocardial perfusion. However, the imaging of deficiencies in myocardial perfusion does not specifically define regions of myocardial necrosis and (Tc-99m) technetium-stannous pyrophosphate overestimates the size of such regions (extent of the infarct). The detection of myocardial necrosis can be based on the accompanying loss of integrity of the myocardial cell (myocyte) membrane and exposure of the cell contents to the extracellular fluid. The diagnostic appearance of creatine kinase in the blood is based on this phenomenon.

the concept of
Myoscint

Myocardial cell myosin is exposed to the extracellular fluid upon myocardial cell membrane disruption resulting from the myocyte necrosis. Because myosin is a large, insoluble protein it remains within the necrotic cell.

Detection of the location of the exposed cardiac myosin therefore defines the location of necrotic (irreversibly damaged) myocyte and hence defines the extent of the myocardial infarct. As mentioned above, antibodies have an inherent and exquisite specificity for the antigen against which they are raised. Antibodies to cardiac myosin bind to the exposed myosin in regions of myocardial infarction but do not bind to healthy myocytes as the antibodies do not cross the undamaged cell membrane.

model system

Polyclonal antibodies to canine cardiac myosin have been prepared, radiolabelled with radioactive iodine and shown to localise specifically in infarcted myocardium in a canine model. This same study used purified, radioiodinated polyclonal F(ab')₂ and found an increased uptake compared to IgG. Fab fragments (ie monovalent antibody fragments) of polyclonal antimyosin antibodies were subsequently prepared and it was observed that the radioiodinated Fab fragments showed higher infarct-to-normal myocardium ratios than the F(ab')₂ in a canine acute infarct model.

Production of antimyosin antibodies was improved by the preparation of a murine monoclonal hybridoma cell line by fusion of a murine myeloma cell line with a murine splenocyte raised against human cardiac myosin. This cell line (coded as R11D10) allows the preparation of large quantities of pure antimyosin antibody.

Π   What properties should the radionucleotide have which will be used to label the antibodies? (Think about how the radiation will be detected, safety of the recipient and how the radionucleotide will be attached to the antibody).

Indium III

Although most of the early studies using polyclonal antimyosin antibodies were performed using I-125 and I-131, these radionuclides are not optimal for gamma scintigraphy. I-125 has too low an emission energy (35 keV), while I-131 has a gamma emission energy (364 keV) that is too high for good resolution and efficient detection. Moreover, I-131 is a beta-emitter which adds significantly to patient-absorbed radiation dose. I-123 has a gamma emission of appropriate energy (159 keV), but is not widely available. Tc-99m is widely available and has a gamma emission optimal (140 keV) for gamma scintigraphy, but reliable methods for producing a stable Tc-99m labelled antibody were not available at the time of the development of Myoscint.

The monoclonal antibody produced called Mifarmonab had a kappa type light chain and with a particular type of heavy chain (isotype IgG2a). See Figure 9.2.

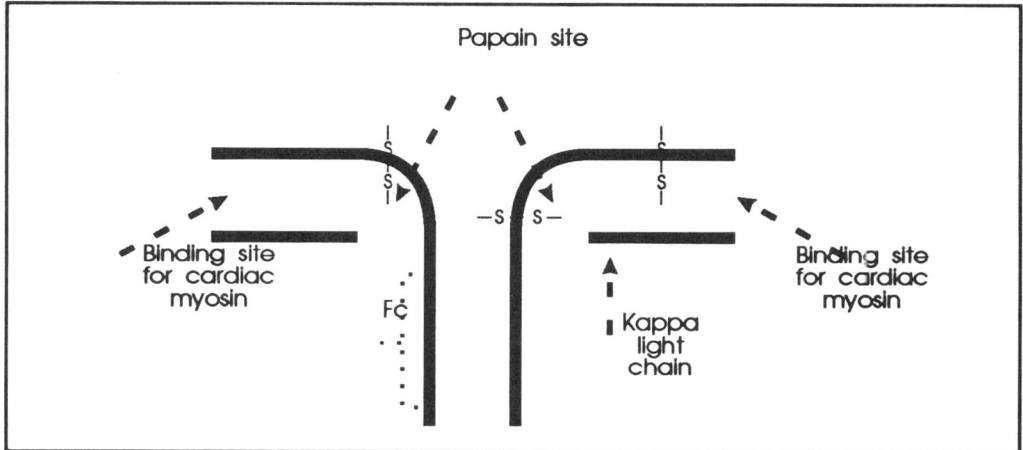

Figure 9.2. Stylized structure of Mifarmonab
Note: Fc or heavy chain constant region of Mifarmonab is of the α2a type, so this antibody belongs to the isotype IgG2a. The sites for papain - Papain is a proteolytic enzyme which will cut the two heavy chains of IgG at the positions indicated to release an Fc fragment and two monovalent Fab fragments. The structure drawn is highly stylized and many of the -S-S- bridges have been omitted for simplicity

Indium-111 was chosen as the radionuclide because of its half-life (2.81 days), the two medium energy gamma emission peaks (172 keV and 247 keV) and the availability of a reliable method of radiolabelling proteins through the chelator diethylenetriamine penta-acetic acid (DTPA). Diethylenetriamine penta-acetic acid, by virtue of its carboxylic acid and amino groups is a good chelating agent which strongly chelates and retains Indium. The half-life of Indium-111 matches the anticipated time required to obtain an appropriate contrast between target and non target tissue. By use of its bicyclic anhydride derivative, diethylenetriamine penta-acetic acid (DTPA) can easily be coupled to the antibody forming a stable amide bond with accessible amino groups. This coupling method was shown to be efficient (up to 70%), fast (minutes) and can be carried out under aqueous conditions at near neutral pH (7.0 - 8.5). The sequence is shown in Figure 9.3. The indium-DTPA complex has an equilibrium constant estimated to be 10^{28}. Using a competitive binding assay, it was shown that at least 2.75 DTPAs could be attached to each Mifarmonab Fab fragment with essentially no loss of immunoreactivity.

formulation of a kit

It was anticipated that the preparation of the diagnostic radiopharmaceutical indium-111 Mifarmonab Fab-DTPA would be performed by the end user (clinic) using a Mifarmonab Fab-DTPA kit (Myoscint) and (In-111) indium chloride supplied commercially. This required that a radiolabelling procedure be developed that was suitable for reliable, routine operation in the field. Solutions of indium(III) must be kept

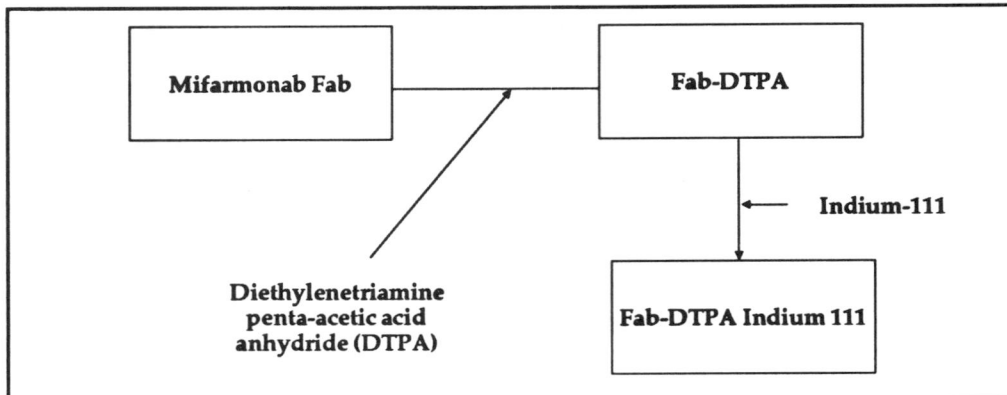

Figure 9.3. The labelling of Mifarmonab Fab fragments with Indium-111

acidic (pH 3.4) in order to prevent the formation of indium hydroxides, which may form colloids or precipitate out of solution. However, the presence of organic anions such as citrate or acetate can complex with indium in solution and prevent precipitation at somewhat higher pH. In the presence of 0.1 M acetate, less than 1% precipitation was observed up to pH 4.9: with citrate, similar results were reported with the pH as high as 7.0. This meant that mild conditions could be used, such that indium-111 could rapidly and reproducibly form a complex with the DTPA attached to protein, and thus avoid the potential for protein denaturation. Thus it was decided that Myoscint would be based on the Fab fragment of Mifarmonab and that it would be radiolabelled with indium-111 in citrate buffer using the chelator DTPA conjugated to the protein by means of its bicyclic anhydride.

SAQ 9.1.

In selecting a radionucleotide to use with Mifarmonab Fab, which of the following criteria were deemed important in the final selection of indium-111?

1) Indium-111 has two emission peaks

2) The energy of radiation of indium-111 is close to the maximum sensitivity of gamma scintigraphy equipment

3) The energy of radiation of indium-111 is sufficiently high to be detected by gamma scintigraphy

4) Indium-111 is highly reactive

5) Stable conjugates between Mifarmonab and other potential radionucleotides were not available

9.2.1. Development of the manufacture of Myoscint

The manufacture of Myoscint has involved meeting and overcoming several key challenges. First was the challenge of growing the mammalian hybridoma cell-line *in vitro* reproducibly, efficiently and economically, such that an adequate and reliable production of Mifarmonab was realised. Second was the challenge to develop a series of processing steps that would isolate and purify the whole immunoglobulin from the cell culture supernatant, convert it to this Fab fragment, purify the Fab fragment, couple the indium-111 chelator DTPA conjugate in a manner that was efficient, reproducible and that gave purified Fab-DTPA suitable for injection. Third was the challenge to formulate the Mifarmonab Fab-DTP in a kit form that would allow the end user to prepare reliably and reproducibly the diagnostic radiopharmaceutical indium-111 Mifarmonab Fab-DTPA (In-111 Myoscint).

The next section describes in detail how these challenges were met. Conceptually the scheme is as outlined in Figure 9.4. The main points to notice are:

- continuous-perfusion fermentation (or cell-culture) was selected as the optimum means of allowing the R11D10 hybridoma cells to grow and secrete Mifarmonab reproducibly on a commercially practical scale;

- downstream processing of the cell-culture supernatant was designed to use state-of-the-art bioprocess chromatography and tangential flow ultrafiltration methods to isolate the Mifarmonab, purify its Fab fragment following enzymatic digestion and prepare and purify the Fab-DTPA conjugate. Key components to the isolation/purification sequence were the use of Protein A-Sepharose affinity chromatography and the scrupulous avoidance of trace-metal contamination after the DTPA coupling step;

- extensive formulation development resulted in the selection of a two-vial kit comprising one vial containing Mifarmonab Fab-DTPA in phosphate-buffered saline containing 10% maltose and a second vial containing citrate buffer to facilitate the radiolabelling with indium-111.

9.3. Manufacture of Myoscint

9.3.1. General manufacturing conditions

dedicated facilities

restricted access

Myoscint was developed by Centocor. It is worthwhile examining the manufacturing facilities used by this company. You will recall from chapter 3 that a description of the facilities is an essential part of the licensing file. Centocor has designed and built a manufacturing plant at Centocor B.V., Leiden, The Netherlands that is dedicated to the manufacture of monoclonal antibody-based pharmaceuticals. The plant is equipped with an extensive and elaborate air-handling system that allows the maintenance of ultra-clean conditions in all parts of the plant. Access to the plant is restricted to authorised manufacturing personnel who wear plant-dedicated overgarments, shoes or shoe-coverings and hair-coverings.

Good Manufacturing Practice

Standard Operating Procedures

For sensitive operations such as vial filling, personnel wear complete-covering disposable production garments. All manufacturing personnel are highly trained to operate strictly according to current Good Manufacturing Practice (GMP), all processes are performed using documented Standard Operating Procedures (SOPs) and all stages of manufacture are performed according to, and recorded on, batch records.

All equipment is carefully maintained, all use of equipment is logged and it is frequently tested to ensure correct performance. Equipment that comes into contact with the product is scrupulously cleaned and, where possible, sanitised before use. Equipment eluate is tested before use to ensure minimal endotoxin burden. All incoming raw materials are controlled and tested by the Quality Control Department (QC) as is the

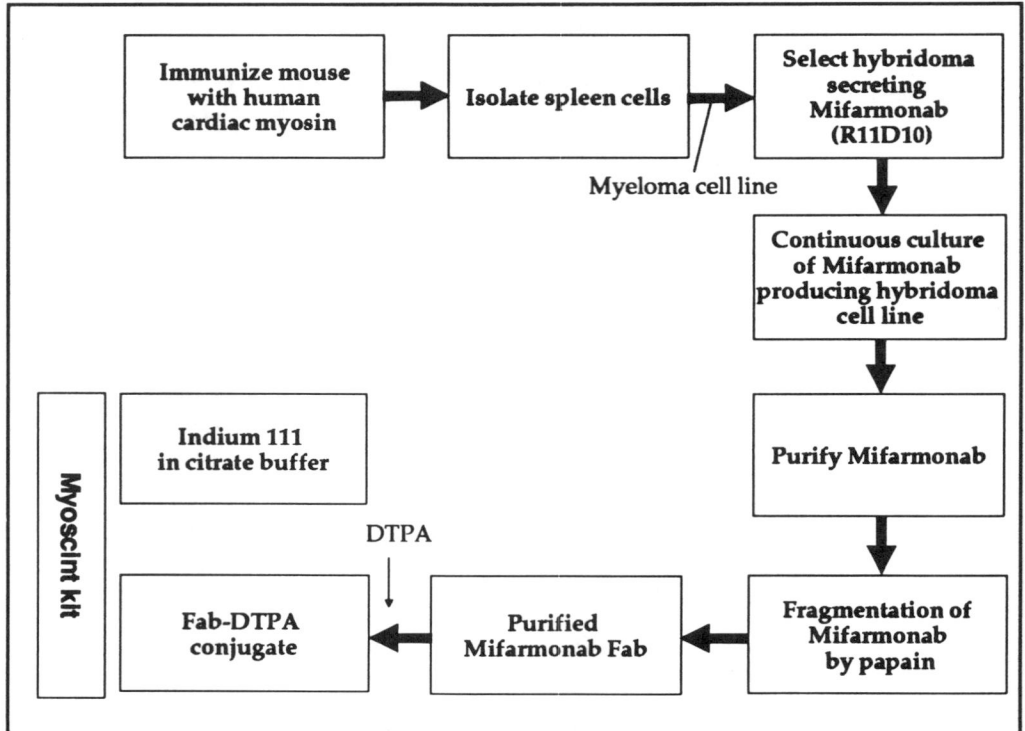

Figure 9.4. An over-view of the stages in the manufacture of Myoscint

in-house 'Water For Injection' (WFI) system which produces the WFI used in all stages of manufacture.

9.3.2. Cell bank system

Master Cell Bank

Manufacturer's Working Cell Bank

To assure a reliable and reproducible source of cells of the cell-line R11D10 that secretes murine antimyosin antibody, Mifarmonab, a cell bank system has been established comprising a Master Cell Bank (MCB) and a Manufacturer's Working Cell Bank (MWCB). The MCB comprises the cells of an individual passage of the cell line that has been aliquoted into multiple vials which are stored frozen in liquid nitrogen. The MCB is the genesis from which all future cells of this cell line will originate. As will be described below under Quality control, the MCB has been extensively characterised and tested for undesirable contaminants (viruses, mycoplasma, etc).

One vial of the MCB is thawed, the cells are cultured and the culture is expanded and aliquoted into multiple vials to prepare a MWCB which is also stored in a liquid nitrogen freezer. It is the cells of the MWCB that are used to start each production fermentation. The MWCB is also extensively tested for possible contaminants.

Extreme care is taken during the establishment of the cell banks to avoid contamination with cells of other cell-lines or with adventitious organisms. The combination of testing of the MCB and of the MWCB provides high assurance that undesirable contaminants are not harboured by the cells and will not be co-cultivated when the hybridoma cell are grown in large scale production fermentation.

9.3.3. Fermentation

pre-culture

An individual vial of MWCB is selected and expanded in small scale cell culture, termed pre-culture.

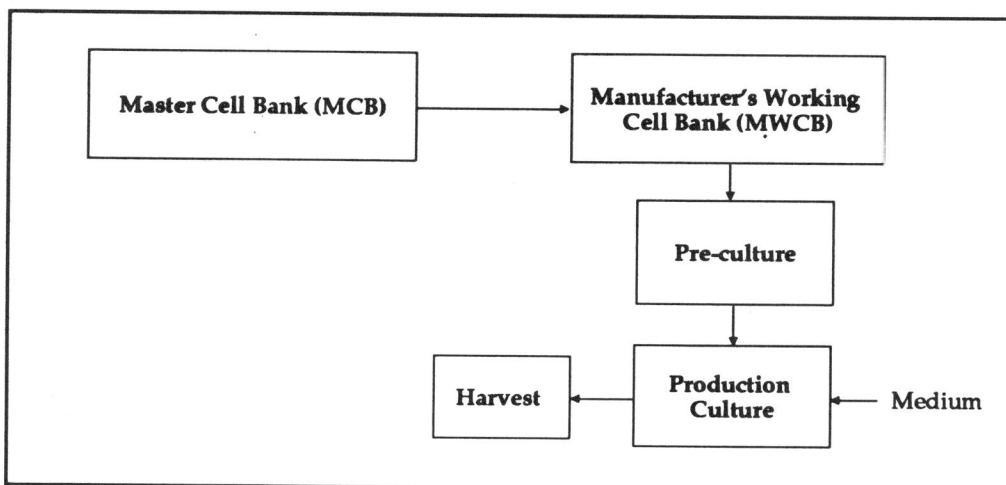

Figure 9.5. The scale up of the R11D10 cell line

perfusion culture

When the cell culture has been expanded sufficiently it is then used to inoculate a production fermenter charged with fresh culture medium. When the cell culture has expanded to a cell density adequate to support production, culture medium is continuously added (perfusion culture) and cell culture supernatant, termed Harvest, is continuously collected at the same rate. The fermentation is automatically and continuously monitored for several key parameters including pH, temperature, dissolved oxygen, and rate of perfusion. Daily aseptic sampling allows the monitoring of cell density, cell viability and antibody concentration. Every 30 days, samples are also taken for testing for murine retrovirus and mycoplasma.

During pre-culture and fermentation extreme care is taken to prevent adventitious contamination of the cell culture by other organisms. The conditions in the culture are such that many micro-organisms and viruses will grow very well. If contamination occurs organisms could grow rapidly and render the culture unfit for further use. The solutions added to the cell culture must therefore be sterile and must be added aseptically.

The harvest is clarified of cells and cell debris and concentrated by a factor of approximately 20 using ultrafiltration devices. This concentration allows more manageable downstream processing. The clarified, concentrated cell supernatant is tested to assure the absence of microbial and viral contamination.

9.3.4. Downstream processing

The downstream processing stage in the manufacture of Myoscint involves a sequence of chromatographic purifications and diafiltrations designed to remove contaminants, such as foreign proteins, DNA, any microbial or viral contamination and substances used during fermentation and downstream processing. It includes the conversion of the intact IgG (Mifarmonab) to the antibody fragment, Mifarmonab Fab-DTPA. At the end of the process a purified, formulated Mifarmonab Fab-DTPA Final Bulk is obtained.

During the process extreme care is taken to avoid microbial contamination. All equipment is scrupulously cleaned and, where possible, sanitised with sodium hydroxide solution. The effluent from each piece of equipment is monitored for endotoxin levels prior to use. The product is $0.22\mu m$ filtered after each step to control potential microbial contamination.

The different steps of the downstream processing are depicted in the flowchart (Figure 9.6.).

Use Figure 9.6. to follow the description of the important steps.

Protein A affinity chromatography

Mifarmonab is isolated from the clarified, concentrated cell supernatant and purified using Protein A affinity chromatography. Protein A is a protein derived from *Staphylococcus aureus* that specifically binds the Fc region of immunoglobulin. In this step, the clarified, concentrated cell supernatant is loaded on the Protein A Sepharose chromatography column. The immunoglobulin is bound and the bulk of the concentrate and its components flow through the column to waste. The bound immunoglobulin is then eluted off the column using a low pH buffer. Eluted Mifarmonab immunoglobulin is greater than 90% pure by gel filtration high performance liquid chromatography (GF-HPLC).

papain

Purified Mifarmonab is then diafiltered (a procedure equivalent to dialysis but using a tangential flow ultrafiltration device) into an appropriate buffer for digestion with papain. Papain is a proteolytic enzyme derived from papaya that requires a free sulphydryl for activity, and which cleaves gammaglobulins to give the monovalent antigen-binding Fab fragments. Pre-activated (using reducing conditions) papain is combined with the purified Mifarmonab. The digestion is monitored by analysing aliquots by GF-HPLC.

iodoacetamide inhibition

When the digestion is complete, iodoacetamide is added to stop (block the activity of) the papain and also to block (prevent disulphide bond formation of) any free sulphydryls exposed on the antibody Fab fragment during or as a result of digestion. The Mifarmonab Fab fragments are isolated and partially purified from the digest again using Protein A affinity chromatography. In this step the digest is loaded onto the Protein A affinity chromatography column and the Mifarmonab Fab and other digestion components flow through the column and are collected. The column binds, and hence removes, residual whole immunoglobulin and Fc-fragments generated by digestion.

anion exchange chromatography

This partially purified Mifarmonab Fab fragment is then further purified by anion exchange chromatography. This step removes residual DNA and endotoxin from the product.

DTPA

The purified Mifarmonab Fab fragment is treated with DTPA bicyclic anhydride to give the Mifarmonab Fab-DTPA conjugate. A sample of the reaction mixture is taken for analysis for number of DTPA's per Fab and the bulk is concentrated using a tangential flow ultrafiltration device. From this stage forward, extreme care is exercised to minimise exposure of the Mifarmonab Fab-DTPA conjugate to trace metals. All buffers that come into contact with the product from this stage forward are treated with ion exchange resin to remove trace-metals.

gel filtration Final Bulk

The concentrated, crude Mifarmonab Fab-DTPA is loaded onto a gel filtration chromatography column and subsequently eluted. Selected fractions are analysed by GF-HPLC. The fractions corresponding to pure, monomeric Fab-DTPA, free of $F(ab')_2$

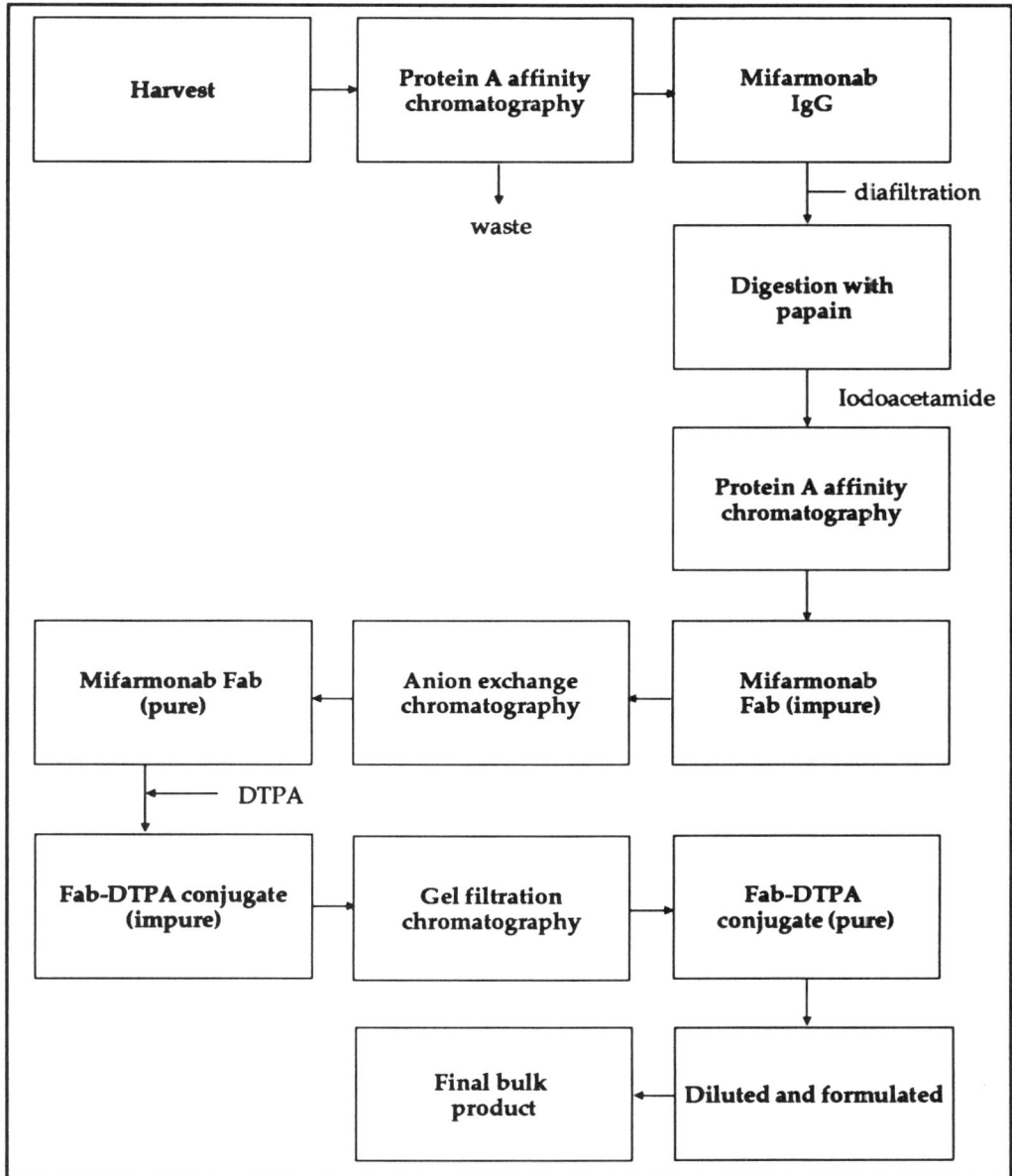

Figure 9.6. Downstream processing of Mifarmonab Fab-DTPA

dimer and free of smaller molecular weight proteins are pooled as purified Mifarmonab Fab-DTPA. The pool of purified Mifarmonab Fab-DTPA is diluted and formulated. The result is Mifarmonab Fab-DTPA Final Bulk containing Mifarmonab Fab-DTPA at a concentration of 0.5 mg/ml in the final formulation of 10% w/v maltose, 0.01 M sodium phosphate, 0.145 M sodium chloride, pH 6.5.

In parallel to the production of Mifarmonab Fab-DTPA Final Bulk, a solution of 0.2 M sodium citrate, pH 5.0 Final Bulk is prepared (from QC tested and released sodium citrate, citric acid and water for injections).

As is described below, many of the final release tests are performed on samples of Mifarmonab Fab DTPA Final Bulk and 0.2 M citrate, pH 5.0 Final Bulk.

9.3.5. Pharmaceutical manufacturing

In this final stage in the manufacture of Myoscint, the Mifarmonab Fab-DTPA Final Bulk solution undergoes a final 0.22μm sterile filtration and is then filled aseptically, 1 ml per vial, into sterilised, depyrogenated DIN 4R Type I glass vials which are closed with sterilised Teflon-faced grey butyl rubber stoppers and sealed with blue flip-off crimp seals.

All filled vials are inspected for defects and for particulate matter and are then stored under quarantine at 2 to 8°C. Samples representing beginning, middle and end of filling are taken for Quality Control testing. Upon successful QC testing the vials are labelled, Myoscint Vial 1, and stored at 2 to 8°C pending final kit assembly.

In parallel, 0.2 M Citrate, pH 5.0 Final Bulk solution is similarly aseptically dispensed into DIN 8R vials, QC tested and labelled 'Myoscint Vial 2'. It is to the contents of these vials that In 111 will be added just prior to use.

Myoscint kit

In the final step, a selected lot of Mifarmonab Fab-DTPA 'Myoscint Vial 1' and a selected lot of 0.2 M citrate, pH 5.0 'Myoscint Vial 2' are packaged as a kit along with the appropriate package insert. The Myoscint kit is stored at 2 to 8°C until being shipped to the customer.

9.4. Quality control

9.4.1. Cell banks

The MCB and MWCB are both characterised for antibody production (secretion), antibody isotype and antibody-antigen binding. These cell banks are also tested for bacterial, fungal, mycoplasmal and viral contaminants which could have arisen from the original cell line or from components of the cell-culture. This testing is extensive, and expensive, but provides assurance that no undesirable contaminants will be introduced with and grow along with the hybridoma cells during production fermentation.

9.4.2. In process

QC of starting materials

As mentioned above, all raw materials including buffer components, chromatography packing and cell culture medium are rigorously tested by QC before being released for use in production.

QC in process

During the actual process, multiple test are performed to control for bioburden, endotoxins, retrovirus and adventitious viruses, mycoplasma and DNA. Protein purity is assessed at several steps using gel-filtration HPLC and antibody isotype is ascertained by immunoelectrophoresis on the Mifarmonab prior to papain digestion. The amount of DTPA coupled to the Fab fragment is also assessed by an in-process test.

9.4.3. Final release tests

The final release tests for Myoscint are compiled from the tests performed on the final vialed products and selected tests performed on the Final Bulks as well as certain tests performed during manufacture. The list of tests, their specifications and stage of testing is presented in Table 9.1.

Test	When test is performed	Specification
Vial 1		
On solution:		
colour	[1]	colourless
appearance	[1]	clear liquid
pH	[1], [2]	6.4 - 6.6
particulate matter	[1]	essentially free from particles
volume uniformity	[1]	mean 1.10 ml; range 1.05 - 1.20 ml
On active constituent:		
protein concentration by OD 280	[1], [2]	0.45 - 0.55 mg/ml
indium-111 incorporation	[1], [2]	≥90% in 10 minutes
immunoreactivity by affinity column assay	[1], [2]	≥84%
purity by gel-filtration HPLC	[2]	≥98% Fab-DTPA
SDS-PAGE	[2]	conforms to standard
isoelectric focusing	[1], [2]	conforms to standard
immunoelectrophoresis	[2]	reacts with anti-Fab and anti-whole mouse serum
DTPA/Fab	[4]	1.25 - 2.25 DTPA/Fab
On other constituents:		
identity: sodium	[2]	conforms
phosphate	[2]	conforms
chloride	[2]	conforms
maltose	[2]	conforms
assay: chloride	[2]	0.130 - 0.160 M
phosphate	[2]	9 - 11 mM
maltose	[2]	9 - 11%
Impurities		
retrovirus (S+L), (XC)	[5]	retrovirus free
mycoplasma	[5]	mycoplasma free
adventitious virus	[6]	virus free
sterility	[1], [2]	no growth
pyrogens*	[1]	non-pyrogenic
general safety*	[1]	passes test
endotoxins by QCLAL	[2]	<20 EU/mg
gentamicin	[2]	non detectable
DNA	[3]	<pg/mg

Test	When test is performed	Specification
Vial 2		
colour	[1]	colourless
appearance	[1]	clear liquid
particulate matter	[1]	essentially free from particles
volume uniformity	[1]	mean 0.95-1.05 ml, (to fall within range 0.90-1.10)
identity: citrate	[1], [2]	conforms
sodium	[1]	conforms
assay: citrate	[1], [2]	0.180-0.220 M
Indium-111 incorporation	[1]	≥95% compared to standard
pH	[2]	4.9-5.1
sterility	[1], [2]	no growth
pyrogens*	[1]	non-pyrogenic
general safety*	[1]	passes test

Table 9.1. Product release tests
* pyrogen and general safety testing are performed on a mixture of equal volumes of the contents of Vial 1 and Vial 2. [1] vialed product [2] Final Bulk [3] half-formulated Final Bulk [4] on DTPA coupling mixture [5] fermenter samples [6] clarified, concentrated cell supernatant

∏ Use this table to list the tests carried out on:

- the contents of the fermenter;

- the supernatant produced after the hybridoma cells are removed;

- on the final bulk products;

- on the vialed products.

As you list them, try to work out the rationale for including particular tests at the different stages of Myoscint preparation.

9.4.4. Process validation

Validation studies form an important part of the development of a pharmaceutical product. Systematically every step has to be validated.

stability of cell line
With regard to the fermentation it is important to assess the stability of the cell line. During fermentation the genetic properties of the cells could change, resulting in the formation of different substances.

late extended cell banks
From several of the first few fermentations, cells were removed at different stages in the fermentation and used to prepare Late Extended Cell Banks (LECBs). The LECBs were

extensively tested to assure that the cell line did not change and that no initially undetected contaminant was observed as the fermentation proceeded. No cell line change nor observation of any previously undetected contaminant was recorded. On the basis of these studies with the LECBs a maximum period of fermentation is defined.

reduction factors

During the manufacture of Myoscint various process components are used whose presence in the final product is undesirable (viruses, DNA, components of the culture medium etc). In order to demonstrate that the manufacturing sequence reduces such process components to minimal (often undetectable) levels in the final product, validation studies were performed, either by testing samples from different stages in the process and showing sequential reduction in the concentration of a component or by spiking a scaled-down version of the individual steps with the component and showing removal.

By using such procedures, it has been demonstrated that bioburden and endotoxin levels of product are controlled and minimised and that components of the cell culture medium and components of the process are removed to sub-part per million levels by the process. For each step of the process the reduction factor can be assessed. Spiking-type validation studies have also been used to show that the downstream process is capable of reducing a virus burden to extremely low levels. For example murine retrovirus is reduced by a least 10^{16} by the downstream process.

∏ What can we conclude from these studies?

The combination of strict control of starting materials, cell bank testing and in-process testing in addition to the testing of both Final Bulk and Final Vialed Product assure the safety, strength, purity and potency of Myoscint. This is backed up by extensive process validation studies that have demonstrated the efficacy of the manufacturing process particularly with regard to removal of unwanted contaminants.

SAQ 9.2.

1) Why are tests for viruses included in the evaluation of the products of hybridoma technology?

2) Which of the following strategies were adopted to ensure that Myoscint was free of virus?

a) Filtration of the Final Bulk product through a 0.22μm sterile filter

b) Chemical sterilization of the equipment

c) By conducting spiking studies to show that processing steps reduced the possibility of viral transfer into the final product

d) Specific tests for viruses by incubation with suitable cell cultures or animals

e) Determination of DNA in the final product

SAQ 9.3.

Below is an abbreviated flow diagram of the manufacture of mifarmonab Fab-DTPA.

Indicate on the diagram where the tests listed below might be conducted.

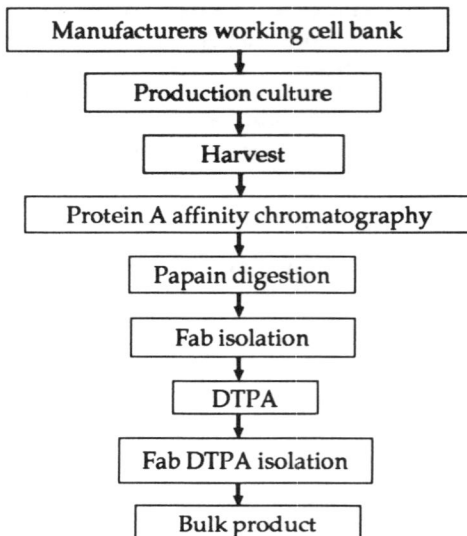

```
┌─────────────────────────────────────┐
│  Manufacturers working cell bank     │
└─────────────────────────────────────┘
                    │
          ┌──────────────────────┐
          │  Production culture   │
          └──────────────────────┘
                    │
            ┌──────────────┐
            │   Harvest     │
            └──────────────┘
                    │
      ┌─────────────────────────────────┐
      │ Protein A affinity chromatography │
      └─────────────────────────────────┘
                    │
           ┌──────────────────┐
           │  Papain digestion │
           └──────────────────┘
                    │
            ┌──────────────┐
            │ Fab isolation │
            └──────────────┘
                    │
              ┌──────────┐
              │   DTPA    │
              └──────────┘
                    │
          ┌────────────────────┐
          │ Fab DTPA isolation  │
          └────────────────────┘
                    │
            ┌──────────────┐
            │ Bulk product  │
            └──────────────┘
```

Tests

1) DNA determination

2) Presence of viruses

3) Immunoelectrophoresis

4) SDS-PAGE

5) Gel filtration HPLC

6) Presence of bacteria

9.5. Toxicological and pharmacological tests

∏ What tests are necessary? (Use your experience to write a list).

At an early stage of the development of a new biotechnology-derived agent, toxicological and pharmacological tests will be conducted to gather information on the safety and the efficacy of the product. Regulatory authorities have laid down the requirements for toxicological and pharmacological testing of medicinal products. Normally industry will carry out the required tests. However, the array of tests, required for xenobiotic drugs will, in most cases, not be appropriate for biotechnology-derived products. Some tests may be meaningless because the effects of a product are species specific. In cases where the product is identical to a human native molecule some tests may be redundant. It is therefore necessary that for biotechnology derived products the relevant studies should be determined on a case-by-case basis.

9.5.1. Toxicology

For the antimyosin murine monoclonal antibody some of the usual tests have been performed to get an impression of the intrinsic toxicity of the product.

single dose i.v. toxicity

Two single-dose intravenous toxicology studies were performed. One in dogs and the other in rats. In these studies antimyosin Fab-DTPA was used. In the dog study the dosage levels approximated 10 times and 200 times the human dose. In the rat study 300-400 times the human dose was used. No toxic effects were seen.

multiple dose i.v. toxicity

To assess the toxicity after repeated administration of the product multiple-dose (14 days) intravenous toxicity studies were performed in Cynomolgus monkeys and Beagle dogs. Because the product is radiolabelled with indium-111 prior to administration to humans, the studies were performed on antimyosin Fab-DTPA conjugated with a mixture of indium and cadmium. Cadmium was used because indium-111 decays to cadmium.

⊓ Why is cadmium of special interest?

Cadmium, in high doses, is known to have deleterious effects on biological tissues. In the experiments the indium-cadmium ratio used was 1:20, which approximates the ratio of these metals used under clinical conditions. Each mg of protein was associated with 0.16μg of the metal combination. The dose employed was 0.1 and 1.0 mg/kg per day, which corresponds to 10 and 100 times the human dose, respectively. No mortality or clinical signs of overt toxicity were observed in any of the animals in the antimyosin or the control group.

antigenicity of Fab-DTPA

In the monkey multiple dose toxicity study tests were included to determine whether antimyosin Fab-DTPA induces antibody formation. This is an important issue because if human-anti-mouse-antibodies (HAMAs) were formed in man, it could be a potential hazard, which would limit the use of Myoscint. In the study some of the animals showed a weak antigenic response. The question was whether similar effects would be seen in man. Studies in humans were therefore required to assess the immunogenic properties of the product in the clinical situation. We will consider these tests a little later.

cross reactivity

Another important concern in considering the toxicological potential of a monoclonal antibody in man is cross reactivity: the presence of the antigenic determinant on cells or tissues other than the intended target tissue. Cross reactivity could present the potential for an undesirable localisation at a non-target tissue site. Therefore, cross reactivity studies have been performed with Mifarmonab Fab-DTPA on a number of human tissue specimens. The results of these studies demonstrated high specificity for cardiac and skeletal muscle only.

foetal toxicity

Foetal toxicity and fertility studies were not deemed necessary. The potential hazard of radiation on developing foetuses is well known and the product should therefore only be used during pregnancy if the potential benefit outweighs the potential risk to the foetus.

mutagenicity study

Since the purified antibody is not expected to possess any mutagenic potential, mutagenicity studies were initially not performed. However because the authorities in one country considered these studies necessary, they have been performed and, as expected no mutagenic effect was found.

Carcinogenicity studies were not deemed necessary in view of the nature and the acute use of the product.

9.5.2. Pharmacodynamics

Fortunately there is sufficient similarity between human myosin and the myosin in certain animals (eg the dog and the rabbit) to enable antimyosin, as used in Myoscint, to react with myosin of those animals. Studies in animals can therefore be done to develop the pharmacological basis for the utilisation of antimyosin as an imaging agent for myocardial infarction. Studies have been designed to evaluate:

- that antimyosin binds specifically to irreversibly injured myocardial cells;

- that, when radiolabelled, it allows the accurate detection of myocardial infarction.

Myocardial cells exposed to extremely low levels of oxygen, as would occur following the occlusion of a coronary artery, become irreversibly damaged. This damage results in a marked increase in cell permeability due to the formation of gaps in the cell membrane.

antimyosin binding *in situ*

The first component of the hypothesis is that these membrane defects are sufficiently large to allow antimyosin to bind to the heavy chain of cardiac myosin contained within the cell. The key experimental evidence to support this hypothesis was obtained in studies which used scanning electron microscopy to observe directly the binding of fluorescent polystyrene sphere, coated with antimyosin, to cultured mouse myocytes, exposed to anoxic conditions.

Micrographs demonstrated dense accumulation of the spheres around ruptures in the membrane. Furthermore, when the cells were sorted based on their ability to bind antimyosin, only those which did not bind antimyosin continued to grow in cell culture. This finding provided physiological evidence that the ability of antimyosin to bind to injured myocytes was a marker of cell death.

antimyosin binding proportional to damage

The second component of the antimyosin hypothesis is that antimyosin accumulates in myocardial tissue in proportion to the severity of the damage. The severity of damage to the myocardium has a strong dependence on the magnitude of reduction in blood flow. Greater uptake of antimyosin should occur in the regions of lower myocardial flow. The close correlation between reduction in myocardial blood flow and increase in myocardial uptake of antimyosin has been demonstrated in canine models of myocardial infarction that have used persistent coronary artery occlusion or temporary occlusion of the coronary artery followed by reperfusion. This relationship unveils the high specificity of antimyosin for infarcted myocardium in that antimyosin concentrates in infarcted tissue despite limitation in its delivery caused by deprivation of myocardial blood flow.

detection of bound antimyosin

The third component of the Myoscint hypothesis is that sufficient amounts of radiolabelled antimyosin are sequestered in necrotic tissue to allow for scintigraphic detection and that this localisation is sufficiently discrete to discriminate necrotic from normal myocardium. Experimental evidence to evaluate this component has come from studies comparing infarct size as evaluated by standard histopathological techniques to that detected by Myoscint. These studies using triphenyltetrazolium chloride (TTC) staining to map infarct size, have shown that infarct size determined with radiolabelled antimyosin was virtually identical to that determined with TTC staining. When normal cells are incubated with TTC, the colour of the tissue changes to a deep red, due to the reduction of the agent by the cytoplasmic dehydrogenase. Injured cells, which have lost their cytoplasmic enzymes do not change colour. The result of the staining is a sharply contrasted boundary between normal tissue (coloured red) and infarcted tissue which remains a lighter dull brown colour.

9.5.3. Pharmacokinetics

blood clearance

The blood clearance of radiolabelled antimyosin in dogs showed an exponential loss of activity from the blood over about 72 hours. The initial rapid loss of blood activity is probably due to distribution of the antibody into the extravascular space. The activity is excreted by the kidneys.

∏ Make a list of the conclusions you can draw from the toxicological, pharmacodynamic and pharmacokinetic studies on Myoscint. When you have completed this, check your list against ours provided below.

The conclusions of the preclinical studies performed on antimyosin can be summarised in the following way:

- the antibody is not toxic to animals;

- there is no cross reactivity with other tissues (except skeletal muscle myosin);

- the efficacy data generated in canines provide the pharmacological basis for the use of antimyosin as an imaging agent for myocardial infarction;

- in a few animals antibodies against the antimyosin antibody were formed, therefore studies in man to assess the formation of similar antibodies. These antibodies are often referred to as HAMAs (human anti-murine antibodies).

9.6. The clinical trials with Myoscint

9.6.1. Introduction

confirmation of diagnosis

The diagnosis of acute myocardial infarction can be established in most patients by history, physical examination, electrocardiogram and serum cardiac enzymic changes. However in many patients the diagnosis may not be definitive. In these patients additional procedures are needed to establish or confirm the diagnosis. A major goal therefore is to identify the high and low risk patients and to determine the appropriate treatment. Patients at high risk generally should undergo early cardiac catheterisation, since many of them may be candidates for percutaneous transluminal coronary angioplasty or coronary artery bypass surgery.

If the diagnosis is uncertain, physicians tend to admit the patient to the hospital rather than risk an inappropriate discharge. If the diagnosis of acute myocardial infarction could be improved, resources could be more effectively allocated to the appropriate care of both high and low risk patients.

risk
stratification The assumption was that Myoscint could improve the diagnosis of acute myocardial infarction and the early risk stratification of the patients. To demonstrate that this assumption is correct, a number of clinical trials have been conducted.

It should be emphasised that the clinical trial programme of the biotechnology-derived product Myoscint does not differ fundamentally from that of other radiopharmaceutical imaging agents: the question to answer is whether the product is a safe and accurate diagnostic agent.

Some of the elements of the clinical programme will be discussed below.

9.6.2. The choice of the dose

The choice of the dose for antimyosin Fab-DTPA and indium-111 chloride was an important problem to be resolved. A number of criteria were set which had to be met:

- sufficient antibody must be administered to effect adequate localisation of the radiolabelled product at the target site;

- sufficient radioactivity must be present to produce a good image in a short enough time;

- the radioactive dose should be acceptable from a safety point of view.

For the antibody fragment a dose of 0.5 mg was chosen. This dose is fairly low but studies have demonstrated that it allows adequate target localisation. For the indium-111 chloride a dose of 74 MBq (2.0 mCi) was chosen. 24 - 28 hours after the injection this dose gives enough counts to make high quality images. For the patient this absorbed radiation dose is acceptable. Dosimetry studies have shown that the kidneys and the bladder wall will receive the highest doses (average dose in Rads: 9.5, 6.9 respectively). These are comparable with doses obtained with other radiopharmaceutical diagnostic agents.

9.6.3. The timing of injection and imaging

The timing of indium-111 antimyosin injection and imaging had to established. Not only the time between the injection and gamma camera imaging, but also the time between the onset of the chest pain and the injection.

The time between the injection and the imaging must be long enough to allow sufficient antibody to be fixed to the myosin in the necrotic tissue and to get sufficient contrast between target and the blood. However, radioactivity should still be high enough to obtain high quality images.

The pharmacokinetic studies showed a two phase curve of radionuclide in the blood stream. The first phase (α phase) had a half-life of 1.9 hours and a second phase (β phase) had a half-life of 19 hours. This means that after 24 hours 19% of the initial radioactivity remains in the blood, compared with 10% after 48 hours. This blood clearance profile allows imaging after 24 and 48 hours without the activity in the blood obscuring the images. In the clinical studies imaging after 24 hours was recommended. Only if there was still a lot of activity in the blood pool, were images taken after 48 hours.

In the various studies Myoscint was administered at variable times after the onset of the chest pain: from as early as 3 hours up to several days. Positive images were obtained at both extremes.

The conclusion of the studies is that Myoscint can be administered as early as immediately after the onset of chest pain and that 24 hours thereafter the diagnostic results should be available.

9.6.4. The diagnostic accuracy

multicentre studies

The diagnostic accuracy of Myoscint has been established in a number of clinical trials, involving more than 1000 patients. The pivotal trial was a multicentre trial in almost 600 patients with chest pain considered to be due to myocardial ischemia. The set up of the trial was such that the patient population that was studied was representative of the patients that would be considered for diagnostic assessment with Myoscint. By including 25 centres with different gamma camera and nuclear medicine computer systems, the results derived from this study could be applied to other hospitals.

Using the normal diagnostic procedures (physical examination, ECG, determination of enzymes etc) the patients were evaluated and the diagnosis was made. In addition, all clinical records were evaluated 'blind' by a cardiologist not involved in the trial.

The patients received Myoscint and images were made after 24 and 48 hours. The images were evaluated by a 'blind' panel of experts, ie the experts did not receive any information on the patients, the diagnosis was solely made on the valuation of the images. By comparing the results of both assessments the *specificity* and *sensitivity* of Myoscint could be established.

specificity sensitivity

The specificity is defined as the number of patients with a definitive diagnosis of chest pain without ischemia or necrosis with a negative scan divided by the total number of definitive myocardial infarction patients. The sensitivity is defined as the number of definite myocardial infarction patients with a positive Myoscint scan divided by the total number of definitive myocardial infarction patients. The specificity of Myoscint was 95% (95% confidence intervals, 85 to 92%), the sensitivity in all patients with infarction 88% (95% confidence intervals, 84 to 91%). These results clearly show that Myoscint is an accurate diagnostic agent for myocardial infarction.

prognostic significance

The Myoscint images also enable a physician to define the area with necrotic tissue and to establish extent of the necrosis. Since the extent of an infarction is associated with an increased risk for a future cardiac event, Myoscint could be a tool to stratify patients with an increased risk. To assess the prognostic significance of the extent of Myoscint uptake, the relationship of Myoscint uptake to post-infarct cardiac events was evaluated. This analysis involved evaluation of the patients during a follow-up period of 60 days. A number of patients had a major cardiac event (ie cardiac death or non-fatal myocardial infarction) during this period. By quantifying the Myoscint uptake an association between the extent of Myoscint uptake and a major cardiac event during the follow-up period could be demonstrated. Thus Myoscint can also be used in risk stratification of patients with a positive Myoscint scan.

Necrotic myocardial cells are formed, not only after myocardial infarction, but they can also be caused by other conditions eg myocarditis and rejection of a transplanted heart. Preliminary clinical trials have shown that positive Myoscint scans are obtained in these conditions. Particularly the use in monitoring heart transplant patients seems a useful application of Myoscint.

9.6.5. The safety of Myoscint

The overall clinical experience demonstrates that Myoscint is safe. In the clinical trials no patient had any adverse experience definitely attributable to the product. In a few cases adverse experiences were probably or possibly related to antimyosin. These include injection site pain and fever. With administration of any murine protein, there always is concern about the potential immunogenicity of the product and the possibility of human antimurine antibody responses that could sensitise patients to subsequent administration of murine monoclonal antibodies. Using a sensitive enzyme linked immunosorbant assay (ELISA,) no dose-related human antimurine antibodies (HAMA) response could be detected in serum collected from more than 1000 patients. Several patients had received two or three doses of Myoscint. Apparently the Fab antibody fragment, used in Myoscint, is *not* immunogenic.

∏   What conclusions can be drawn from the clinical trials?

- Myoscint is an accurate diagnostic agent for the detection of myocardial infarction. In cases where the findings obtained with traditional diagnostic tools are ambiguous a definitive diagnosis can be made;

- Myoscint can be used for the risk stratification of patients with myocardial infarction;

- Myoscint is a safe product. Adverse experiences are virtually absent. A dose related HAMA response is not observed. The radiation exposure to critical organs is acceptable.

Summary and objectives

This case study has focused on the downstream processing and the evaluation of a monoclonal antibody system used as an *in vivo* diagnostic reagent. Now that you have completed this chapter you should be able to:

- explain the overall strategy for using monoclonal antibodies or Fab fragments as an aid to *in vivo* diagnosis using radionuclide tagging;

- define the requirements of radionuclides to be used in *in vivo* diagnosis;

- explain the need for quality control testing of materials at the various stages of production of monoclonal antibodies and Fab fractions and the criteria used in selecting the tests to be undertaken;

- explain why the testing of monoclonal antibodies or Fab fragments against human proteins in animals is not always appropriate or applicable.

Responses to SAQs

Responses to Chapter 2 SAQs

2.1. Virtually all of the indications are that you should advise your company not to become involved in producing this vaccine.

1. Although it may be morally right to attempt to cure any disease, no organisation can simply pursue every moral issue. Doing so would soon lead to financial disaster and the organisation would cease to function.

2. If the market is so small that it is unattractive to competitors, it is unlikely to be attractive to your own company. In other words, if competitors see it as a non-viable proposition, it is also likely to be so for your organisation, unless you have some special skills which gives you substantial competitive advantage.

3. Generally a large, growing market offers the most likely commercial success. Statement 3 is valid. It could also be added that the disease is a relatively mild one. There are many other more damaging diseases with a much greater number of sufferers which could offer greater commercial and social benefit for the same kind of investment.

4. This is also true and would indicate that the project should not be undertaken. It is not even certain if the disease is caused by a virus. Before any specific plans for producing a vaccine can be developed, questions such as: How is the virus to be collected? How (can) it be cultivated in the laboratory? Is its genome accessible to restriction enzyme digestion? Can the gene for its coat protein be isolated? Is it expressed in bacterial cells? Will the coat protein, when injected into humans, provide immunity against the infection? There are far too many fundamental questions left outstanding and a tight strategy for developing the vaccine cannot be developed.

5. Although this is true, the reasons given in (4) far outweigh the reason given for (5).

6. This is true and is perhaps one of the few positive reasons that can be given for proceeding with the project. It must however be remembered that the implementation of this project will mean that resources (staff, equipment etc) will not be available for other projects.

7. Very dangerous attitude to take. Irrespective of the remoteness of the market, it cannot be sound business sense to rely on dodging the inspectorate or for not doing good science. Apart from that it is totally immoral. Companies operating in the medical/pharmaceutical areas depend on displaying a high ethical and moral profile, without which the products become unacceptable and their employees face the prospect of fines or imprisonment.

8. Although it is true that the remoteness of the market will probably deter rivals, it should also deter your organisation. Do you have, for example, sufficient resources

for the organisation to collect samples and carry out clinical trials in such a remote place?

2.2 You should have been able to identify many questions that need answering since we gave you very little information. Let us see if we can identify some of them. Do not be despondent if you have not asked exactly the same kind of question as we have detailed below. We will go through the various stages.

First we need to have sufficient viruses to extract DNA for subsequent manipulation. Where are such viruses going to be obtained? From patients? (unlikely). Can the virus be cultivated *in vitro* in animal culture? (unknown). If so can it be purified from host materials? (probably).

If we can obtain sufficient viruses and obtain DNA, what restriction enzymes will be needed to 'cut out' the coat protein gene. We have no knowledge of this. Finding an appropriate one (if one exists) would either involve an empirical approach (ie try many different enzymes in the hope that one works) or carry out an expensive gene mapping/restriction enzyme mapping of the virus genome. This latter approach makes many assumptions about the ease of cultivating the virus.

If we assume that we can cut out the appropriate fragment, can we ligate this into an expression vector? In all probability the answer is yes. It might mean however that we need to use lined sequences to join the gene into the vector (we will meet an example of this in the insulin case study). We cannot however answer this question until we have obtained a suitable fragment. The next major question relates to identifying the clone(s) of host cells which carry and express the appropriate gene. How do we identify such a clone? We could use a variety of strategies. The most likely to succeed is to use antibodies against the coat protein to identify of the appropriate clone(s). This raises other questions such as are antibodies against WBC coat proteins available?

In this SAQ, we have tried not to be dogmatic about the response we anticipate from you. What we intended for this SAQ is to test your ability to use existing knowledge and match it against a plan to identify issues that need resolving.

The next stage of the process would be to prioritise the order in which these problems should be resolved. In the case under discussion, obtaining sufficient viruses would seem to be of greatest importance for without them, little else could be done.

2.3. To be able to answer this question you need to think of the properties of each cell type. Important attributes are: easily cultivated; non-pathogenic and ready containment.

1. Unlikely to be satisfactory for two reasons. The probability is that the laboratory (attenuated) strain of *E. coli* was originally selected because it was unlikely to survive for long outside of the laboratory. Transferring mammalian hormone genes to a strain of bacteria capable of growing in the alimentary tracts of mammals, poses substantial risk. Unintentional release of the organism into the environment with subsequent infection of animals, including humans, of such a strain could result in enormous medical problems. A second reason relates to the metabolism of the attenuated and wild type strains of *E. coli*. If one breaks down the hormone, so might the other.

2. Unlikely to be satisfactory. *Corynebacterium diphtheriae* is a pathogen (causes diphtheria). The use of this organism poses risks to researchers and production line

staff and therefore imposes additional costs on the company, to provide additional containment faculties and procedures. In general, the use of pathogens as primary producer, whether gene manipulated or not, should be avoided if possible.

3. This could be a satisfactory system but the researcher would have to recognise that its use would lead to substantial adjustments to subsequent stages of the project. For example, quite a different set of gene vectors would be needed and, on scale up, substantially different growth rates and product yield should be anticipated.

4. Unlikely - this is a strict anaerobe and difficult to cultivate. Furthermore very little genetic manipulation has been attempted using *Clostridia* and therefore the project would become based on a less firm scientific experience.

5. Unlikely. The growth of algae, using light as a source of energy is notoriously expensive and difficult to operate as an industrial process under aseptic conditions.

6. Could be a possibility, but the selection of this option poses quite different issues further down the development pathway. These cells are much more costly to cultivate than, for example, the bacterium *E. coli* or yeast. The use of this type of host system also raises questions about the contamination of the product by latent viruses, oncogene sequences etc., that may be present in such cell lines. It is accepted however that although transformed (cancerous) mammalian cells offer greater risk, they have better growth characteristics than normal mammalian cells. The use of these as host cells will, however, invariably demand more stringent quality control/product quality assurance steps to be taken.

7. Unlikely. Although *Enterobacter aerogenes* is non-pathogenic, it is a normal gut organism found in many mammals. It also has not been the subject of many genetic manipulation studies. Thus for the reasons given for 1 and 2, this is probably not a satisfactory system.

2.4. The straightforward answer to this question is that all of the listed features are important. Without suitable vectors (item 1) being available or there being some expectation that such vectors can be constructed, there is little chance that the project will achieve much unless there are alternative strategies for producing the required cell lines.

Likewise, the nature of the host (item 2) to be used is of great importance. The need for particular containment facilities, the requirements for product quality, the design of the bioreactor and its anticipated yields are all dependent on the nature of the host.

The progress of a project is highly dependent upon the knowledge and skills of those working on the project (item 3). The availability of suitably qualified staff is, therefore, vital. Items (4), (5) and (7) (number of patients, their distribution and the presence of rival products) are also vital, since these will govern the size and accessibility of the market for the product.

The number of animals (item 6) to be used is important, since this represents significant expenditure. There are also ethical issues that need to be considered including the pressure that exists from animal welfare groups.

Although the bioreactor (item 8) would not be built early in the process, some thought would need to be given to this early in the project. At the outset, this may be only a provisional analysis, but the organisation would need to have some notion of the size and likely operating cost of the vessel in order to make reasonable budget provision. Those working on the cell system would also need to have some framework in which to develop their analysis of the influence of various parameters on growth and product yield.

Item 9, the other projects under consideration also have a bearing on the proposals since they must be regarded as competitors for company resources.

This SAQ serves to emphasise that the development of a proposal to apply biotechnology to the production of medicinals, demand many different inputs. It is not just sufficient to think of one step at a time, but to think beyond just producing a cell line capable of making the product. It demands knowledge of bioreactors, downstream processes and the need to respond to the regulatory requirements. It demonstrates that commercial and social (ethical) issues are also of great importance. Success is only likely if all of these issues are considered quite early in the project.

2.5. We were anticipating that your flow diagram would look like this.

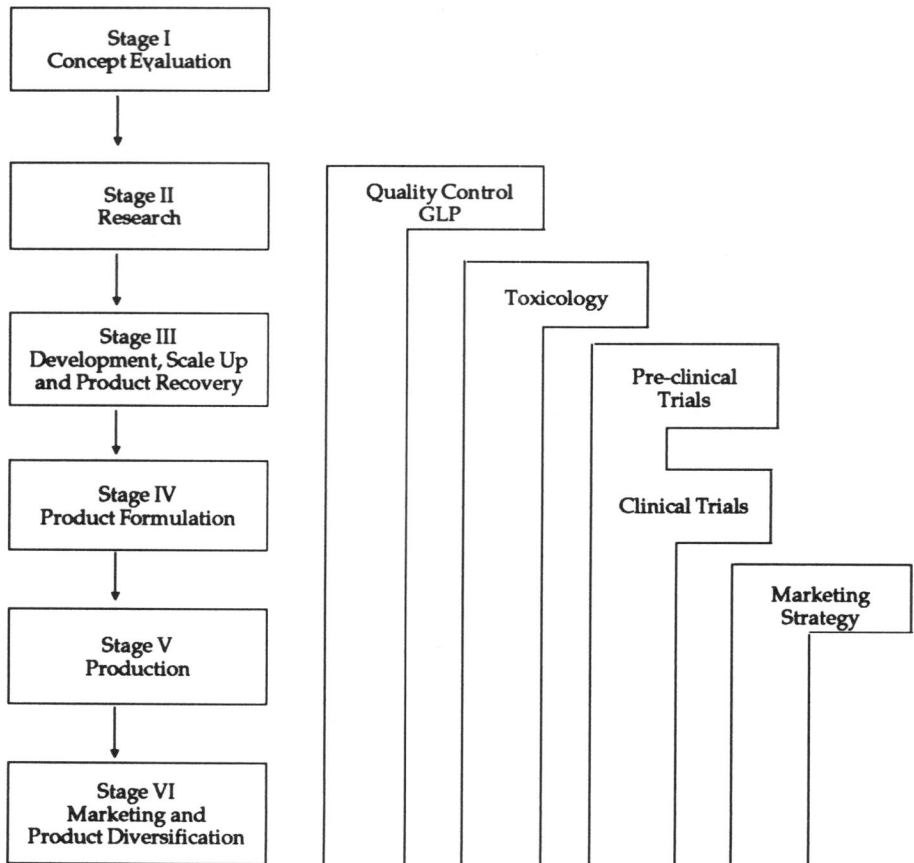

You might however have drawn it quite differently, the key points are that you should display the development of the process and the evaluation of the product being conducted in parallel with each other.

If you have produced a figure similar to that displayed above without returning to examine Figure 2.1, then you have understood the major messages of this chapter.

Responses to Chapter 3 SAQs

3.1.   1) Directive 78/25/EEC.

2) Careful examination of the list will reveal that directives 75/318/EEC; 81/852/EEC; 83/189/EEC; 83/571/EEC and 87/119/EEC directly relate to the analysis, pharmacological, toxicological and clinical evaluation of medicinal products. We could also cite directives which are specific for veterinary medicines such as 81/852/EEC and 87/20/EEC. Some of the remainder of the directives listed in Table 3.1 will make reference to analysis and efficacy and safety evaluations, but these are not the principle concern of these additional directives.

3) Directive 87/18/EEC.

3.2.   1) √   2) √   3)   4)   5) √

Option 1), 2) and 5) are specifically exempt from the EC directive on the use of genetically modified micro-organisms.

3.3.   The correct statements are:

1) Type A operations are operations used for teaching research or non-commercial purposes.

2) Type A operations are operations that are conducted in a total volume of less than 10 litres.

3) Group 1 organisms have a long record of safe use and are considered safe.

4) The procedures for allowing Group IA operations are less stringent than those for Group IIB operations. From the definition of Group I and II organisms and type A and B operations, it follows that Group 1A operations offer the least potential hazard, while Group IIB offer the greatest risk. The table presented in Appendix 2 (section 1.5.2) indicates those operations for which notification of the competent authorities alone is sufficient and those for which specific authorisation is required.

3.4.   The EC directive on the deliberate release of genetically modified organisms is divided into A) - General Provisions; B), Research and Development (except placing on the market); C) Placing the product on the market, D) Final provisions.

3.5.   1) 90 days (section 2.5.4. in Appendix 2)

2) 60 days (section 2.5.5. Appendix 2)

3) 90 days (section 2.4.2. Appendix 2)

4) 30 days (section 2.4.3. Appendix 2)

3.6.   You may have made a drawing like this.

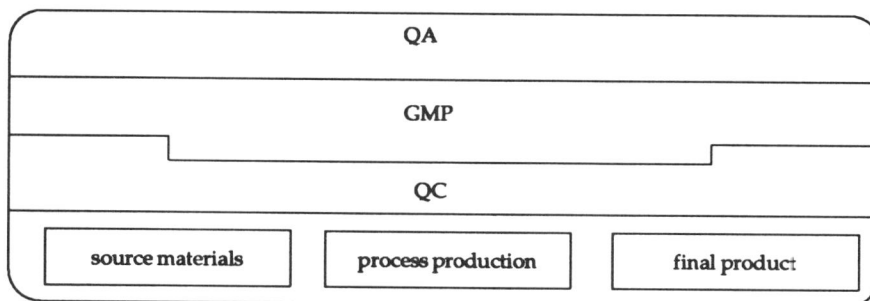

What we hope your drawing depicted was that QA (Quality Assurance) embraces the whole process (production, laboratory work, materials etc) and that GMP also covers the process beginning with the starting materials and ending with the final product. Quality control is also important throughout the process but tends to predominantly focus on ensuring the starting materials are of high quality and that the final product also reaches certain specifications. Also the production process is monitored by having a number of quality control checks at various stages. In our drawing we have tried to indicate this by the thickness of the bar carrying the symbol QC.

3.7.
1) If 1 vial in 10 000 is contaminated then the selection of the single contaminated vial each time a vial is removed is 1/10 000. Since 40 vials are selected, the chances of including the contaminated vial is therefore 40 x 1/10 000 x 100 = 0.4%.

2) If 100 vials in every 10 000 are contaminated, the chances of selecting a contaminated vial each time a vial is removed is 100/10 000. Since 40 vials are selected the chances of including a contaminated vial is therefore approximately

40 x 100/10 000 x 100 = 40%

In these calculations, we have made approximations. In reality, one has to consider this from a probability stand point. Nevertheless the calculations show that this sampling method is unreliable especially when the proportion of contaminated vials is low.

3.8.
1) Since the upper limit of contamination by DNA is 1 ng/ml and each patient is to receive 10 x 0.1 ml doses, then the maximum DNA dose to be received will be 1 ng (or 1000 pg).Since the risk of a tumorigenic event is 1 in 5-20 million per 100 pg DNA received, then the risk from the product is about 10 times greater since the patient may receive 1000 pg of DNA. Therefore the risk is of the order of 1 in 0.5-2 million.

2) It is probably that all these tests need to be included. *Escherichia coli* is an enteric organism and may contain pyrogens. The Limulus Amoebocyte Lysate assay would need to be used to prove that pyrogens were not present. You would also probably need also to include a rabbit pyrogen test.

There is general agreement that protein products should not be contaminated with DNA. This is especially true of products derived from transformed animal cells where the risk from oncogene transfer is greater. It would be essential to check that the culture was pure since this not only influences the amount of product made (and therefore commercial viability) but is required by GMP. The immunoassay and HPLC analysis represent types of chemical analysis that may be used to characterise the product and they could well be used as part of the Quality Control criteria. Likewise, the check for bacterial contamination is necessary to ensure that the product does not provide the vehicle for transmitting infection and is not subjected to biodegradation during storage.

3.9. Your sequence should be (b), (c), (a)/(e), (d), (f). The sequence is not however entirely linear and some overlap may occur. Let us think through the logic of the sequence. The first stage, involved in product assessment, must be to determine whether or not the product is what you think it is. Generally chemical analysis (eg composition, structure) is much more precise, cheaper and more sensitive than bioassays. Of course the precise choice and range of technique used depend on the specific nature of the product. Once the product has been shown to possess the expected chemical characteristics, its properties in living systems (pharmacological, pharmacokinetics) are analysed. These usually are first conducted in animals (item c) but, with some human-specific products such analysis is not always ideal. This analysis should give an indication of the dose size and likely pharmacological effects in humans. The animal tests should also provide an indication of any likely toxic effect.

If all is satisfactory, the product may then be administered to small numbers of healthy volunteers to evaluate its pharmacokinetic and toxicological properties (item a) and to evaluate the dose range required (item e). If this analysis so indicates, a more extensive trial is run with perhaps 100 or so subjects before embarking on a multicentre, large scale study (item d). If this study leads to full market authorisation, it is still necessary to continue monitoring the performance of the product in order to detect any adverse effects which may occur at low frequency (item f).

Responses to Chapter 4 SAQs

4.1.

1) True - In diabetes mellitus fats and proteins are used to a greater extent as sources of energy since the cells cannot take up glucose. These compounds yield considerable quantities of acetyl CoA, but little oxaloacetate. Therefore the acetyl CoA cannot be broken down by the Krebs cycle enzymes. Instead the acetyl CoA is further metabolised to form ketones such as acetone and acetoacetate.

2) True - Glucose absorption by cells is greatly reduced in the presence of low levels of insulin. The glucose remains in the blood until it reaches the kidneys where, if it exceeds a threshold level, it is passed into the urine.

3) False - It is true that sufferers of diabetes mellitus usually drink excessively to recoup the liquid that is excreted as urine. But this drinking does not result in coma. The coma state is caused by the failure of the brain cells to absorb glucose needed as a metabolic fuel. This, in turn, causes changes in the metabolism of proteins and fats. This uncontrolled metabolism of these non-carbohydrates and their metabolic end products are the major causes of coma.

4) False - Absolute deficiency of insulin is typical of Juvenile (Type I) diabetes. The Adult type is more usually caused by a delayed release of endogenous insulin.

5) False - The opposite is true. In diabetes mellitus, the uptake of glucose into cells and its storage as glycogen is severely restricted.

4.2.

This SAQ tests your understanding of the sequence of events which leads to the production of recombinant DNA clones. The order should be B, D, E, G, H, A. Steps C and F are not needed.

Let us briefly consider each option:-

A This technique is involved with identifying the recombinant DNA bearing clones and, therefore, occurs late in the sequence.

B This is the primary source for isolating the human insulin nucleotide sequence and therefore occurs very early in the scheme.

C This is not a very sensible step. If plasmid DNA and human DNA are incubated with a mixture of restriction enzymes, the DNA samples would be broken into many small fragments which would be difficult to analyse and difficult to ensure that the appropriate pieces were joined up to produce the desired human gene:plasmid combination.

D This step is needed fairly early and is the step in which the human mRNA is converted into a DNA form.

E This is needed to cut open specifically the plasmid DNA, ready to receive the cDNA for the insulin gene.

F   This is not specifically required unless, of course, the researcher needs to produce more plasmid material for step E.

G   This is carried out to complete the insertion of the cDNA into the plasmid DNA.

H   The products of G are incubated with an *E.coli* strain which is sensitive to ampicillin. Any *E.coli* cells which take up intact plasmid (pBr322) should acquire resistance to ampicillin. Therefore this step enables identification of plasmid carrying *E.coli*. If such cells are replica plated onto tetracycline containing medium then those cells carrying plasmids which contain inserts in their tetracycline resistance gene will be tetracycline sensitive. Those, however, which are tetracycline resistant will have intact pBR322 plasmids. Thus cells carrying pBr322 plasmids containing cDNA inserts can be identified. Some of these could be carrying cDNA for the insulin gene. These can be identified using an antibody assay for insulin (Step A above).

In this SAQ, we did not include all of the stages in the process. We can, however, summarise the overall process in the following way:-

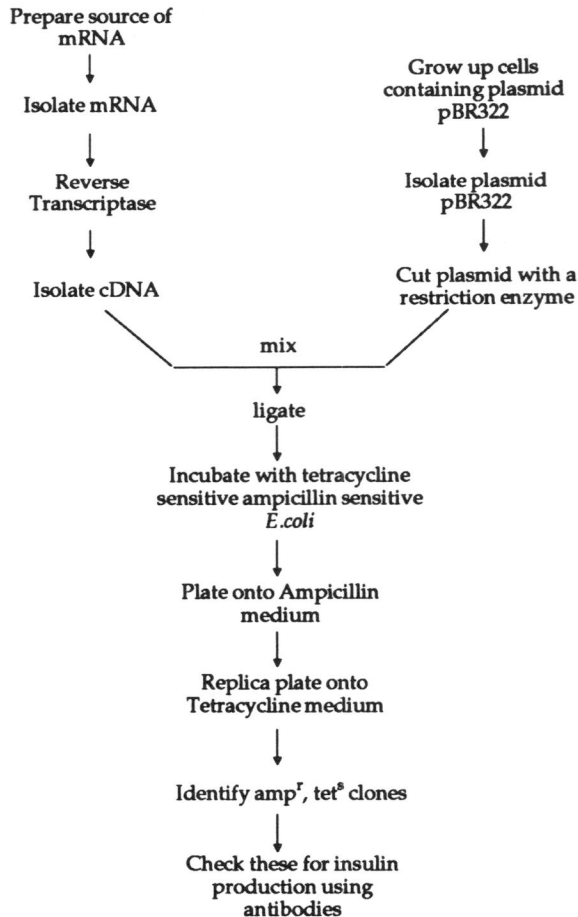

4.3. For each molecule of the chimeric protein (Tryp E-pro-insulin) synthesised, the total number of amino acids that need to be joined together is $191 + 86 = 277$.

But each chimeric Tryp E pro-insulin molecule will need to be modified to produce a shorter insulin molecule (ie 51 amino acids long).

Therefore for each 277 amino acids jointed together each pro-insulin formed will produce one molecule of insulin.

ie 1 molecule of insulin is produced for each 277 amino acids joined together.

The ratio of amino acids is $51:277 = 0.18$.

This ratio is much higher than that obtained when the two chains are produced separately. In the text, we calculated that the ratio was $51:433$ if the chains were produced separately.

4.4. Potentially any $-SSO_3^-$ group could react with any other. Therefore the sorts of structures which could be produced include:

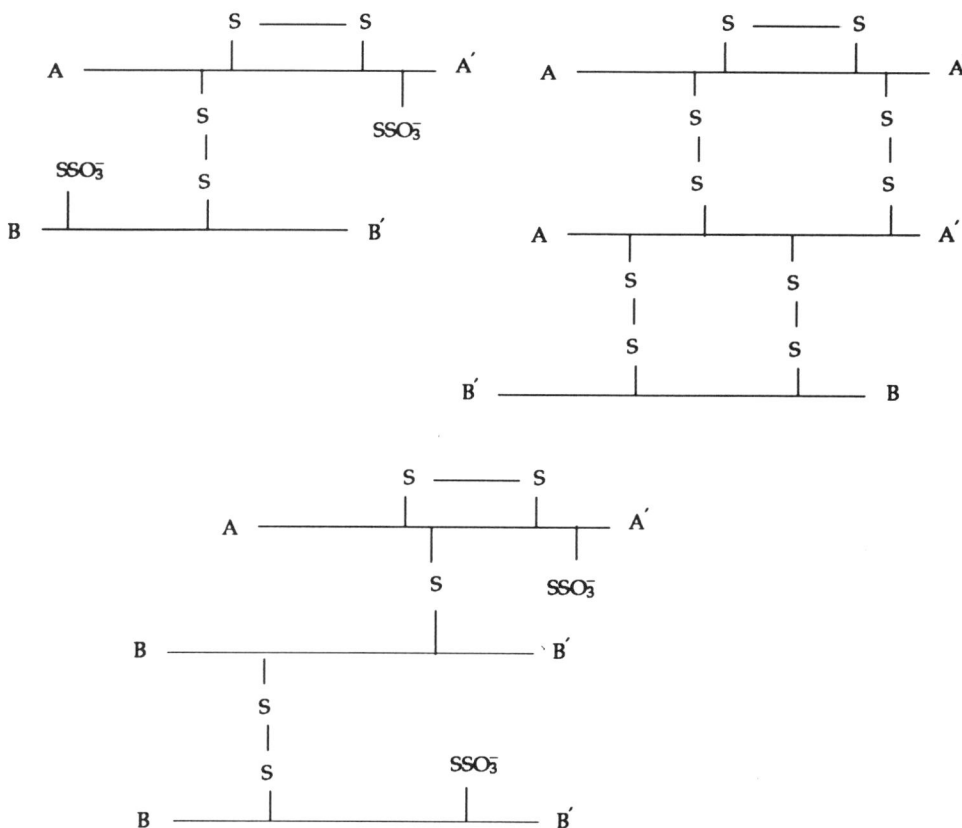

There are in fact an enormous number of such structures that are possible. Although some heterogeneity is derived from the reaction mixture, the main product is insulin indicating that the steric arrangement of -SS0₃- residues on A and B imposes some restrictions on the way -S-S- bridges are made.

4.5.

1) *E.coli K12* is regarded as a safe organism to use. Although strains of *E.coli* commonly grow in animal alimentary tracts, strain K12 is a weak laboratory strain. If the genetically manipulated organism was accidentally released from the laboratory, it is thought that it would only survive for a few hours, therefore, offering little long term risk to health or the environment. The system is also well understood genetically and there are many genetic vectors (plasmids) available for use in this strain.

2) The tryptophan synthetase promoter leads to a greater proportion of insulin production (see Figure 4.12. and associated text).

3) There is a higher yield of insulin (see SAQ 4.3.) produced using this system and there is no need to produce two separate fermentations. The pro-insulin route also enables the C-peptide and pro-insulin to be produced and these are potentially useful products.

4) The methionine allows a specific point for the use of cyanogen bromide to break the desired polypeptide from the promoter protein.

5) The pro-insulin is first oxidised to break disulphide bonds because those that are formed as the pro-insulin is made within the bacterial cells are not the ones to produce insulin in the correct configuration. Once the structure has been opened up by breaking the disulphide bridges, the molecules can be re-folded in the correct configuration and the appropriate disulphide bonds formed.

6) Use is made of the high degree of specificity of the enzymes. Chemical hydrolyses to break peptide bonds would be quite random within the pro-insulin molecule and therefore many different types of molecules would be produced. The enzymes used split the pro-insulin quite specifically producing high yields of insulin.

4.6.

1) H is correct. HPLC devices take perhaps only 1 1/2 - 2 meters of bench space. The bio-assay system takes 144 rabbits for each test. The HPLC can be used many times.

2) H - Experience shows that precision of the HPLC assay is much greater than the rabbit bioassay. Data presented in the text show this very clearly (examine the range and standard deviations of the data in Tables 4.2. and 4.3.).

3) H - The rabbit bioassay is not capable of distinguishing between closely related insulins such as bovine, porcine and human insulins. By carefully selecting the chromatography column to be used and the eluting solvent, the HPLC procedure can be capable of separating these closely related compounds. (See Figure 4.27.).

4) H - Although HPLC devices represent quite a substantial capital investment, it is still much cheaper than the husbandry of the enormous numbers of animals needed in the bioassay procedure.

5) R - Clearly the biological activity can only be measured in a "biological" system.

6) H - In order to measure insulin in rabbits, the level of sugar (glucose) in the rabbit blood needs to be monitored for 3–4 hours after the administration of insulin (ie insulin is measured by its hypoglycaemic effect). It also has to be monitored in several animals in order to get a reasonably precise result. Using HPLC, insulin elutes in a matter of 10–15 minutes. The precise time depends on the column used and the flow rate of eluting solvent. Nevertheless, a result can be obtained within minutes.

On the basis of the criteria cited in this question the HPLC method for measuring insulin is far more attractive than the conventional rabbit bioassay. Of course techniques like HPLC, which are based on chemical and physical features, do not determine the biological activities of the material subject to evaluation.

4.7.   Although many of these tests used gave very similar results for the test sample and authentic human insulin, there are sufficient differences for the test sample to be suspect.

Thus, the HPLC profiles a) were very similar. It would, however, have been useful to run a mixture. Even slight differences in the two materials may well have been picked up as a double or asymmetric peak. Samples run separately does not allow distinction to be made for those samples which elute close together, slight differences in each run leads to slight variations in elution times.

The amino acid analysis indicates that the test sample had one less phenylalanine moiety per molecule (2.00 compared with 2.99). This shows up with the HPLC profiles of the protease digests d), in which a slightly different result is achieved with the two samples.

The circular dichroic spectra of the two samples are very similar, indicating that they have similar conformations.

This SAQ serves to illustrate that it is important not to rely on a single type of determination in confirming similarities between complex molecules like insulin. It is important to make comparisons of a variety of properties of such molecules.

4.8.   Although true or false can be written for each of the statements given, the answers are not always clear cut. We have chosen to enlarge on this.

1) This is probably true but has yet to be proven. Certainly some patients suffer from loss of glycaemic control and this may be a reflection of the increased rates of adsorption observed in pharmacological studies.

2) Not really true. The sera were examined to ensure that the patients were not becoming sensitised to antigens derived from E.coli. If they had become sensitised, then it would appear that they had been receiving E.coli peptides. Of course, if they had not been receiving such antigens, it is unlikely that they would have received pyrogens, but this was not the primary purpose of the test.

3) This is basically true. The studies showed that the transfer was followed by some decrease in the circulating anti-insulin antibodies. At the same time these antibodies have lower affinity for rDNA human insulin. Therefore, there was a much lowered tendency to exhibit immunological insulin resistance.

4) This is basically true - since it should be anticipated that human insulin should work best in humans.

5) This is false. Although it may be true that human insulin will work best in humans and that rDNA human insulin appears to be chemically and structurally the same, there are some important differences. Firstly the rDNA human insulin is made in an alternative host and may be contaminated with host materials which may have some deleterious effects (see 3).

Secondly rDNA human insulin is not delivered to the body in the same way as natural insulin. This may have some important pharmacological effects.

Thirdly, the rDNA human insulin is administered in a mixture. This may have some important consequences especially in terms of reaction by the immune system.

Fourthly, although rDNA human insulin looks identical to normal human insulin, there may be subtle differences which have long term side effects.

It is therefore essentially to conduct proper clinical trials and to continue to monitor the performance of the material long after its initial market launch. This post-market monitoring is, of course, demanded as part of the licence application (see Chapter 3).

6) Although many might argue that this statement is true, we do not know sufficiently about the release of insulin from insulin:antibody complexes to make a firm judgement. Many would equally argue that the binding of insulin by antibodies makes the treatment of diabetes less reliable because the release of insulin from an insulin:antibody reservoir might lead to unwanted hypoglycaemia. It should be realised that the removal of insulin from circulation is a complicated process. Insulin complexed with antibodies is likely to be removed from circulation by phagocytotic cells including both circulating cells and cells lining the reticuloendothelial system.

Responses to Chapter 5 SAQs

5.1   All of the reasons could be cited as benefits.

1.   Blood transfusion is an expensive operation involving the need for many donors, blood storage and typing as well as medical and clinical staff. Of course, whether or not an alternative is cheaper depends on the cost of the alternative!

2.   A correction of blood deficiencies by non-transfusion techniques lowers the risk of infection from donors.

3.   The protection of patients against blood deficiency by non-transfusion procedures would obviously not produce transfusion reactions.

4.   Iron intoxication is a consequence of repeated blood transfusion as iron is released from the breakdown of administered erythrocytes.

5.   Transfusion is regarded as a potentially uncomfortable process. Alternative procedures could reduce this discomfort.

5.2   1.   It is true that RBCs, have no nucleus and therefore do not contain the genetic information that is needed to code for new or replacement cell constituents.

2.   This is untrue. Red blood cells actively metabolise sugars mainly by the Embden-Myerhoff glycolytic pathway. About 10% of the sugar is metabolised by the hexose monophosphate shunt. This latter pathway produces reduced nicotinamide adenine dinucleotide phosphate (NADPH) which, in turn, reduces glutathione (GSH) which prevents peroxidation of haemoglobin.

3.   This is true. Although young RBCs contain ribosomes, these are gradually lost during the maturation processes. Ribosomes are essential for protein synthesis.

4.   This is untrue. Although mature RBCs do not have mitochondria, they can generate ATP anaerobically using the Embden-Myerhoff glycolytic pathway. It is perhaps paradoxical that these cells which carry oxygen should carry out anaerobic metabolism. These cells however contain much reduced glutathione to provide a reducing environment and the oxygen present in these cells is bound to haemoglobin. Red blood cells require ATP to carry out some maintenance functions such as in retaining a correct osmotic balance.

5.   This is untrue. Although red blood cells contain substantial amounts of iron, this is chemically bound up in haemoglobin. There is no evidence that the iron in red blood cells would inhibit protein synthesis.

5.3a   1.   Haemorrhage is a loss of blood by bleeding and therefore can result in anaemia.

2.   Iron deficiency will result in restricted haemoglobin production and thus impair RBC production leading to anaemia.

3. Low oxygen tension in the kidneys leads to erythropoietin production which stimulates RBC development and is, therefore, unlikely to lead to anaemia. It is more likely to produce the opposite disorder - polycythaemia.

4. Folic acid is needed for thymidylate synthesis. A deficiency of this vitamin leads to restricted RBC production and those cells which are produced are larger and are more quickly destroyed than normal RBC thereby producing anaemia.

5. The consequences of a kidney infection depend on the causes of the infection. If the infection causes renal loss of blood or reduced erythropoietin production this will result in anaemia. If the main result of the infection is to reduce blood flow and oxygen tension in the kidneys, this may results in increased erythropoietin production which will lead to enhanced RBC production.

6. Living at high altitudes results in lower blood oxygen tensions. This in turn causes more erythropoietin to be produced. Therefore RBC production is stimulated (ie the reverse of anaemia = polycythemia).

7. Vitamin B_{12} is needed for DNA synthesis. A deficiency of B_{12} gives a similar effect to that described for folic acid deficiency as described in 4, ie it will result in anaemia.

8. Pyridoxine (Vit B_6) is needed to make a precursor of haemoglobin. A deficiency of this vitamin will result in reduced haemoglobin production and, therefore, result in anaemia.

9. Intrinsic factor is needed for the adsorption of vitamin B_{12}. Therefore a failure to produce this factor gives the same result as vitamin B_{12} deficiency described in 7.

10. Androgens stimulate erythropoietin (EPO) production. Since EPO stimulates RBC production, an overproduction of androgens is unlikely to lead to anaemia. More probable is that it would lead to polycythaemia.

5.3b   None of the conditions of anaemia caused by dietary deficiency would be effectively treated by administrating erythropoietin.

Thus anaemia induced by iron deficiency (2), folic acid deficiency (4), vitamin B_{12} deficiency (7) pyridoxine (vitamin B_6) deficiency (8) and failure to produce intrinsic factor (9) would not respond to erythropoietin treatment.

We might, however, anticipate erythropoietin to be effective in anaemia produced as a result of haemorrhage (1) or kidney infection (5).

5.4   a) The logical way to work out this sequence is to first convert the amino acid sequence into a mRMA nucleotide sequence.

Thus

Meth	Meth	Tryp	Ile	Tyr
AUG	AUG	UUG	AUU	UAU
xx	xx	xx	AUC	UAC
xx	xx	xx	AUA	xx

Remember that codons read from the 5' end of the messenger. Therefore a possible mRNA for this sequence is:-

5'AUGAUGUUG(AUU)(UAU)3'

alternatively the (AUU) sequence can be replaced by AUG or AUA and the (UAU) by UAC.

These mRNA are transcribed against DNA in an antiparallel manner, ie RNA is synthesised from the 5'end reading DNA in the 3' —> 5' direction.

Since A pairs with T, U with A, C with G then the DNA sequence could be

3'TACTACAAC(TAA)(ATA)5'

Note however that we have no evidence which of the three possible codes for isoleucine is used in the natural system. Thus (TAA) could be replaced by TAG or TAT. Likewise (ATA) could be replaced by ATG.

Thus the nucleotide sequence which could code for the amino acid sequence in the genome could be

3'TAC	TAC	AAC	TAA	ATA 5'
xx	xx	xx	TAG	ATG
xx	xx	xx	TAT	xx

To produce a probe which will hybridise with this we should produce a complementary nucleotide sequence

5'ATG	ATG	TTG	ATT	TAT
xx	xx	xx	ATC	TAC
xx	xx	xx	ATA	xx

Thus suitable probes would be those detailed as sequences 3 and 5 in the question.

b) The chosen amino acid sequence is quite good because most of the amino acids have single codons. (ie Meth = AUG; Tryp. = UUG) which means that probes for these can be synthesised with certainty. Sequences which contain amino acids with several possible codons increases the number of possible oligonucleotide probes that would need to be produced to ensure hybridisation with the target gene.

5.5

1) Normal CHO cells are not sensitive to ampicillin. Incorporation of the SV40 vector into these cells will have no effect on this lack of sensitivity and therefore, ampicillin cannot be used as a selective device for distinguishing between 'normal' and 'infected' CHO cells.

2) There is no evidence for SV40 vector directed coat protein being produced on the surface of CHO cells - therefore this is not the criteria.

3) CHO cells do not contain dihydrofolate reductase and cannot produce certain amino acids. If these amino acids are omitted from the medium then only those CHO cells which have been transfected with the SV40 vector carrying the gene for this enzyme will be able to grow. This was the selective device used to identify the transfected CHO cells.

4) The assay is difficult to carry out and can only be detected after the cells have grown. This assay of the enzyme was not used to select transfected CHO cells. Demonstration of this enzyme does however, confirm the presence of an active SV40 expression vector.

5) Once cells carrying the SV40 expression vector have been selected for by option 3), it is quite important to demonstrate that they produce EPO. An immunoassay could in principle be used to detect such production.

5.6

EPO is a relatively large molecule and would be capable of inducing antibody production if introduced into an alternative animal species (ie it is immunogenic). The likelihood is that EPOs from different animals are sufficiently different to each other that the introduction of, for example human EPO into dogs would stimulate the recipient's immune system which would show an immune response to subsequent introduction of human EPO. It might be anticipated that the introduction of exogenously produced human EPO into humans would not cause such an immune reaction. It is important however to be cautious. We learnt in the insulin case study that rDNA human insulin does induce an immune response in some cases when it is introduced into humans. This probably arises because the insulin is not delivered into the circulation system in exactly the same way and environment as occurs naturally through its release from the Islets of Langerhans cells. An analogous circumstance could arise with the administration of exogenously produced EPO.

One final point about the results described in the question. A great difficulty posed by the production of human proteins is the suitability of tests conducted in animals. It could well be argued, as it was for insulin, that tests in animals pose problems (ie immunological reactions) which are irrelevant to the use of the material in humans. The manufacturer is faced with something of a dilemma. On the one hand, past practice has demanded that pre-clinical tests be done on animals, while on the other hand such tests are likely not to generally reflect the pharmacological and toxicological properties of the material in human. Thus in the case of EPO, contra-indications in animal tests must be treated critically and need not prove to be sufficient to abandon the product. It should also be noted that many of the tests were conducted using doses much greater than those anticipated for clinical use.

5.7

1) No. No clinician would administer a new drug to a volunteer to find out how much is needed to poison him! The wide range was used because the therapeutic dose had not been established. The doses were certainly kept below the levels at which they might be toxic.

2) Yes. Erythropoietin has to stimulate cell division and development of red blood cells. This is not an instant process and the half-life of administered erythropoietin is only a matter of about 5 hours.

3) Yes. Reticulocytes are maturing red blood cells. High levels of reticulocytes therefore signify increased red blood cell development.

4) The answer "no" is a little too definite. But healthy volunteers are not the best candidates because they already have sufficient erythropoietin and, therefore red blood cells (ie they have a high basal level upon which to detect increased red blood cell numbers). It would be expected that those with depressed erythropoietin production would show a greater response. This was shown in other trials.

5) Yes. If the volunteers diets were depleted of any of the requirements for erythrocyte production (eg iron, Vitamin B_{12}, B_6 etc) then erythropoietin would not be anticipated to stimulate this process.

6) No. The pharmacokinetic studies certainly showed that the method of administration (i.v, s.c) greatly influenced the levels and turnover rates of erythropoietin, but it did not seem to influence erythropoiesis.

5.8

The answer is that they are all true.

The list is important because these were the conclusions that have been drawn from the clinical studies. If you did not get this question right, we suggest you write out the list of conclusions given in SAQ 5.8 and re-read the pre-clinical and clinical testing sections again. This time, however, see which of the conclusions can be drawn from each study. Now return to the text to find out what else needs to be learnt of Eprex®.

Responses to Chapter 6 SAQs

6.1. Hepatitis B can be a devastating infectious disease. It has been estimated that there are over 25 million chronic carriers worldwide. Of these approximately one in five will die of liver cirrhosis, and one in 20 will die of liver cancer. As there is no cure once a person is infected, a vaccine is needed to prevent such infections.

6.2. We can cite three main reasons: plasma-derived vaccine is in limited supply; the cost of processing this vaccine is high and there are many adverse reactions to this vaccine.

You should also note that as soon as the program, initiated by the WHO, for world-wide vaccination started, the number of carriers decreased. In future this number will probably decrease significantly. This means that the plasma available from carriers, and thus the supply of vaccine, will be limited.

Because the vaccine is prepared from the plasma of carriers, extensive processing and safety testing are necessary to ensure that the vaccine is pure and free of any extraneous live material.

The acceptance of a vaccine derived from human plasma by persons eligible for vaccination, is less than expected given its safety and efficacy.

6.3. All hepatitis B subtypes have determinant 'a' in common.

6.4. Antibodies produced in response to an antigen may cross react with a different antigen, because they share determinants (eg the a determinant). Thus vaccination with subtype adw will protect against subtype adr or ayw. It would not, however, protect against a subtype which carried no shared determinants with the adw subtype.

6.5. There are several differences but the main ones are:

the HBsAg produced by the yeast cell remains in the cell, whereas the mammalian cell secretes the antigen;

all yeast-derived HBsAg is non-glycosylated versus approximately 75% glycosylation of the mammalian-derived HBsAg.

6.6. There are several reasons for selecting yeast mainly connected with the ease of culture and the safety of the product. Yeast cell cultures can be readily scaled up and the molecular stability of recombinant yeast cells during fermentation can be controlled more precisely than with mammalian cell cultures.

Mammalian cells have one or more properties associated with *in vivo* tumorigenicity or *in vitro* transformation, and may harbour endogenous proto-oncogenes and retroviruses. It is also impossible to ensure total removal of host cell DNA from the product.

6.7.   You should have produced a sequence similar to that described below. The stages are the fragmentation of hepatitis B viral DNA and its insertion into the plasmid pBR325 (identified in *E.coli* by its sensitivity/resistance to ampicillin and chloramphenicol) and its subsequent insertion into pBR322 and PC1/1 and cultivation in yeast.

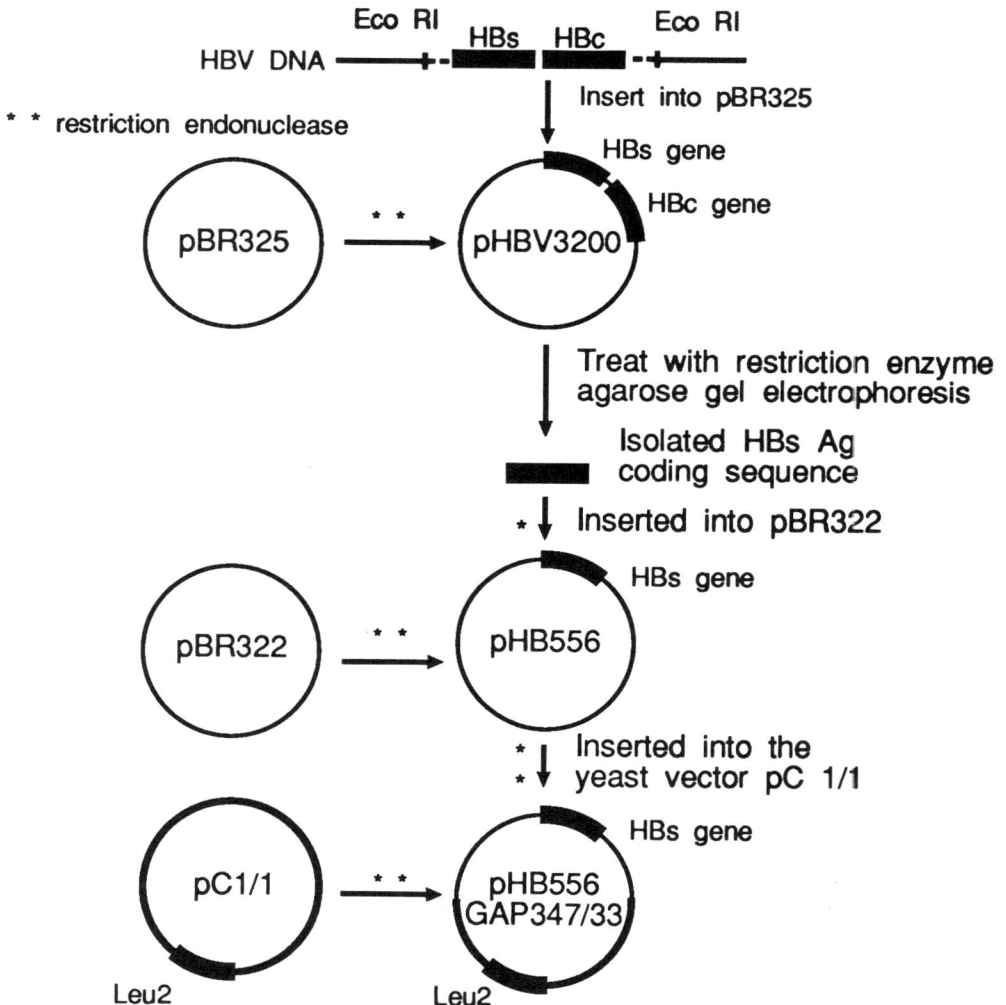

6.8.1.   i)   Cells which contain the plasmid, also have the Leu-2 gene and can grow in the absence of leucine in the media. Cells which do not carry the plasmid are unable to synthesise leucine and therefore need it supplied in the growth medium. B is the correct answer.

ii) This can be achieved by getting a measure of the amount of plasmid DNA present in a known number of cells. Technique G is designed especially for this purpose. G is, therefore, the correct answer.

iii) Here we are concerned with the fine detail of the structural gene. This can be determined by sequencing the nucleotides present. Technique A is the correct answer.

iv) It is much too complex to try to sequence the whole plasmid. Treating the plasmid DNA with restriction endonucleases and analysing the fragments by their electrophoretic mobility provides a suitable way of comparing plasmids. This technique will not, however, detect small changes in the plasmid. Choice D is the correct answer.

6.8.2. The product is a protein. The most straightforward techniques to use to compare proteins are Radioimmunoassay (F) and Western Blotting (E). Circular dichroic spectroscopy (C) gives quite detailed information about the structural configuration of protein and could be used.

6.9.

1) Pyrogenicity test is for safety, therefore should have been labelled with a C.

2) SDS-PAGE depends on the structure/size of a protein and is used to measure stability - it is also carried out *in vitro*. This test should have been labelled A and B.

3) Mouse potency test is used to measure the efficacy of a vaccine and to measure its stability. It is carried out *in vivo*. Therefore it should be labelled A and D.

4) Microbiological sterility test is carried out to ensure that the vaccine does not provide a vehicle for the transmission of disease. It is therefore a safety test (C).

5) Monoclonal antibodies identify and react with specific structures. In this case they are used to identify and assay hepatitis B antigen. They are therefore used as a test of antigen (vaccine) stability. The test is carried out *in vitro* therefore you should have labelled this as A and B.

6) The protective value tests give a measure of how good the vaccine is in protecting an animal against a particular infection. The answer is therefore D.

6.10 First you should have taken logarithms of the titres ie

Titre	Log Titre
16	1.2041
300	2.4771
420	2.6232

100	2.0000
7	0.8451
90	1.9542
400	2.6021
10	1.0000
100	2.0000
100	2.0000
Add these together	**18.7057**

Divide by number of samples to give a mean value = 1.87057. Take the antilog of this number = 74.2. This is the GMT value.

6.11. This was a little bit of a trick question. The correct response was a), c) and f).

In safety assessment of a vaccine study, the emphasis is on discomfort at the injection site and possible rise in temperature in addition to the person's general well-being. The other features listed are concerned with the efficiency of the vaccine not with its safety.

6.12. If individuals are seronegative it means that they have not been infected by the virus. Therefore the ability of the vaccine to stimulate antibody production can be measured. When a subject is not seronegative for one of the HBV serologic markers, this means that he or she has been infected with the virus. When you vaccinate an HBsAg-positive subject, you just add more HBsAg to the pool of HBsAg being produced in the body. When a subject is anti-HBs-positive, you may however boost this subject by vaccination, resulting in an increasing anti-HBs titre.

Note: It is meaningless to vaccinate persons for a disease they have already experienced as they have acquired immunity unless you want to increase their immunity further.

6.13. A seroconversion rate reflects the percentage of vaccinated individuals who developed antibodies in response to the vaccine administered.

6.14. A responder is a vaccinee whose anti-HBs titre equals or exceeds 10 IU/l. A hyporesponder has a anti-HBs titre which lies below 10 IU/l.

If no anti-Hbs antibodies can be demonstrated in the serum than the individual is a non-responder.

6.15. Although a vaccine may be potent (ie elicit an anti-HBs titre) it may not be effective. The elicited antibodies may not recognise the natural antigen. In other words, the potency of a vaccine does not say anything about the efficacy of a vaccine. To demonstrate efficacy in man, one needs to carry out a vaccination study in a group in whom the incidence of the infection is high. As over 90% of newborns born to HBs-positive mothers become chronic carriers, a study was carried out in children born to these HBs-positive mothers. The result was that 4.8%, instead of the predicted 90%, became chronic carriers, thereby proving the efficacy of recombinant hepatitis B vaccine.

Responses to Chapter 7 SAQs

7.1.

1) False - Class II MHC antigens are only found on leucocytes. Class I MHC antigens are found on all nucleated cells (Figure 7.1.).

2) True - If the antibodies react with the Class II antigens on the surface of lymphocytes, then these lymphocytes will become sensitive to the addition of complement. This is the basis of the cytotoxic antibody matching test carried out just prior to transplantation (Figure 7.3.).

3) True - See Figure 7.4.

4) True - See Figure 7.4.

5) False - Interleukin 2 predominantly acts by stimulating the proliferation of T-cells. Activated T-Helper cells stimulate B-cell proliferation (and therefore antibody production) using different lymphokines (eg B-cell growth factor). There may however be some minor reduction in antibody production since interleukin 2 can also stimulate T-Helper cells (See Figure 7.4.).

6) False - There is dual recognition between T-Helper and Antigen Presenting Cells - this recognition involves both the antigen and the Class II surface antigens. Thus T-Helper cells will recognise and respond to APCs carrying the appropriate (ie self) Class II antigen as well as the foreign antigen.

7) False - Both T-Helper and T-Suppressor cells carry the surface antigen CD3. Therefore the antigen cannot be used to distinguish between them. CD4 is only found on T-Helper cells, so this can be used to identify this sub-population of T-cells.

7.2.

(Refer to Figure 7.7. and Table 7.1.)

1) The most selective agent of those listed for stopping interleukin 1 activity are corticosteroids. The other agents will interfere with interleukin 1 but only in an indirect manner.

2) Although antimetabolites such as azathioprine and alkylating agents such as cyclophosphamide will inhibit cell proliferation, they are not selective. The most selective way to inhibit T-cell proliferation is to inhibit interleukin 2 production and release by T-Helper cells. The best agent of those listed is cyclosporin.

3) The antimetabolites azathioprine and the alkylating agent cyclophosphamide will inhibit both B- and T-cell proliferation. Note, however, that they also inhibit the proliferation of other cells and their use therefore invariably produces a large number of side effects. If we used polyclonal antibodies against lymphocytes (ALS) these would **remove** B and T cells.

4) Administration of polyclonal antibodies against lymphocytes (ALS) or T cells (ATG) will result in the removal of T-cells from the blood stream. Their use, however, often results in undesirable side effects because such preparations contain such a variety of antibodies against a number of host antigens. Monoclonal antibodies against the T-cell specific CD3 antigen are much more selective. They will only cause the removal of cells carrying CD3 surface components. Therefore monoclonal antibodies against CD3 antigen would be the reagent of choice.

7.3. A more selective immune suppressor to OKT3 would be option b) - a MoAb against surface antigen CD4. This is because CD4 is confined to T-Helper cells. After the induction of the rejection process by the APC, the T_H-cell is the starting point for IL-2 secretion and proliferation. By removing T_H-cells, there should be an interruption of this cascade more specifically than just removing all CD3-positive cells. Indeed, an anti CD4 antibody would remove T_H cells with a sparing of the T-Suppressor (CD8-positive) subset.

7.4. 1) The answer is c) lymph cells. The CD3 is an antigen found on the surface of T-cells. It is not produced by the other cell types listed.

2) In principle a) or b) could be used although in practice b) is the preferred method. CD3 is not a simple molecule and it would be difficult to prepare it in a radioactive form, thus it would be difficult to use technique a). On the other hand, T-cells contain CD3 as a surface component. Therefore, hybridoma cells producing antibodies against CD3 would bind T-(lymphocyte) cells. Platelets do not produce CD3 and therefore would not be bound by hybridomas producing antibodies against CD3, therefore c) is incorrect. Usually in selecting anti-CD3 producing hybridomas a blood platelet incubation is included as a negative control. There is no reason to suspect that hybridomas producing anti-CD3 antibodies will especially inactivate viruses, therefore option d) is wrong.

3) a) and b) are correct. Filters with a pore size of 0.22μm will remove bacteria and other metabolising microbes (eg protozoans, fungi, yeast). Therefore after filtration they should be free of these microbes. Since these filters remove all organisms capable of metabolism, they should prevent the spoilage of the product by such organisms.

Such membranes, however, will not exclude viruses. These are smaller than the pore sizes and will pass through into the product. Thus answer c) is incorrect.

Antibodies are quite water soluble and are readily dispersed in aqueous solution. Filtration will not aid this process and therefore answer d) is incorrect.

7.5. 1) For the purified antibody tests a), c), d), e), f) and g) should be done. Tests a) and g) are examples of tests to prove the product contains no infective agents. Test c) proves the product is negligibly contaminated with murine derived genetic material. Test d) proves there are no pyrogenic toxins present and tests e) and f) confirm molecular integrity and prove the identity of the product.

2) For confirming the safety of the cell banks essentially we have to examine them for infective agents, thus tests a) and g) are carried out.

3) To test the status of the mice used to produce ascites fluid we again seek evidence for the presence of infective agents. After quarantine, evidence of the viral-free status of the mice is sought by looking for anti-viral antibodies in mice sera. Thus tests b)

and g) would be conducted. Also evidence for the presence of mycoplasmas would be sought (test a)

Note the tests listed in this question are only a few examples of the many test which have to be conducted. Further tests are provided in Tables 7.2. - 7.7. in the text.

Responses to Chapter 8 SAQs

8.1. Use the diagram below to help you follow this answer. You should have placed:

the Us label around the region gX to gI;

the short repeats at the left hand end;

the long repeat between genes gH and gX (gX is in the Us region), ie gH is separated from the Us region by a long repeat;

the inverted repeat at the right hand end;

the thymidine kinase (TK) could have been placed anywhere between the short repeat and long repeat since all we have told you is that it is found in the 65 kbp region. Actually it is found between gII and gH;

the segment between the short repeats and long repeat is 65 kbp long;

the segment between the long repeat and the inverted repeat is 6 kbp long;

the long repeat and inverted repeat are 9.9 kbp long.

Thus you should have written:

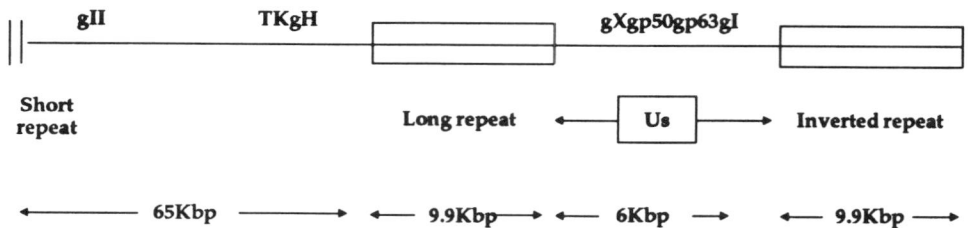

Could you have exchanged the positions of the labels of 'long repeat' and 'inverted repeat'?

The answer is yes because both segments are equivalent to each other and each are inverts of each other!

8.2. 1) A - Adjuvants may be added with attenuated viruses but they are not usually necessary. With inactivated viruses and purified antigens, adjuvants are usually necessary to stimulate a greater immune response.

2) B and C - This tends to favour inactivated (dead) viruses or purified antigens since these will not revert into virulent form. Attenuated viruses may be restored to virulent forms by the acquisition of new genes or by repair of modified genes.

3) A - Live vaccines create a greater immune response than inactive ones.

4) Favours A and, especially B. If knowledge of the specific antigens produced by a virus is lacking, it is difficult to visualise how such antigens can be selected to produce purified antigen based vaccines. It must also be borne in mind that most viral diseases are caused by a 'family' of slightly different viruses each with a slightly different array of antigens. It is important to choose an antigen present on all forms of the viruses to provide full protection. If you recall the hepatitis B vaccine, 'a' antigenic determinants were present in all viruses whereas other determinants (eg 'd' 'w') were not.

5) B and C - Vaccines based on inactivated viruses and purified antigens are more stable than live, attenuated viral suspensions.

6) C - Purified antigens are chemically and physically simpler than complete viral particles. It is therefore easier to characterise them more thoroughly.

8.3.

1) *Hind* III cuts PRV DNA in the inverted repeats on either side of the Us region.

2) A combination of Ml (*Mlu*I) and Bg (*Bgl* II) will cut out the major part of the sequence.

3) The enzymes Sp (*Sph*I), Bg (*Bgl* II) and Bl (*Bal* 1) will each cut the sequence in a single place as the Us region only contains a single recognition site for these three restriction enzymes. In the case of producing the deletion mutants of PRV, Bl was selected.

4) *Bal* 31 is an exonuclease, it therefore hydrolyses DNA from the ends of the molecule. *Bam*HI is a restriction enzyme which will cut DNA at its recognition sites. T4 polymerase is a DNA synthesising enzyme. T4 ligase joins the ends of DNA molecules. *Hind* III is a restriction enzyme and cuts DNA at its recognition sites.

5) pHB2.1 - this plasmid carries a *Hind* III fragment of PRV in which 2.1 kbp have been removed. pHB2.4 is similar but with 2.4 kbp removed while pHB2.8 has 2.8 kbp removed.

8.4.

1) The strains which carry an active (intact) thymidine kinase gene would be most sensitive to bromodeoxyuridine (BdUr). Of those listed, only 2.4 N3A TK 783 has an inactive TK gene, ie all of the other strains would be sensitive.

2) Thymidine kinase is needed to phosphorylate ^{14}C-thymidine leading to its incorporation into DNA. Thus strain 2.4 N3A TK 783, which does not contain an intact TK gene, is least likely to facilitate ^{14}C -thymidine incorporation into DNA.

3) NIA3 is a virulent strain of PRV. Inoculation of a single animal will undoubtedly cause illness to the infected animal and the viruses released from this animal are likely to infect other herd members. 2.3 N3A TK 783 is an attenuated strain. The evidence is that it would not cause the symptoms of the disease and the animal would develop antibodies (immunity) against PRV. Other animals in the herd would not be affected.

8.5.

Strategy I - This strategy is very similar to that used to produce a vaccine for hepatitis B. In this case, however, it is unlikely to work. Although Gly 1 is produced by all of the viruses tested, it does not seem to be part of the virus. Thus although a vaccine using

Gly 1 would stimulate antibodies against Gly 1, these antibodies would not inactivate this molecule since it is produced and used within the host cells.

Strategy II - This is experimentally similar to Strategy I except that capsomer proteins are used. Vaccines based on this product are unlikely to work because of the epitopic differences between capsomers from different viruses. Thus although raising antibodies against capsomers from one virus may provide protection against the virus producing the capsomers, it is unlikely to protect against related viruses with different capsomers. It should also be noted that the presence of envelopes would also reduce the efficacy of antibody binding to the capsomers.

Strategy III - This looks quite promising. Since incubation of viral DNA with EcoRI results in only 3 fragments, one of which carries Gly 1, then in principle it should be relatively easy to produce viruses with this gene deleted.

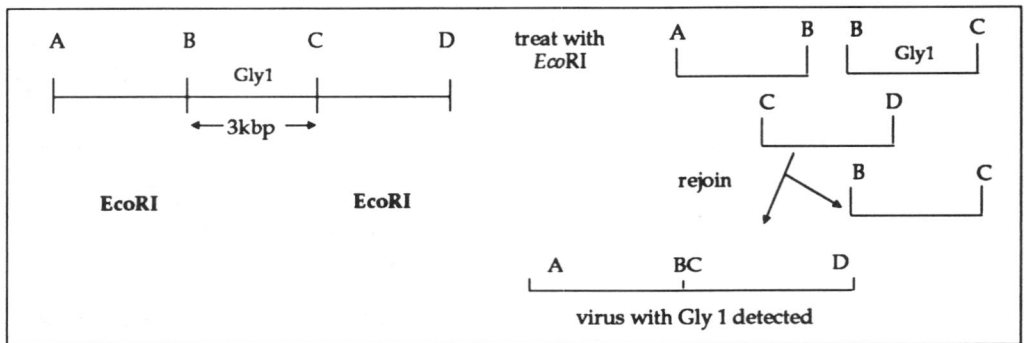

Such a virus would have many antigenic determinants and might offer universal protection against the viruses causing the disease in much the same way as 783. provides protection against Aujeszky's disease. Since Gly 1 is needed for maturation, the strain would almost certainly be attenuated. Whether or not it would be attenuated too much or too little would require thorough testing.

Responses to Chapter 9 SAQs

9.1.
1) It was not particularly important that indium-111 had two emission peaks. What was important was that the energy of these emission peaks was close to the energy levels maximally detected by the gamma scintigraphy equipment

2) See 1) - Although other radionuclides could be bound to the Fab fragments, the energy of their emissions were not optimal for gamma scintigraphy. This is especially true for I-125 and I-131

3) In some ways this is rather superfluous. If the radionuclide gamma emissions are in the sensitivity range of the gamma scintigraphy equipment, then the radiation energy would be sufficient to pass through the body tissues. Gamma radiation is relatively non-interactive with biological material compared to beta and alpha radiation. This has two major consequences. Firstly, it readily penetrates biological tissues and secondly, because of its relatively low adsorption, it causes less radiation damage

4) Indium-111 is not particularly reactive. This is important because it is not desirable to use a highly reactive species which may react and bind indiscriminately with body constituents.

5) This was an important consideration. On energetic consideration technetium (Tc) - 99m would appear to be a better radionuclide to use but no reliable method of binding this to the antibody preparation could be found. The very strong binding of indium by DTPA (equilibrium constant 10^{28}) means that once the indium is bound to DTPA attached to the antibody, it will not be readily released.

9.2.
1) We can cite several reasons for testing for viruses in monoclonal antibody preparations. Since these products have been produced from cancerous (myeloma) animal cells they may contain either the agent which caused the transformation of the animal cells or latent viruses carried by these cells. Since the product is to be administered to humans, it is not desirable to use a product which may contain viruses some of which may carry oncogenes.

2) a) The answers to this are not quite straightforward. 0.22μm filters will not remove viruses. Such filtration is used to remove metabolically active cells particularly bacteria. Such filtration removes the chances of transmitting bacterial infections and decrease the chances of biological degradation of the product

2) b) Chemical sterilisation of equipment is predominantly designed to remove metabolically active cells (e.g. contaminating bacteria, mycoplasmas etc). It will, of course, also reduce the chances of viruses being transmitted.

2) c) Spiking studies were undertaken to demonstrate that, even if viruses were present in the production processes, the downstream process would significantly reduce their numbers in the final product. This was part of the process validation analysis and was important in demonstrating that Myoscint was unlikely to contain viruses.

2) d) Such tests are part of the quality control process and are carried out on cell lines as well as with the final product.

2e) Although DNA analysis of the final product will show that it is not heavily contaminated by DNA viruses, this test is not specifically designed to demonstrate the absence of viruses. It is designed to show that the protein product is not heavily contaminated by DNA from any source but especially by DNA from the hybridoma cells.

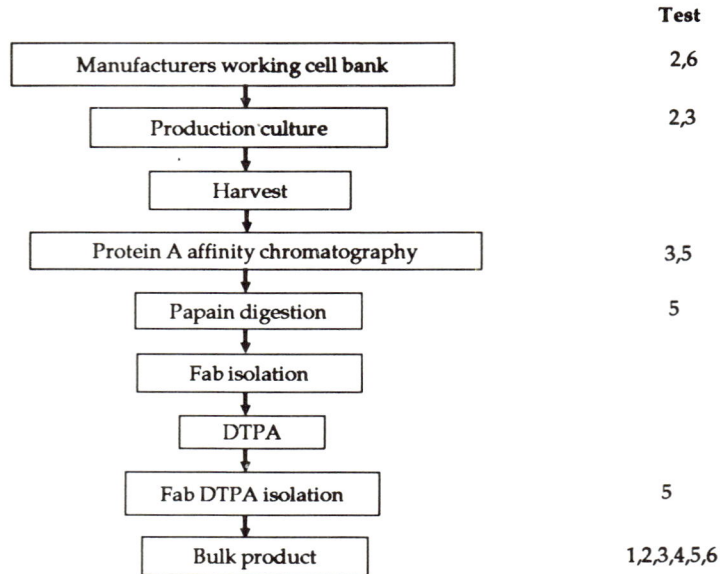

	Test
Manufacturers working cell bank	2,6
Production culture	2,3
Harvest	
Protein A affinity chromatography	3,5
Papain digestion	5
Fab isolation	
DTPA	
Fab DTPA isolation	5
Bulk product	1,2,3,4,5,6

9.3. We have chosen not to include all the test that could (and are) conducted at each stage in order to convey particular messages. The messages we would like to convey are:

at the stages of production where cells are involved, the primary concern is to ensure that cultures are microbiologically clean. Of particular concern is the presence of viruses mycoplasmas and bacteria. At later stages in the process, when we are attempting to isolate specific molecular species, we become more concerned with the identity and purity of this species. Therefore the tests undertaken become predominantly chemical in nature. In this case we have indicated Gel-Filtration HPLC as the technique of choice, but other techniques could be used such as SDS-PAGE or immunoelectrophoresis. The choice of method(s) used is made on the basis of the confidence that can be placed on the data produced as well as on issues of cost and speed;

in the final phase of the process, we are concerned with both its composition and the possibility of chemical and biological contamination. Thus a large number of tests need to be undertaken to ensure that the product meets the stringent specifications of potency and safety.

You may well have indicated slightly different tests than those we have included on our diagram. For example, in monitoring papain digestion, you may have selected SDS-PAGE. In principle, this could be used to monitor the release of Fab fragments from IgG but in practice this is rather slow and less precise than the data derived from gel-filtration HPLC. Re-examine your selection of tests and see if you can justify your selections by addressing the following questions:

Is the test I have selected testing for the presences of microbial or chemical contamination, or is it concerned with the chemical identity of the product?

Is the test I have chosen appropriate for particular stage(s) in the process?

Is the technique I have chosen the best I could have made in terms of precision/sensitivity/reliability and in terms of speed and cost?

Are there particular tests that the regulations demand must be carried out at particular stages?

Appendix 1

Names and addresses of competent authorities

Belgium:

Ministere de la Sante publique
Inspection generale de la
Pharmacie
Cite Administrative, Quartier
Vesale
B-1010 Bruxelles
Ministerie van Volksgezondheid
Farmaceutishe Inspektie
Vesalius Gebouw, B-1010 Brussel
Tel: (32) (2) 210 49 00 and 210 49
01
Telex: 25768 MVGSPF B
Telefax (32) (2) 210 48 80

Spain:

99 Ministerio de Sanidad y
Consumo Direccion General de
Farmacia y
Productos Sanitarios
Paseo del Prado, 18-20
E-28014 Madrid
Tel: (34) (1) 467 34 28, Telex:
22608 MSASS Telefax: (34) (1)

Germany:

Institue fur Arzneimittel des
Bundesgesundehitsamtes
Seestr 10
D-1000 Berlin 65
Tel: (49) (30) 45
Telex: (2627) (17) 308062 BGESA D
Telefax: (49) (30) 4502207

For sera, vaccines and allergens:

Paul-Ehrlich-Insitut
Bundesamt fur Sera und Impfstoffe
Paul-Ehrlich-Str 42-44
Postfach 700810
D-6000 Frankfurt/M 70
Tel: (49) (69) 634402
Telex/Teletex: 6990716
Telefax: (49) (69) 63 44 02

Denmark:

Sundhedsstyrelsen
Lagemiddelafdelingen
Frederikssundsvej 378
DK-2700 Bronshoj
Tel: (45) (2) 94 36 77
Telex: 35333 IPHARM DK
Telefax: (45) (2) 84 70 77

France:

Ministere de la Solidarite
de la Sante et de la Protection sociale
Direction de la Pharmacie et du
Medicament 1, place de Fontenoy
F-75700 Paris
Tel: (33) (1) 40 56 60 00
Telex: 250011 SANTSEC F
Telefax: (33) (1) 405 65 355

Greece:

E.O.F. (National Drug
Organisation)
Voulis Str 4
Athens 10562
Tel: (30) (1) 323 0911, Telex:
223514
Telefax: (30) (1) 323 86 81

Ireland:

National Drugs Advisory Board
63-64 Adelaide Road
Dublin 2
Tel: (353) (1) 76 49 71 - 7, Telex:
90542 Telefax: (353) (1) 78 60 74

Italy:

Ministero della Sanita, Servizio
Farmaceutico Viale della Civilta
Romana, 7
I,00144 Roma, EUR
Tel: (39) (6) 592 58 63
Telex: 625205 MINSAN I
Telefax: (39) (6) 592 58 24

Netherlands:

College ter beoordeling van
geneesmiddelen P.O. Box 5811
NL-2280 HV Rijswijk
Tel: (31) (70) 40 70 08
Telex: 31680 WVCRW NL
Telefax: (31) (70) 40 50 48

United Kingdom:

Department of Health and Social
Security Medicines Division
Market Towers
1 Nine Elms Lane
London SW8 5NQ
Tel: (44) (1) 720 2188
Telex: 883669 DHSSHQ G
Telefax: (44) (1) 720 5647

For delivery of the dossier:

Department of Health
Medicines Division
Britannia House
7 Trinity Street
London SE1 1DA
Tel: (44) (1) 407 5522

Luxembourg:

Direction de la Sante
Division de la Pharmacie et des
Medicaments 10, rue C. M. Spoo
L-2546 Luxembourg
Tel: (352) 4 08 01, Telex: 2546 SANTE
LU

Portugal:

Ministerio da Saude
Direccao Geral dos Assuntos
Farmaceuticos Av Estados Unidos
da America, 37
P-1700 Lisboa
Tel: (351) (1) 80 41 31
Telex: 15655 MAS P
Telefax: (351) (1) 88 03 31

Address of the Secretariat of the
Committee for Proprietary
Medicinal Products:

DG III B 6 "Pharmaceuticals,
veterinary medicines", Commission
of the European Communities, rue
de la Loi 200, B-1049 Brussels,
Telephone: 236 03 32/235 69 35,
Telex: 21877 COMEU B

Appendix 2

1. EC directive on the contained use of genetically modified micro-organisms

1.1. Purpose of the directive

Article 1 gives the purpose of the directive as:

"to lay down common measures for the contained use of genetically modified micro-organisms with a view to protecting human health and the environment".

1.2. Scope

The scope of every regulation depends on the definitions and the exemptions.

This directive covers contained use of genetically modified micro-organisms. In the following paragraphs, therefore the definitions of the terms genetically modified, micro-organisms, contained use, and exemptions are explained.

1.3. Definitions

1.3.1. Micro-organisms

Micro-organisms are defined as:

"any microbiological entity, cellular or non-cellular, capable of replication or transferring genetic material".

It should be noted that this directive only covers micro-organisms, and not all organisms. It was argued that including all organisms in this directive would given an unacceptable delay in the implementation of the proposals. The Commission undertook to keep the whole biotechnology sector under review and to make appropriate proposals to extend the scope of this directive to genetically modified organisms.

The national Member States may maintain and adopt national measures for the contained use of organisms other than micro-organisms.

Genetically modified micro-organisms are in this document abbreviated as GMMOs.

1.3.2. Genetically modified micro-organism

Genetically modified micro-organisms are defined as:

"micro-organisms in which the genetic material has been altered in a way that does not occur naturally by mating or by recombination".

In order to specify this, a list of techniques is given by which genetic modification can occur.

This non-limitative list is Annex Ia, of the Directive and contains:

- recombinant DNA techniques using vector systems;

- techniques involving the direct introduction of heritable material prepared outside the micro-organism, such as micro-injection;

- cell fusion and hybridisation techniques by means or methods that do not occur naturally.

It should be noted here that the scope of the regelation is not limited to recombinant DNA techniques. This gives recognition to the fact that a variety of molecular genetic transformation techniques, including recombinant DNA, are widely used and may have similar safety considerations.

1.3.3. Contained use

Contained use means:

"any operation in which micro-organisms are genetically modified or in which such organisms are cultured, stored, used, transported, destroyed or disposed of and for which physical barriers together with chemical and/or biological barriers, are used to limit their contact with the general population and the environment".

The key term here is the physical barrier.

1.4. Exemptions

The directive does not apply where genetic modification is obtained through the use of certain techniques (Article 1a). These techniques are:

- mutagenesis;

- construction and use of somatic animal hybridoma cells;

- cell fusion of plants which can also be produced by traditional breeding methods;

- self cloning of certain non pathogenic naturally occurring micro-organisms.

The directive does furthermore not apply to:

- the transport of GMMOs;

- GMMOs which have been placed on the market under Community legislation (Article 3)

1.5. System of the directive

1.5.1. Group I and Group II organiss/Type A Type B operations

Group I and Group II organisms

For the purpose of the directive, micro-organisms are classified in two groups (Article 2).

Group I - those satisfying certain criteria

Group II - those other than group I

Group I organisms have a long record of safe use and are considered to be safe when used under specific conditions. The criteria for group I are given in Annex II of the directive, which gives criteria for the recipient and parental organisms (non pathogenic etc), to the vector and the insert used, and to the final GMMO. This Annex is based on the criteria for GILSP (Good Industrial Large Scale Practice) set up by the OECD.

For GMMOs of Group I, principles for good microbiological practice and of good occupational safety and hygiene shall apply. These principles are also based on the OECD report of 1986.

In addition to these principles, certain containment measures set out in an Annex shall be applied to ensure a high level of safety for GMMOs of Group II (Article 5).

Type A and Type B operations

For the purpose of the directive, a distinction is made between Type A and Type B operations.

Type A operations are operations used for teaching, research and development, or non-industrial or non-commercial purposes and which are of a small scale (e.g. 10 litres volume or less).

Type B operations are operations other than operations of type A.

1.5.2. System of the directive

The regulatory system of this directive contains two sorts of procedures:

- activities for which a notification is required;

- activities for which an authorisation is required.

The two distinctions can be combined to four possible activities:

- Type A operations with Group I - IA operations;

- Type B operations with Group I - IB operations;

- Type A operations with Group II - IIA operations;

- Type B operations with Group II- IIB operations.

In addition to this, the first use of an installation for an operation involving GMMOs is considered to be an activity for which a procedure is required.

Articles 6, 7 and 8 of the Directive assign specific procedures to each of the possible activities mentioned above and form the basis of the procedures.

The possible procedures are:

- to keep records of the work carried out and make them available to the competent authority (Article 7.1.);

- a notification within a reasonable period before commencing the use (Article 6);

- a notification and a waiting period (Article 7.2. and 8.1.);

- an authorisation (Article 8.2.).

The possibilities are presented below:

1.5.3. Optional stricter measures

Article 9.5. gives the competent authorities the possibility to require that for operations notified under Articles 6, 7.2. and 8.1., its consent must be given.

1.6. Additional provisions

Articles 11 to 14 lay down some specific obligations for the Member States:

- to ensure that, where necessary, before an operation commences an emergency plan is drawn up and information on safety measures is available (Article 11);

- to ensure that, in case of an accident, the proper measures and steps will be taken (Article 12);

- consult, when necessary, with other Member States and inform the Commission (Articles 13 and 14);

- to ensure that inspections are carried out (Article 13a).

1.7. Confidentiality

Article 14a:

"The Commission and the competent authorities shall not divulge to third parties, any confidential information notified or exchanged under this directive and shall protect the intellectual property rights relating to the data received".

The notifier indicates what information needs to be kept confidential. However, it is the competent authority which decides, after consultation with the notifier, which information shall be kept confidential.

2. EC Directive on the deliberate release of genetically modified organisms

2.1. Parts of the Directive

This directive consists of four parts:

Part A - General provisions

Part B - Research and Development (R & D) and other introductions into the environment than placing on the market

Part C - Placing of products on market

Type of operation	procedure	article
First use in an installation	notification within reasonable period in advance	6
IA operations	keep records	7.1.
IB operations	notification and waiting period of 60 days and additional 60 days at request of competent authority	7.2.
IIA operations	notification and waiting period of 60 days and additional 60 days at request of competent authority	8.1.
IIB operations	authorisation to be communicated within 60 days and additional 60 days at request of competent authority	8.2.

Part D - Final provisions

2.2. Part A: General provisions

2.2.1. Purpose of the directive

The purpose of this directive is laid down in Article 1:

"to approximate the laws, regulations and administrative provisions of the Member State and to protect human health and the environment when carrying out a deliberate release or placing on the market of genetically modified organisms".

In order to gain a better understanding of the purpose and background of this directive, the considerations in the preamble should also be studied.

In addition to the purpose of this directive Article 4 emphasises in general terms the obligations of member states in accomplishing this purpose.

2.2.2. Scope of the directive-definitions

Repeating what was explained earlier: the scope of every regulation depends on the definitions and exemptions.

This directive covers the deliberate release of genetically modified organisms. In the following paragraphs the definitions of the directive are explained. These definitions are summed up in Article 2.

Organism

Organism is defined in this directive as:

"any biological entity, capable of replication or of transferring genetic material".

Since the term 'biological entity' is open for multiple interpretation, an explanation is given in the statements for inclusion in the Council's minutes:

"This definition covers: micro-organisms, including viruses and viroids; plants and animals; including ova, seeds, pollen, cell cultures and tissue cultures from plants and animals".

Hereafter, the term genetically modified organisms is abbreviated to GMO.

Modified organism

The definition of a genetically modified organism is analogous to the definition of a genetically modified micro-organisms, provided that the term 'micro-organism' is replaced by the term 'organism'.

Deliberate release

Deliberate release is defined in paragraph 3 of Article 2 as:

"any intentional introduction into the environment of a GMO or a combination of GMOs without provisions for containment such as physical barriers or a combination of physical barriers together with chemical and/or biological barriers used to limit their contact with the general population and the environment".

A further clarification is given in the Council's Statements.:

"the introduction by whatever means, directly or indirectly, by using, storing, disposing, or making available to a third party".

By using the terms "without provisions for containment such as..." as a cross reference to the directive for contained use, a complementary system is achieved.

In other words: every activity that is not a contained use is regarded as a deliberate release.

2.2.3. Exemptions

The exemptions of the scope of this directive are laid down in Article 3:

"This directive shall not apply to organisms obtained through the techniques of genetic modification listed in Annex Ib". These include:

- mutagenesis;

- cell fusion of plant cells when the plant can also be produced by traditional methods.

The background of these exemptions is that these specific applications have been used in a number of applications and have a long safety record.

2.3. System

The system of this directive is based on two notions:

- the release of a GMO into the environment can have adverse effects on the environment which may be irreversible;

- GMOs, as well as other organisms, are not stopped by national frontiers.

These two notions led to the choice of a system whereby:

- every introduction of a GMO into the environment is subject to an authorisation by the competent authority of the country where the introduction takes place;

- before an authorisation is given, the competent authority consults the other Member States of the Community.

In addition to this, a distinction between Research and Development (R & D) and placing on the market is made.

2.3.1. R & D and placing on the market

In this directive, a distinction is made between:

- Research and Development (R & D) and other introductions into the environment than placing on the market (part B of the directive) ;

- Placing on the market (part C of the directive).

The result of this distinction and the system of the directive is that placing on the market involves a system of international consultation whereby the competent authority can not take a decision without the agreement of the other Member States.

All other introductions into the environment (Part B, which basically consists of R & D introductions) are subject to an authorisation of the competent authority which may give its decision without the approval of other Member States, though be it that these introductions are also notified to the other Member States who may give comments.

The reason for this distinction, is found in the numbers and the spread of a GMO connected with placing on the market. When a product containing GMOs or consisting of a GMO is placed on the market, it will be spread all over Europe in, possibly, vast numbers under uncontrolled circumstances. Whereas R & D introductions are normally small scale introductions of a limited number of GMOs and under controlled circumstances.

2.4. Part B: Research and Development (R & D) and introductions into the environment other than placing on the market

The basis of part B of this directive is laid down in the combination of the Articles 5, paragraph 1 and Article 6, paragraph 4.

Article 5, paragraph 1 states:

"Any person before undertaking a deliberate release of a GMO for the purpose of research and development or for any other purpose than placing on the market, must submit a notification to the competent authority of the Member State within whose territory the release is to take place".

Article 6, paragraph 4, states:

"The notifier may proceed with the release only when he has received the written consent of the competent authority, in conformity with any conditions required in this consent".

2.4.1. The notification

Article 5, paragraph 2, gives the requirements of a notification under part B of the directive:

"The notification shall include the information specified in Annex II".

Annex II is an indicative list of points of information set out under 5 headings:

I   General information

II   Information related to the GMO

III   Information relating to the conditions of release and the receiving environment

IV   Information relating to the GMO and the environment

V   Information on monitoring, control, waste treatment and emergency response plans

This Annex II is based on the OECD report of 1986. It is essential to realise that this Annex contains an indicative list, and that not all the points included will apply to every case.

2.4.2. Authorisation

The authorisation procedure is laid down in article 6.

Paragraph 1: "On receipt and after acknowledgement of the notification the competent authority shall examine the conformity of the notification with the requirements of this directive".

Paragraph 2: "The competent authority, having considered where appropriate, any comments by other Member States, shall respond in writing to the notifier within 90 days by indicating either:

- that the release may proceed;

- that the release does not fulfil the conditions of this directive and the notification is therefore rejected.

For calculating the waiting period of 90 days, the period needed for the notifier to supply further information and the period in which a public inquiry is carried out, shall not be taken into account.

The third paragraph of Article 6 gives the steps to be taken when new information becomes available with regard to the risk of the product. In that case the notifier shall revise the information, inform the competent authority and take the necessary measures to protect human health and the environment.

2.4.3. International consultation

Within 30 days after the receipt of a notification, the competent authority shall send to the Commission a summary of the notification. The Commission shall immediately forward these summaries to the other Member States which may, within 30 days, present observations. It should be stressed here that these observations are not binding to the original competent authority.

2.5. Part C: Placing on the market products containing genetically modified organisms

2.5.1. General provisions

Part C of this directive stars with Article 10, which gives in paragraph 1, a set of general conditions before any product can be placed on the market.

These conditions are that:

- consent has been given under part B of the directive, meaning that no GMO can be placed on the market without a proper R & D stage;

- the product should comply with this directive and relevant product legislation.

Paragraph 2 of Article 10 indicates that the procedure for placing a product on the market shall not apply to products covered by Community legislation which includes a specific environmental risk assessment similar to that provided in this directive. The background of this provision is that:

- it is desirable to have only one procedure for placing the products on the market;

- product legislation already contains procedures for placing on the market.

2.5.2. System

The same system of part B is found in part C.

Article 11, paragraph 1:

"before a GMO or a combination of GMOs are placed on the market as such or in a product, the manufacturer or the importer to the Community shall submit a notification to the competent authority of the Member State where they are placed on the market for the first time".

Article 11, paragraph 5:

"the notifier may only proceed when he has received a written consent".

2.5.3. The notification

The first paragraph of Article 11 says that the notification shall include the information of Annex II, information obtained from R & D releases and specific product information laid down in annex III (use, labelling, packaging etc).

The final paragraph of Article 11 gives the steps to be taken when new information becomes available with regard to the risk of the product, analogous to Article 6.

2.5.4. Authorisation

The authorisation procedure of placing on the market is in fact a two step procedure. The first step is given by article 12:

Paragraph 1: "On receipt and after acknowledgement of the notification the competent authority shall examine the conformity of the notification with the requirements of this directive".

Paragraph 2: "The competent authority shall respond within 90 days by either:

- forwarding the dossier to the Commission with a favourable opinion;

- informing the notifier that the release does not fulfil the conditions of this directive and the notification is therefore rejected;

For calculating the waiting period of 90 days the period needed for the notifier to supply further information shall not be taken into account.

2.5.5. International consultation

The second step of the procedure is laid down in Article 13:

"The Commission shall immediately forward the dossier to the other Member States which may, within 60 days, present observations are received from the other Member States".

When an objection is received and the competent authorities concerned cannot reach an agreement within these 60 days, the commission shall take a decision in accordance to a specific procedure.

2.5.6. Placing on the market: Community wide

One of the key articles of this part C is Article 15:

"A Member State may not restrict or impede, on grounds relating to the notification and written consent of a release under this directive, the placing on the market of product containing or consisting of GMOs which comply with the requirements of this directive".

This means that when a product has received a consent after the procedure of part C, no Member State may restrict the placing on the market on grounds of protecting human health or the environment.

When a Member State has justifiable reasons (e.g. new information) that such a product constitutes a risk, it may provisionally restrict the product, after which the Commission shall take a decision in accordance to a specific procedure (Article 16).

The commission shall publish a list of products which received consent under this directive (Article 18).

2.6. Part D: Final provisions

2.6.1. Confidentiality

Article 19:

"The Commission and the competent authorities shall not divulge to third parties any confidential information notified or exchanged under this directive and shall protect the intellectual property rights relating to the data received".

The notifier indicates what information he wants to be kept confidential, though be it that certain information cannot be kept confidential, like the name and address of the notifier and a description of the GMO, methods for monitoring and the evaluation of foreseeable effects.

It is the competent authority which decides, after consultation with the notifier, which information shall be kept confidential.

2.6.2. Commission procedure

The specific procedure mentioned before is explained in Article 21, which in general terms says that the Commission will be assisted by a Committee which votes by qualified majority. If measures envisaged by the Commission are not in accordance with the opinion of the committee, it will be submitted to the Council.

Appendix 3

Units of measurement

For historical reasons a number of different units of measurement have evolved. The literature reflects these different systems. In the 1960s many international scientific bodies recommended the standardisation of names and symbols and a universally accepted set of units. These units, SI units (Systeme Internationale de Unites) were based on the definition of: metre (m), kilogram (kg); second (s); ampare (A); mole (mol) and candela (cd). Although, in the intervening period, these units have been widely adopted, their adoption has not been universal. This is especially true in the biological sciences.

It is, therefore, necessary to know both the SI units and the older systems and to be able to interconvert between both sets.

The BIOTOL series of texts predominantly uses SI units. However, in areas of activity where their use is not common, other units have been used. Tables 1 and 2 below provides some alternative methods of expressing various physical quantities. Table 3 provides prefixes which are commonly used.

Mass (S1 unit: kg)	Length (S1 unit: m)	Volume (S1 unit: m^3)	Energy (S1 unit: $J = kg\ m^2\ s^{-2}$)
$g = 10^{-3}\ kg$	$cm = 10^{-2}\ m$	$l = dm^3 = 10^{-3}\ m^3$	$cal = 4.184\ J$
$mg = 10^{-3}\ g = 10^{-6}\ kg$	$Å = 10^{-10}\ m$	$dl = 100\ ml = 100\ cm^3$	$erg = 10^{-7}\ J$
$\mu g = 10^{-6}\ g = 10^{-9}\ kg$	$nm = 10^{-9}\ m = 10Å$	$ml = cm^3 = 10^{-6}\ m^3$	$eV = 1.602 \times 10^{-19}\ J$
	$pm = 10^{-12}\ m = 10^{-2}\ Å$	$\mu l = 10^{-3}\ cm^3$	

Table 1 Units for physical quantities

Concentration (SI units: mol m^{-3})

a) $M = mol\ l^{-1} = mol\ dm^{-3} = 10^3\ mol\ m^{-3}$

b) $mg\ l^{-1} = \mu g\ cm^{-3} = ppm = 10^{-3}\ g\ dm^{-3}$

c) $\mu g\ g^{-1} = ppm = 10^{-6}\ g\ g^{-1}$

d) $ng\ cm^{-3} = 10^{-6}\ g\ dm^{-3}$

e) $ng\ dm^{-3} = pg\ cm^{-3}$

f) $pg\ g^{-1} = ppb = 10^{-12}\ g\ g^{-1}$

g) $mg\% = 10^{-2}\ g\ dm^{-3}$

h) $\mu g\% = 10^{-5} g\ dm^{-3}$

Table 2 Units for concentration

Fraction	Prefix	Symbol	Multiple	Prefix	Symbol
10^{-1}	deci	d	10	deka	da
10^{-2}	centi	c	10^2	hecto	h
10^{-3}	milli	m	10^3	kilo	k
10^{-6}	micro	μ	10^6	mega	M
10^{-9}	nano	n	10^9	giga	G
10^{-12}	pico	p	10^{12}	tera	T
10^{-15}	femto	f	10^{15}	peta	P
10^{-18}	atto	a	10^{18}	exa	E

Table 3 Prefixes for S1 units

Appendix 4

Chemical Nomenclature

Chemical nomenclature is quite a difficult issue especially in dealing with the complex chemicals of biological systems. To rigidly adhere to a strict systematic naming of compounds such as that of the International Union of Pure and Applied Chemistry (IUPAC) would lead to a cumbersome and overly complex text. BIOTOL has adopted a pragmatic approach by predominantly using the names or acronyms of chemicals most widely used in biologically-based activities. It is recognised however that there remains some potential for confusion amongst readers of different background. For example the simple structure CH_3COOH can be described as ethanoic acid or acetic acid depending on the environment or industry in which the compound is produced or used. To reduce such confusion, the BIOTOL series makes every effort to provide synonyms for compounds when they are first mentioned and to provide chemical structures where clarity and context demand.

Appendix 5

Abbreviations used for the common amino acids

Amino acid	Three-letter abbreviation	One-letter symbol
Alanine	Ala	A
Arginine	Arg	R
Asparagine	Asn	N
Aspartic acid	Asp	D
Asparagine or aspartic acid	Asx	B
Cysteine	Cys	C
Glutamine	Gln	Q
Glutamic acid	Glu	E
Glutamine or glutamic acid	Glx	Z
Glycine	Gly	G
Histidine	His	H
Isoleucine	Ile	I
Leucine	Leu	L
Lsyine	Lys	K
Methionine	Met	M
Phenylalanine	Phe	F
Proline	Pro	P
Serine	Ser	S
Threonine	Thr	T
Tryptophan	Trp	W
Tyrosine	Tyr	Y
Valine	Val	V

Index

A

ADH-1 terminator, 148
alanine aminotransferase
(ALT), 155
ALG, 176
allograft, 167
ALS, 176
ALT, 155
amplification, 148
anaemia, 102 , 109 , 174
and renal disease, 102
pernicious, 105
antibodies
affinity for target, 227
against EPO, 124
against gI, 223
against myosin, 241
anti-idiotype, 181
anti-OKT3, 196
anti-yeast, 158
avidity constants, 157
cardiac myosin, 228
cross absorption, 157
human anti-mouse, 241
monoclonal, 4
to PRV, 222
antibody
cross reactivity, 227
antibody response, 155 , 160
antigen
recognition of, 171
antigen presenting cells, 171
antigen stripping, 174
antigen-antibody complex, 180
internalisation, 180
antigenemia, 155
antigenic response, 241
antigenicity
of Fab-DTPA, 241
antigens
Class I, 168
Class II, 168 , 169
antimyosin
and infarction, 242
binding, 242
blood clearance, 243
radiolabelled, 242
uptake, 242
antimyosin injection timing, 244
antisera
against lymphocytes, 176

ATG, 177
ascites, 182 , 184
aseptic filling, 188
aspartate aminotransferase
(AST), 155
AST
(aspartate aminotransferase), 155
ATG, 176
Aujeszky, 200
Aujeszky's disease
epidemiology, 200
history, 200
authorisation, 292
to market and GMOs, 293
authorities
national, 31
autograft, 167
Avery, 4
azathioprine
immunosuppressive agent, 175

B

B cell, 166
B cell growth factor, 171
B-cells, 181
b-gal
promoter, 20
b-galactosidase
promoter, 20 , 72
B2-microglobulin, 168
Bal 31, 211
bilirubin, 106
bioassay
insulin, 21
pyrogens, 153
bioburden, 239
biochemistry, 2 , 3
bioreactor, 22
bioreactors, 5
blood
glucose, 57
blood transfusions, 102
and AIDS, 103
and EPO, 132
and haemochromatosis, 102
bromodeoxyuridine, 217 , 218
bulk
final, 46
product, 46

C

C-region
> of HBV, 143
CAF1 mice, 184 , 185
capsid of PRV, 202
carcinogenicity
> myoscint, 242
CD1, 173
CD2, 173
CD3, 173 , 180 , 181 , 182
CD4, 173
CD5, 173
CD8, 173
cDNA, 64 , 67
> production of, 62 , 63
cell banks
> characterisation, 189
> Quality Control, 236
cell death
> marker, 242
Centocor, 231
chemotactants, 175
chimeric
> proteins, 70
Chinese hamster ovary cells, 115
> as host, 20
Class I antigens, 168
Class II antigens, 168 , 169
clinical
> studies, 22 , 23 , 117
> trial, 51
> trials, 50
clinical
> testing, 50
clinical studies
> phase I, 158
> phase II, 158
> phase III, 158
clinical trials
> OKT3, 194
clone selection, 183
cloning
> of human insulin gene, 67
coat antigens
> See HBsAg
Codes
> of Federal Regulations, 32 , 42
coinfection, 222
commercial
> return, 15
compassionate approval, 195
competent
> authorities, addresses of, 282

authority and GMOs, 292
complement cascade, 180
complex
> shedding, 180
concept evaluation, 12
confidentiality, 30
> and EC directives, 287 , 294
contamination
> by viruses, 47
> reduction of, 44
> viruses, 115 , 189
copy DNA
> See cDNA
copy number, 150
core of PRV, 202
costs
> of development, 14
> of manufacture, 14
CPMP, 31
creatine kinase, 228
creation
> manipulation of, 7
CRF, 32
cross reactivity
> Myoscint, 241
Cushing Syndrome, 176
cyclophosphamide, 174
cyclosporin, 176 , 196
cytoxic antibody matching test, 169
cytoxic T cells, 171

D

dedicated facilities, 231
deletion mutants, 206 , 208 , 222
development
> cost of, 14
DHFR
> dihydrofolate reductase, 115
> promoter, 115
diabetes, 176
> mellitus, 56 , 57
diafilter, 234
diagnosis
> myocardial infarction, 228 , 243
diagnosis precision
> of Myoscint, 245
diagnostic tools, 6
> in vivo, 226
dialysis patients
> numbers in Europe, 128
diethylenetriamine pentaacetic acid, 229
dihydrofolate reductase
> See DHFR

directives
EC, 32 , 284
disease eradication, 201
diuresis, 57
DNA
residual, 47
test, 47
dose-range studies, 158
double blind
trials, 158
downstream processing, 5
Mifarmonab, 235
Myoscint, 233
downstream processing , 19
DTPA, 234
DTPA conjugates, 231

E

E.coli
and HBsAg as a source, 144
as host, 20
E.coli peptides, 90
EC
See also CPMP
committee for proprietary medicines, 31
directives, 32 , 33
recommendation, 35
EC
guidelines, 32
*Eco*RIsite
site in TK gene, 216
ECPs, 90
effective dose
ED50, 153
efficacy
and safety, 14 , 20
of PRV vaccine, 220
proof of, 14
of a drug, 158
ELISA, 47 , 246
endonucleases
restriction, 4
endotoxin burden, 231 , 239
envelope of PRV, 202
enzyme
reactor, 3
epitope, 153
EPO
and blood pressure, 131 , 132
and blood transfusion, 132
and CAPD patients, 127 , 136
and chronic renal failure, 127
and epilepsy, 136

and Grand mal, 129 , 136
and hypertension, 136
and non-responders, 132
and thrombosis, 131
anti-EPO antibodies, 124
clinical studies, 128
dose titration of haematocrit, 132
haematocrit, 133
haematocrit in trials, 129
haemodialysis patients, 129 , 131
LAL test, 120
master cell bank, 115
pharmacodynamics, 121
pharmacokinetics, 120 , 123
pharmacological studies, 118
phlebotomy, 129
placebo-controlled trials, 122
pre-clinical studies, 117
production of, 112
reticulocyte count, 122
therapeutic value, 138
toxicological studies, 119
Eprex
See EPO
production of, 112
erythron
transferrin, 123
erythropoiesis, 104
regulation of, 106
erythropoietin
amino acid sequence, 114
See EPO
Escherichia coli
and HBsAg, 144
as host, 20
ethical
issues, 12
Eurocollins, 166
Eurotransplant, 167
evaluation
pharmacological, 12
toxicological, 12
exonuclease, 211
expiry
time, 48
expression system, 146
extrinsic impurities, 152

F

$F(ab^1)_2$ fragments, 227
Fab fragment, 227
facility
dedicated, 189

FDA, 32 , 181
fermenter, 45 , 233
fertility studies, 241
filling
        lot, 46
final release tests, 236
first line treatment, 195
fistula, 133
Fleming, 2
foetal toxicity, 241
folic acid
        deficiency, 106
Food
        and Drug Administration, 32
        See alsoFDA
formulation
        general, 23

G

GAPDH promoter gene, 148
genetic
        engineering, 13
        manipulation, 18
        manipulation and release, 12
genetic engineering, 4
genetically modified micro-organisms
        (GMMO), 284
        See also GMMO
genetically modified organisms, 37 , 288
        See also GMO
geometric mean titre, 160
        (GMT), 153
gI, 205
gI test kit, 222
GILSP
        (Good Industrial Large Scale Practice), 286
GLP, 24 , 39 , 41 , 42 , 43
        See also Good Laboratory Practice
glucogenic
        amino acids, 57
glucosuria, 57
glyceraldehyde 3-phosphate dehydrogenase
        GAPDH promoter, 147
glycoproteins of PRV, 205
GMMO, 37 , 284 , 285
GMMOs, 286
GMO, 37 , 289 , 290 , 291 , 293
GMP, 24 , 39 , 40 , 41 , 42 , 43
        See also Good Manufacturing Practice
        myoscint, 231
        OKT3, 188
Good Industrial Large Scale Practice
        (GILSP), 286

Good Laboratory Practice, 8 , 24 , 39
        See also GLP
Good Manufacturing Practice, 7 , 24 , 40
        See also GMP
gp50, 205
gp63, 205
Group I
        organisms, 286
Group II
        organisms, 286
growth arrest, 216
guidelines
        EC, 32 , 35
gX, 205

H

haematocrit, 116 , 119
haemochromatosis, 102
HB Vax
        See also Hepatitis vaccine
HBsAg
        and vaccinia, 145
        as a vaccine, 142
        gene, vector construct, 148
        Hepatitis surface antigen, 144
        isolation, 151
        selectable marker, 148
HBsAg coding sequence, 147
HBV
        coat antigens, 142
        Dane particle, 142
        envelope protein, 143
        first vaccine, 142
        gene cloning, 146
        genetic composition, 143
        incubation period, 142
        structure, 143
        subtypes, 143
        transmission, 142
heart transplant
        heart, 245
Hepatitis B virus
        See HBV
Hepatitis vaccine, 148
        adjuvant, 151
        analysis, 152
        clinical evaluation, 158
        clinical studies, 158
        effective dose, 162
        hyporesponders, 163
        immunogenicity, 159
        non-responders, 163
        potency, 153 , 154

pre-clinical studies, 153
preservative, 151
protective efficacy, 163
purity, 152
responders, 163
safety, 152
stability, 153
tolerance to, 159
hepatotoxicity, 174
Herpes virus, 200
Herpesviridae, 202
high risk patients, 243
Hind III fragments, 210 , 212
histiocytes, 180
HLA, 168
HLA-antigen, 171
homologous recombination, 222
host system
choice of, 63
HPLC
insulin, 21
HPLC
chemical potency tests, 47
human leucocyte antigen
See HLA
hybrid
cells, 18
hybridoma, 4 , 182 , 184 , 226 , 230
production of, 182
technology, 13
hybridoma cells, 231
hybridoma technology, 18 , 166 , 182
hydrocortisone, 175
hyperglycaemia, 57
hypersplenism, 132
hypertension, 176
hypochromia, 105

I

iatrogenic, 176
igG, 227
IgM, 227
immobilised
enzyme reactor, 3
immunogenicity, 153
of myoscint, 246
immunosuppressive agents, 174
alkylating agents, 174
antimetabolites, 174
corticosteroids, 175
immunosuppressive drugs, 170
indicator of cell death, 242
indium-111, 226 , 229

inflammatory reaction, 175
insulin
amino acid composition, 85
amino acid sequence, 56 , 59
and clone selection, 65
and glucose, 57
and human cadaver pancreata, 60
and hyperglycaemia, 57 , 58
and hypertriglyceridemia, 58
and lipolysis, 58
and proteolysis, 58
and the diabetic syndrome, 58
as a chimeric protein, 70
bioassay, 81
biological characterisation, 88
biological effects, 95
carboxypeptidase cleavage of pro-insulin, 78
case reports, 98
chemical analysis, 80
circular dichroic spectra, 85 , 86
clinical experience, 98
clinical studies, 94
clinical trials, 97
commercial, 60
dependent diabetes, 56
enzymic conversion of pro-insulin, 79
fat and protein metabolism, 57
from different sources, 60
gene isolation, 63
HPLC analysis, 76 , 81 , 83 , 84 , 90
hypoglycaemia, 60
immunological analysis, 90
immunological effects, 96
pharmacological studies, 94
produced as separate chains, 74
production by pro-insulin route, 78
production of, 60 , 69
promoter systems, 69
purification, 75 , 80
semisynthetic, 61
toxicity tests, 88
trypsin cleavage of pro-insulin, 78
V8 protease, 86
interleukin, 171
interleukin-1, 181
interleukin-2, 171 , 176 , 181
interleukin-2 receptors, 171
intrinsic factor, 105
inverted repeats, 204
iodine-123, 229
iodine-125, 228
iodine-131, 226 , 228
iodoacetamide, 234
iron deficiency, 105

islets
        of Langerhans, 59
isoenzyme profile, 189

K

karyotype, 189
ketone
        bodies, 57
Kohler and Milstein, 181
Kupffer cells, 180

L

LAL
        OKT3, 188
        test, 92 , 120
        tests, 48
latency of PRV, 218
latent state, 216
law and the export of pigs, 222
leucopenia, 174 , 176
licence
        dossier, 43
licencing
        file contents, 30
licencing
        file, 30
licencing file
        See licencing dossier
Limulus
        amoebocyte lysate, 48 , 120
        See also LAL
linker, 74
linkers
        for DNA constructs, 65
lyophilising
        lyophylisation, 21

M

macrocytosis, 106
macrophage activating factor, 175
macrophages, 171 , 180
maintenance therapy, 176
manufacture
        cost of, 14
mapping infarction, 243
market
        authorisation, 32 , 42
        needs, 13
        size, 6 , 15
marketing, 25
        and GMOs, 292

        authorisation, 40
master cell banks
        OKT3, 184 , 189
        Myoscint, 232
master seed
        hepatitis vaccine, 149
mean time to death, 214
mercaptopurine
        immunosuppressive agent, 175
metabolism, 3
methylprednisolone, 175
mice
        containing viral genomes, 191
        pathogen free, 184 , 190
        protection during transit, 190
        quarantine, 190
        sentinal, 190
microcytosis, 105
Mifarmonab, 229 , 230 , 233
        purification, 234
mitogenic
        OKT3, 181
monoclonal antibodies, 4 , 166 , 226
monoclonal antibody
        concentration, 187
        crude preparation, 184
        preparation, 185
        production, 181
        purification, 186
monocytes, 180
mRNA
        and cDNA, 62
multi-state
        procedure, 31
multicentre study
        OKT3, 195
mutagenicity
        Myoscint, 241
myeloma, 166
myeloma cells, 182
myelophibrosis, 109
myocardial cell, 228
myocardial infarction, 242
Myoscint, 229 , 231
        kit, 236
        purification, 234
        scans, 245
        sensitivity, 245
        specificity, 245
myosin, 228

N

national
    authorities, 31
necrotic cells, 228
nephrotoxicity, 176
neutral
    protamine hagedorn (NPH), 94
NIA3 strain of PRV, 208
notification, 287
    and release, 291
nuclear medicine, 226
nucleotide
    sequencing, 21

O

obesity, 176
Official
    journal, 32
OKT series, 173
OKT3, 166 , 180
oligonucleotide
    See also probes
    synthetic, 114
operations
    Type A , 286
    Type B, 286
opsonisation, 180
optimisation
    of processes, 11
organ transplantation
    See renal transplantation
organisms
    Group I, 286
    Group II, 286
osteoporosis, 176

P

P-region
    of HBV, 143
PAB, 32
papain, 234
Pasteur, 2
patenting, 8
pathogens
    elimination of, 45
patients at risk, 245
pBR322, 210 , 211
penicillin, 2
perfusion culture, 233
peritoneal cavities, 182
pernicious

anaemia, 105
personnel
    authorised, 231
    responsibilities of, 44
phagocytosis, 180
Pharmaceutical
    Affairs Bureau, 32
    See also PAB
pharmacokinetic
    studies, 50
pharmacokinetics of myoscint, 243
pharmacological
    evaluation, 12
pharmacological tests
    myoscint, 240
pharmacology
    OKT3, 193
pHB2.1, 212
pHB2.4, 212
pHB2.8, 212
physiological barriers, 180
planning, 13
plasma-exchange, 169 , 170
plasmids
    as vectors, 63
    number, 150
pN3B7, 211
pN3HB, 210
polyclonal antibodies, 226
    canine myosin, 228
polycythaemia, 107
polyuria, 57
post-marketing
    surveillance, 51
potency, 48
    challenge tests, 47
    competition assay, 188
    test, 46
pre-clinical
    studies, 22 , 23 , 117
    testing, 49
    trials, 7
pre-culture, 233
prednisolone, 175
premises
    design of, 44
pristane, 184
probes, 114
process
    bioreactor, 3
process engineering, 3
process optimisation, 11
process validation, 43 , 44 , 238

product
    bulk, 46
    diversification, 25
production
    general, 24
    optimisation, 11
    procedures, 43 , 46
prognostic significance of myoscint, 245
promoters, 69
prophylactic treatment, 195
protective clothing, 189
protective efficacy, 154
protein A affinity chromatography, 234
proto-oncogenes, 146
PRV, 200
    2.4 N3A, 214
    2.8 N3A, 214
    absence of gI gene, 218
    absence of thymidine kinase, 218
    deletion mutants, 214
    glycoproteins, 205
    Hind III sites, 209
    latency, 218
    molecular features, 202
    reactivation, 218
    restriction enzyme sites, 210
    spreading, 218
    strain NIA3, 208
    TK mutant, 217
    virulence, 218
PRV eradication programme, 223
pseudorabies virus
    See PRV
purity
    of desired product, 19
pyrogens, 48 , 92 , 153

Q

QA, 41 , 42
    See also Quality Assurance
QC, 24 , 39 , 40 , 43
    See also Quality Control
Quality
    See also QC
    reproducibility, 48
    reproduction of, 45
Quality Assurance, 24 , 41 , 188
    OKT3, 193
Quality Control, 6 , 8 , 24 , 40 , 41 , 48 , 192 ,
231
    myoscint, 236
quarantine, 190

R

radioimmunoassay, 226
RBCs
    and dietary deficiency, 105
    bilirubin, 106
    breakdown, 106
    chronic inflammatory disease, 110
    decreased erythropoietin output, 109
    deformation, 106
    hypochromia, 105 , 106
    macrocytosis, 106
    microcytosis, 105
reactor
    immobilised enzyme, 3
recombinant
    viruses, 145
recombinant vaccines, 207
red blood cell
    physiology, 103
    production, 104
    See also RBCs
regulation, 8
regulatory
    agency, 51
    issues, 37
    obligations, 21
rejection
    cellular, 171
    humoral, 167
    hyperacute, 169
    renal allograft, 195
    reversal rate, 196
release
    deliberate, 288
    of genetically manipulated organisms, 12
renal
    failure and artificial kidneys, 138
renal transplantation, 167
repeats, 204
reprocessing, 48
    main stream, 48
    side stream, 48
reproducibility
    of quality, 48
research
    and development, 18
resources
    and investment, 10
restricted access, 231
restriction
    endonucleases, 4

restriction enzyme analysis
    of PRV, 218
restriction enzyme mapping, 150
restriction enzyme sites of PRV, 210
resuce therapy, 195
reticuloendothelial system, 180
reversion, 219
RIA, 226

S

S gene, 148
S-gene
    sequence of nucleotides, 150
S-region
    of HBV, 143
Saccharomyces
    as host, 20
    HBsAg gene expression, 145
safety
    and efficacy, 14 , 20 , 31
    hepatitis vaccine, 152
    long term, 158
    of genetically engineered vaccines, 220
    of vaccines, 219
    proof of, 14
    tests, 48
safety and clinical pharmacology, 158
scintigraphic imaging, 227
SDS-PAGE, 152
seed lot, 45
    hepatitis vaccine, 149
segregation
    asymmetry, 150
selection
    of plasmid, 146
sentinal mice, 190
sequenced
    plan, 13
sera
    comparison of, 156
seroconversion, 160
seronegative, 160
Shope, 200
Single Community Act, 201
source
    materials, 45
Southern blot hybridisation, 150
spiking, 45
spiking studies
    myoscint, 239
    viruses, 192
spreading of PRV, 218
stability, 48

Hepatitis vaccine, 153
    hepatitis vaccine constructs, 150
standard
    international preparations, 48
standard operating procedures, 231
sterile filtration
    Myoscint, 236
sterilisation, 188
stock seed
    hepatitis vaccine, 149
suppressor T cells, 171

T

T-cell replacing factor, 171
T-cell subsets, 173
T-helper cell, 171
Tc cells, 171
technetium-99m, 226 , 229
technetium-stannous pyrophosphate, 228
tegument of PRV, 202
thallous chloride, 228
thrombocytopenia, 176
thymidine [14C], 218
thymidine kinase, 204 , 216
toxicological
    evaluation, 12
toxicological tests
    Myoscint, 240
transgenic
    organisms, 5
transplant, 245
triphenyltetrazolium chloride, 242
tryp E
    promoter, 20 , 72
tryptophan
    synthetase promoter, 72
tryptophan E
    promoter, 20
Ts cells, 171
tumorigenicity, 6
Type A
    operations, 286
Type B
    operations, 286

U

Unique short region
    See Us region
Us region, 209
UW rinsing solutions, 166

V

vaccination, 206
vaccines
    conventional, 206
    inactivated, 206
    live, attenuated, 206
    peptide, 207
    recombinant, 207
    safety, 6 , 220
    subunit, 207
    testing, 219
validation
    of a process, 18
    virus removal, 191
validation studies
    Myoscint, 238
vector, 18
vector promoter
    DHFR, 115
vectors, 63
virulence, 206
virulence of PRV, 218
virus
    assay in oropharyngeal fluid, 216
    free, 115
    inactivated, 210
    reconstructed, 214
    reservoir, 201
    tests, 47
    transmission, 201

W

waste water, 3
water
    waste, 3
Watson and Crick, 4
working cell bank
    OKT3, 184
working cell banks
    myoscint, 232
    OKT3, 189

X

X-region
    of HBV, 143
xenograft, 167